BASIC INDUSTRIAL ELECTRICITY

A Training and Maintenance Manual

BASIC INDUSTRIAL ELECTRICITY

A Training and Maintenance Manual

Kenneth G. Oliver

INDUSTRIAL PRESS INC.

Library of Congress Cataloging-in-Publication Data

Oliver, Kenneth G.
    Basic industrial electricity : a training and maintenance manual /
Kenneth G. Oliver. — 1st ed.
      p. 21.6 × 27.9 cm.
    Includes bibliographical references and index.
    ISBN 0-8311-3006-7
    1. Electric engineering. 2. Factories—Electric equipment.
3. Factories—Electric equipment—Maintenance and repair.
I. Title.
TK146.0395 1990
621.319′24—dc20   90-38038
                                             CIP

I N D U S T R I A L P R E S S I N C.
200 Madison Avenue
New York, N.Y. 10016-4078

First Edition

Basic Industrial Electricity: A Training and Maintenance
Manual

6 8 9 7 5 3

DEDICATION

With the expectation that we shall vastly improve their operating efficiency and effectiveness, we dedicate this book to the thousands of part-time and apprentice electrical maintenance men in the industrial plants of this country, most of whom perform their duties with a minimum of training.

FOREWORD

Parker Schools for Industry has been engaged in the training of industrial plant maintenance personnel since 1968 with gratifying success. As a result of its success, a questionaire was sent out to the many plant managers in the Southern California area, to ascertain their needs for further courses to help them train their plant operators and maintenance people. One of the highest demands reported was for a basic course in electrical maintenance, which would assist the part-time electrician, who was usually a mechanic in the department, to perform the many small routine electrical maintenance tasks required to keep the plant operating on a day-to-day basis.

Discussions with several plant engineers resulted in the establishment of the level of technical material required, and a further search of the technical book sources did not uncover any existing textbooks that could be used for this level of training. Consequently, this book was written, aimed at explaining only the barest essentials in the principles involved in the operation of the electrical equipment in the average medium-sized industrial plant. We have endeavored to show the maintenance electrician how his equipment performs its functions, and what is required to maintain it.

Therefore, the reader will not find many of the complicated mathematics usually found in electrical textbooks, but will find only that basic mathematics which describes the simple functions of the majority of the apparatus under his care. A pocket calculator is all that will be required to handle the mathematics in this book.

In short, this textbook, and the courses for which it was designed, is intended to be an adjunct to a "hands-on" approach.

The author of this textbook does not in any way intend to replace the mass of instructional material supplied by the electrical equipment manufacturers. The student and the reader are advised to read such material over very carefully, and follow it closely. The material we have presented here is to serve as a general guide only, and to provide further instructions that may have been omitted or glossed over by the equipment suppliers.

It is fortunate that the basic laws upon which the plant electrical apparatus is designed do not permit too many variations in the design of the equipment built to apply those laws. The differences between the various suppliers of apparatus will be mainly in the arrangement of the components and the details of manufacture. The rules and principles are the same throughout the industry, probably to a greater extent than in many other engineering fields.

Basic Industrial Electricity was written based on codes, products, and practices existing and in common use at the time of writing, all of which are subject to change from time to time. However, the subjects covered in this book probably represent about 95% of the existing codes, products, and practices in use at the time of writing. A product may no longer be available from the supplier, but the item in question may be installed and giving good service in thousands of plants, and, therefore, is a legitimate subject for discussion in this book.

It is the responsibility of the industrial plant operating personnel continually to keep abreast of current changes in those governing codes pertinent to their plant and also to trends in operating practices.

Also, the fact that your plant was designed based on codes, products, or practices no longer in existence does not necessarily mean that you are working in unsafe conditions, but it is to your advantage to check into the latest changes, and determine for yourself whether or not your equipment should be updated. Your local building and safety inspector will be able to help you in this respect.

As there are literally thousands of combinations of motors, controllers, panels, and electrical distribution equipment in today's industrial plant, we can only stress generalities. As it is impossible to cover all of the safety factors necessary to the operation and maintenance of that equipment, we can assume no responsibility for the safety of any industrial plant being operated or maintained by any person who has either attended one of our schools, or who has read any of the textbooks, supplements, or miscellaneous handout materials associated with the school.

Consequently, we stress the point that the ultimate responsibility for the plant safety is in the hands of others who are involved in the operation of such plant. It is the responsibility of the plant managers, the insuring agencies, and, in the final analysis, the operators and servicers of the plant equipment to safeguard the plant and personnel under their jurisdiction.

With that, we leave you to your search for knowledge, happy to know that we have contributed in a small way to your advancement up the ladder to a successful and rewarding occupation maintaining one of this country's great industrial complexes.

ACKNOWLEDGMENTS

Any textbook of this nature, covering the use, maintenance and philosophy of manufactured equipment, must, of necessity refer considerably to those products of specific suppliers in our society. Those suppliers have been most helpful in the assembly of material for this book, and we wish to give them all our heartiest thanks. We trust that those who read this book or take the course for which it was originally produced, will remember the following firms when it comes time to install new equipment in the plant:

Allen-Bradley Co.	GNB Batteries, Inc.
American Petroleum Institute	GTE Products Corp.
Amprobe Instrument Co.	Panalarm Div., Ametek, Inc.
BEHA Div., Greenlee Tool Co.	Phillips Lighting Co.
Cornell-Dublier Electronics	Pass & Seymour, Inc.
Crouse-Hinds Co.	Sangamor-Western, Inc.
Emergi-Lite Inc.	Simpson Electric Co.
General Electric Co.	Square D Manufacturing Co.

The author has borrowed heavily from the various training manuals produced by the U.S. Government Printing Office, under the auspices of various Government agencies. These are listed in the Bibliography.

The author was assisted considerably by two men who have kindly given of their time and expertise on a technical basis in the preparation of this book. Mr. Elliot Gray, the instructor in the course for which the prototype of this book was produced, not only provided additional input by means of his lectures, but also reviewed those portions that gave the author some difficulty. William Greninger, an electrical engineer whom the author has known for several years, was also very generous with his time, advice and final review. To both I am deeply indebted and extend my gratitude.

CONTENTS

C H A P T E R 1
TREAT IT WITH RESPECT

A. Introduction

Because of the invisible, mysterious nature of electricity, and its inherent power and energy, it is imperative that any study course or textbook on the subject begin by stressing the safety aspects involved in handling it. In fact, before any person, however gifted or mechanically inclined, commences to do any work with electricity or the equipment for its use, that person must, for his own safety, understand the rudiments of it. Therefore, this course of study will begin with a complete discussion of proper safety techniques and procedures in the subject of electrical maintenance, troubleshooting, installation, and any other use that may be found in the modern—and not so modern—industrial plant.

Read this section over very slowly and carefully, and make special note of those portions that apply to your work now being performed in conjunction with the electrical services in your plant.

Remember, in accidents involving electricity, the person directly involved in the work is usually the one that suffers most from his mistakes. Not his partner, his supervisor, or others in the plant, but the electrical worker. It is your life that is at stake. Keep that in mind, study the material in this book on the basics of the power you are handling, and you should enjoy a long, happy life in your chosen trade.

During the various discussions throughout this book, there will be other safety matters discussed, usually connected with the specific subject in hand.

In the performance of his normal duties, the electrician is exposed to many potentially dangerous conditions and situations. No training manual, no set rules or regulations, no listing of hazards can make working conditions completely safe. However, it is possible for electricians to complete full careers without serious accident or injury. Attainment of this goal requires that they be aware of the main sources of danger and that they remain constantly alert to those dangers. One must take the proper precautions and practice the basic rules of safety. One must be safety conscious at all times, and this safety consciousness must become second nature.

Safety precautions in this chapter are not intended to replace information given in instructions or maintenance manuals. If at any time there is doubt as to what steps and procedures to follow, consult your supervisor.

B. Effects of Electric Shock

The amount of current that may pass through the body without danger depends on the individual and the current quantity, type, path, and length of contact time.

Body resistance varies from 1000 to 500,000

ohms for unbroken, dry skin. (Resistance and its unit of measurement will be discussed subsequently.) Resistance is lowered by moisture and high voltage and is highest with dry skin and low voltage. Breaks, cuts, or burns may lower body resistance. A current of 1 milliampere can be felt and will cause a person to avoid it. (The term milliampere is discussed subsequently; however, for this discussion it is sufficient to define milliampere as a very small amount of current, $\frac{1}{1000}$th of an ampere.) Current as low as 5 milliamperes can be dangerous. If the palm of the hand makes contact with the conductor, a current of about 12 milliamperes will tend to cause the hand muscles to contract, freezing the body to the conductor. Such a shock may or may not cause serious damage, depending on the contact time and your physical condition, particularly the condition of your heart. A current of only 25 milliamperes has been known to be fatal; 100 milliamperes is likely to be fatal.

Owing to the physiological and chemical nature of the human body, five times more direct current than alternating current is needed to freeze the same body to a conductor. Also, 60-hertz (cycles per second) alternating current is about the most dangerous frequency. This is normally used in residential, commercial, and industrial power.

The damage from shock is also proportional to the number of vital organs transversed, especially the percentage of current that reaches the heart.

Currents between 100 and 200 milliamperes are lethal. Ventricular fibrillation of the heart occurs when the current through the body approaches 100 milliamperes. Ventricular fibrillation is the uncoordinated actions of the walls of the heart's ventricles. This in turn causes the loss of the pumping action of the heart. This fibrillation will usually continue until some force is used to restore the coordination of the heart's actions.

Severe burns and unconsciousness are also produced by currents of 200 milliamperes or higher. These currents usually do not cause death if the victim is given immediate attention. The victim will usually respond if rendered resuscitation in the form of artificial respiration. This is due to the 200 milliamperes of current clamping the heart muscles, which prevents the heart from going into ventricular fibrillation.

When a person is rendered unconscious by a current through the body, it is impossible to tell how much current caused the unconsciousness. Artificial respiration must be applied immediately if breathing has stopped.

C. High Voltage Safety Precautions

Many pieces of electrical equipment use voltages that are dangerous and may be fatal if contacted. Practical safety precautions have been incorporated into electrical systems, but when the most basic rules of safety are ignored, the built-in protection becomes useless.

The following rules are basic and should be followed at all times by all personnel when working with or near high voltage circuits:

1. CONSIDER THE RESULT OF EACH ACT—there is absolutely no reason for an individual to take chances that will endanger his life or the lives of others.

2. KEEP AWAY FROM LIVE CIRCUITS—do not change parts or make adjustments inside the equipment with high voltage turned on.

3. DO NOT SERVICE ALONE—always service equipment in the presence of another person capable of rendering assistance or first aid in an emergency.

4. DO NOT TAMPER WITH INTERLOCKS—do not depend on interlocks for protection; always shut down equipment. Never remove, short circuit, or tamper with interlocks except to repair them.

5. DO NOT GROUND YOURSELF—make sure you are not grounded while adjusting equipment or using measuring equipment. Use only one hand when servicing ener-

gized equipment. Keep the other hand behind you. Wear rubber soled shoes at all times.

6. DO NOT ENERGIZE EQUIPMENT IF THERE IS ANY EVIDENCE OF WATER LEAKAGE—repair the leak, wipe up the water, and dry out the equipment before energizing.

These rules, together with the idea that voltage shows no favoritism and that personal caution is your greatest safeguard, may prevent serious injury or even death.

D. Working on Energized Circuits

Insofar as it is practicable, repair work on energized circuits should not be undertaken. When repairs on operating equipment must be made because of emergency conditions, or when such repairs are considered to be essential, the work should be done only by experienced personnel, and if possible, under the supervision of a safety officer. Every known safety precaution should be observed carefully. Ample light for good illumination should be provided; the worker should be insulated from ground with some suitable nonconducting material such as several layers of dry canvas, dry wood, or a rubber mat of approved construction. The worker should, if possible, use only one hand in accomplishing the necessary repairs. Helpers should be stationed near the main switch or the circuit breaker so that the equipment can be deenergized immediately in case of emergency. A person qualified in first aid for electric shock should stand by during the entire period of the repair.

E. Grounding of Equipment

Because there is no absolutely foolproof method of ensuring that all tools are safely grounded (and because of the tendency of the average electrician to ignore the use of a grounding wire), the old method of using a separate external grounding wire has been discontinued. Instead, a three-wire, standard, color-coded cord with a polarized plug and a ground pin is required. In this manner, the safety ground is made a part of the connecting cord and plug. Since the polarized plug can be connected only to a mating receptacle, the user has little choice but to use the safety ground.

All new tools, properly connected, use the green wire as the safety ground. This wire is attached to the metal case of the tool at one end and to the polarized grounding pin in the connector at the other end. It normally carries no current, but is used only when the tool insulation fails, in which case it short circuits the electricity around the user to ground and protects him from shock. The green lead must never be mixed with the black or white leads, which are the true current-carrying conductors.

Check the resistance of the grounding system with a low-reading ohmmeter to be certain that the grounding is adequate (less than 1 ohm is acceptable). The ohmmeter and its use will be discussed subsequently. If the resistance indicates greater than 1 ohm, use a separate ground strap.

Some old installations are not equipped with receptacles that will accept the grounding plug. In this event, use one of the following methods:

1. Use an adapter fitting that has the ground terminal grounded.

2. Use the old-type plug and bring the green ground wire out and ground it separately.

3. Connect an independent safety ground line. When using the adapter, be sure to connect the ground lead extension to a good ground. (Do not use the center screw that holds the cover plate on the receptable.) Where the separate safety ground leads are externally connected to a ground, be certain to first connect the ground and then plug in the tool. Likewise, when disconnecting the tool, first remove the line

plug and then disconnect the safety ground. The safety ground is always connected first and removed last.

F. Confined Spaces

When personnel are working in confined spaces, adequate ventilation must be provided. This includes oxygen for normal breathing, cooling to prevent heat exhaustion, air movement and exchange to prevent hazardous accumulations of vapors, and an additional or alternate source of ventilation for use in an emergency. When a worker is sent into a confined space for any reason, provisions should be made in advance for his rescue in the event of accident or emergency. These provisions include the use of safety lines for locating the worker and for retrieving him from the space. Some means of communication must be established so that the conditions existing inside and outside the space may be made known to all personnel concerned. A safety person must be provided to keep a constant check on the condition of the space and the worker, and he must be prepared to sound the alarm for additional help or to render assistance to the worker in the confined space, as required.

Fumes tend to collect in confined spaces, so the condition of the space should be checked prior to entry. The worker should also check conditions as he enters and monitor them during the stay. The worker should maintain constant communication with the safety person and inform him of any abnormal conditions that may exist.

Equipment used by personnel working in confined spaces is a matter of considerable importance. Enough light should be provided so that one can see clearly to perform the task. The light provided should be insulated so that it does not present a shock hazard (confined spaces are usually quite warm, and a safety light produces additional heat, so perspiration may become a serious problem). When possible, explosion-proof equipment should be used in confined spaces, and protective gear should be used if toxic fumes are known or suspected to exist within the space.

Caution should be exercised in the use of portable gasoline engine driven generators to prevent concentration of the exhaust in dangerous areas.

G. Fire Prevention and Protection

This is a subject about which a complete volume could be written. It is such an important item in today's industrial plant that we will not attempt to give complete coverage to it in this book. We will, however, make a few suggestions, which we sincerely trust you will follow.

If you have any questions concerning your fire protection, there are several agencies that may be approached for help. One of the better known ones is the National Board of Fire Underwriters, commonly known as the NBFU. They have offices in most large cities, and will either help you directly or refer you to one of their member agencies. They publish several pamphlets, which are handy guides to proper fire protection in the industrial plant.

In the case of fire hazards applying to the electrical equipment, the National Electrical Code®,* together with the Underwriter's Laboratories, have formulated a classification system for designating the degree of hazard for various areas. It is an excellent guide for the plant manager or owner in helping him choose the proper equipment.

The National Fire Protection Association is another agency that has published many codes and guides for use by fire underwriters and insurance agents. Contact them.

There are also various state, county, and city agencies that may be called in to offer help. The local building and safety department, or its equivalent, will have their own set of codes and standards that will have to be followed. Also,

---

*National Electrical Code® and NEC® are Registered Trademarks of the National Fire Protection Association, Inc., Quincy, MA.

the local fire department probably has a man on its staff who is a specialist in spotting potential danger areas, and who will know to which agency to refer you for further help.

There is one thing we will strongly urge right here, and that is that it is a matter of good policy for you to become acquainted with the local fire protection authorities, as it is much better that you cooperate with them than try to avoid them or their rulings. They have seen many more fires than you will ever see, and it is very much to your advantage to make friends with them—or at least do not make enemies of them!

And, above all, do not skip over those fire drills! They are not just "kid stuff."

In most plants, fire protection is under the supervision of either one person or a committee. This does not relieve you, the maintenance technician, of responsibility, however.

As a practical matter, there are several areas in your plant where you should constantly be alert for the possibility of fire damage, or impediments to fire protection, and we shall list some of them here.

As the high voltage supply is one of the main sources of danger, the manual disconnect switch, item 2, Fig. 16-2, should be clearly identified and marked, and the pathway to the enclosure should always be free and clear.

The front of all panels and the area surrounding them should be free of debris at all times. This includes removing all trash containers from the danger area, wiping up all spilled inflammables, and keeping all rags, loose paper, etc., from the area.

The electrical maintenance office, with the records, papers, notes, etc., should be in a relatively danger-free area.

Know where all of the fire extinguishers are located, as well as the fire hoses, hydrants, foam trailers, etc.

Electrical switchgear should be isolated from the rest of the plant, if at all possible, and the correct type of fire extinguisher for fighting electrical fires should be readily in hand. If the switchgear is protected by one of the package, special types of fire fighting systems, such as a CO_2 or Halon system, then this system should

be carefully checked at regular intervals, and all debris or plant equipment spares should be kept away from the systems. Here, again, there should be a clear pathway to the equipment.

General cleanliness of the work area and of electrical apparatus is essential for the prevention of electrical fires. Oil, grease, and carbon dust can be ignited by electrical arcing. Therefore, electrical and electronic equipment should be kept absolutely clean and free of all such deposits.

Wiping rags and other flammable waste material must always be placed in tightly closed metal containers, which must be emptied at the end of the day's work.

Containers holding paints, varnishes, cleaners, or any volatile solvents should be kept tightly closed when not in use. They must be stored in a separate building or in a fire-resisting room that is well ventilated and where they will not be exposed to excessive heat or to the direct rays of the sun.

As the proper selection of fire extinguishers is essential to safe fire fighting, we refer to Tables 1-1 and 1-2, and we strongly recommend that the electrical maintenance worker familiarize himself with the types of fires that may be encountered in the plant, and the proper type of extinguisher to apply. In the case of electrical fires, the cardinal rule is: NEVER USE WATER OR ANY TYPE OF LIQUID THAT WILL TRANSMIT AN ELECTRICAL CURRENT OR PRODUCE A TOXIC VAPOR ON CONTACT WITH HEAT.

Table 1-1. Fire Classifications

Class A	Ordinary combustible materials, such as wood, paper, cloth, and most plastics.
Class B	Flammable liquids, gases, and greases.
Class C	Electrical equipment, including fires in the immediate vicinity of energized electrical equipment.
Class D	Combustible metals, such as magnesium, titanium, zirconium, sodium, and potassium.

Table 1-2. Proper Extinguisher Applications

Class A	Water, multipurpose dry chemical, foam, and Halon 1211.
Class B	Foam, carbon dioxide, dry chemical, and Halon 1211.
Class C	Carbon dioxide, dry chemical, and Halon 1211.
Class D	Dry powder only, as stated on the body of the extinguisher.

H. When Fires and Accidents Happen

You have followed all of the prescribed procedures, safety regulations, and maintenance functions required, and still a fire or accident happens. They do happen, even though you feel they should not. So what do you do then? We shall give you a few general steps to be followed to minimize the impact upon the plant, its personnel, and equipment. These are a guide only; you must modify them to suit your particular installation, equipment, and plant policy.

Troubles experienced in the average industrial plant may consist mostly of fires, accidents, and area-wide catastrophies.

Fires may be broken down into two classifications for our purpose here, those in the electrical system and those not caused by or not in the electrical system, but which endanger the entire plant.

In case of electrical fires, the following steps should be taken:

1. Deenergize the circuit.

2. Call the Fire Department.

3. Control or extinguish the fire, using the correct type of fire extinguisher.

4. Report the fire to the appropriate authority.

For combating electrical fires, use a CO_2 (carbon dioxide) fire extinguisher and direct it toward the base of the flame. Carbon tetrachloride should never be used for firefighting, since it changes to phosgene (a poisonous gas) upon contact with hot metal, and even in open air this creates a hazardous condition. The application of water to electrical fires is dangerous; the foam-type fire extinguishers should not be used, since the foam is electrically conductive; it is also very messy and difficult to clean up from the equipment and the surrounding surfaces.

In case of cable fires in which the inner layers of insulation or insulation covered by armor are burning, the only positive method of preventing the fire from running the length of the cable is to cut the cable and separate the two ends. All power to the cable should be secured, and the cable should be cut with a wooden handled ax or insulated cable cutter. Keep clear of the ends after they have been cut.

When selenium rectifiers (discussed subsequently) burn out, fumes of selenium dioxide are liberated, which cause an overpowering stench. The fumes are poisonous and should not be breathed. If a rectifier burns out, deenergize the equipment immediately and ventilate the area. Allow the damaged rectifier to cool before attempting any repairs. If possible, move the equipment containing it out of doors. Do not touch or handle the defective rectifier while it is hot, since a skin burn might result through which some of the selenium compound could be absorbed.

Accidents in the shop or on the premises involving electric shock to anyone are covered in this chapter, Section L, so will not be discussed further. However, in the case of other types of accidents, we give here the general procedures to be followed. The extent to which you, the electrical technician, will be expected to participate depends entirely on plant organization, where you fit into it, and to what extent you have been assigned particular roles to play in plant problems.

1. Upon arriving at the scene of the accident, determine first if there is any action required concerning the electrical circuits or equipment, and if so, take appropriate action immediately.

2. Take a quick check for casualties and offer any assistance necessary.

3. If necessary, call the local fire and rescue squad, or similar locally available organization. Their official report may help the insurance company in its settlement of any claims.

4. Remove all unauthorized personnel from the area. This includes mostly plant personnel who have gathered to gawk at the carnage, as well as others who may have come in off the street to see what's happening.

5. Do not touch, adjust, or alter any part of the damaged equipment or its accessories, unless essential to safeguard the area, personnel, or the plant.

6. Notify immediately all plant officials necessary, and await their further instructions.

7. Take photographs of all damaged portions or areas, and close-ups of all suspected trouble spots or equipment.

8. Do not attempt to analyze the trouble or find its source at this time. The insurance inspector is the best one to do that, and you may be called upon to assist him, and answer any questions concerning the events leading up to the time of the accident.

9. Make no statements to any outside personnel or unauthorized agencies, firms, etc., until cleared to do so by the plant officials or the insurance underwriter. Damage claims may be involved, and off-the-cuff remarks made to news reporters, gawkers, etc., have a way of getting into the testimony.

In addition to all of the preceding responsibilities, each employee has a responsibility to do his share in protecting or saving the plant, which employs him, in the case of area-wide catastrophies such as block fires, floods, storms, earthquakes, etc. Every well-run, progressive industrial organization has some contingency plan for implementing in case of such major calamities. It really makes sense to accept some responsibility for safeguarding the source of your income, doesn't it?

I. Using Electrical Tools

As a general precaution, be sure that all tools used conform to applicable standards as to quality and type, and use them only for the purposes for which they were intended. All tools in active use should be maintained in good repair, and all damaged or nonworking tools should be turned in for repair or replacement.

When using a portable power drill, grasp it firmly during the operation to prevent it from bucking or breaking loose, thereby causing injury to yourself or damage to the tool.

Use only straight, undamaged, and properly sharpened drills. Tighten the drill securely in the chuck, using the key provided; never with wrenches or pliers. It is important that the drill be set straight and true in the chuck. The work should be firmly clamped and, if it is metal, a center punch should be used to score the material before the drilling operation is started.

In selecting a screwdriver for electrical work, be sure that it has a nonconducting handle. The screwdriver should not be used as a substitute for a punch or a chisel, and care should be taken that one is selected of the proper size to fit the screw.

When using a fuse puller, make certain that it is the proper type and size for the particular fuse being pulled.

The soldering iron is a fire hazard and a potential source of burns. Always assume that a soldering iron is hot; never rest the iron anywhere but on a metal surface or rack provided for that purpose. Keep the iron holder in the open to minimize the danger of fire from accumulated heat. Do not shake the iron to dispose of excess solder—a drop of hot solder may strike someone or strike the equipment and cause a short circuit. Hold small soldering jobs with pliers or clamps.

When cleaning the iron, place the cleaning rag on a suitable surface and wipe the iron across it—do not hold the rag in the hand. Disconnect the iron when leaving the work, even for a short time—the delay may be longer than planned. See Chapter 19 for additional advice on the safe use of soldering equipment.

J. Portable Power Tools

All portable power tools should be inspected carefully before being used to see that they are clean and well-oiled and are in a proper state of repair. The switches should operate normally, and the cords should be clean and free of defects. The case of all electrically driven tools should be grounded. Sparking portable electric tools should not be used in any place where flammable vapors, gases, liquids, or exposed explosives are present.

Be sure that power cords do not come into contact with sharp objects. The cords should not be allowed to kink nor be left where they might be run over. They should not be allowed to come into contact with oil, grease, hot surfaces, or chemicals; and when damaged, they should be replaced instead of being patched with tape. When unplugging power tools from receptacles, grasp the plug, not the cord.

K. Shop Machinery

Daily electrical work requires the use of certain pieces of shop machinery, such as a power grinder or drill press. In addition to the general precautions on the use of tools, there are a few other precautions that should be observed when working with machinery. The most important ones are:

1. Never operate a machine with a guard or cover removed.

2. Never operate mechanical or powered equipment unless thoroughly familiar with the controls. When in doubt, consult the appropriate instruction booklet or ask someone who knows.

3. Always make sure that everyone is clear before starting or operating mechanical equipment.

4. Never try to clear jammed machinery without first cutting off the source of power.

5. When hoisting heavy machinery (or equipment) by a chain fall, always keep everyone clear, and guide the hoist with lines attached to the equipment.

6. Never plug in electric machinery without ensuring that the source voltage is the same as that called for on the nameplate of the machine.

L. Treatment for Shock

Electric shock is a jarring, shaking sensation resulting from contact with electric circuits or from the effects of lightning. The victim usually feels that he has received a sudden blow; if the voltage and resulting current is sufficiently high, the victim may become unconscious. Severe burns may appear on the skin at the place of contact; muscular spasm may occur, causing the victim to clasp the apparatus or wire that caused the shock and be unable to turn it loose.

The following procedure is recommended for rescue and care of shock victims:

1. Remove the victim from electrical contact at once, but do not endanger yourself. This can be accomplished by:

   a. Opening the switch if it is nearby.

   b. Cutting the cable or wire to the apparatus, using an ax with a wooden handle while taking care to protect your eyes from the flash when the wires are severed.

   c. Using a dry stick, rope, belt, coat, blanket, or any other nonconductor of electricity, and dragging or pushing the victim to safety.

2. Determine whether the victim is breathing. Keep the victim lying down in a comfortable position and loosen the clothing about the neck, chest, and abdomen so that he can breathe freely. Protect him from exposure to cold, and watch over the victim carefully.

3. Keep the victim from moving about. In this condition, the heart is very weak, and any sudden muscular effort or activity on the part of the patient may result in heart failure.

4. Do not give stimulants or opiates. Send for a medical officer at once and do not leave the patient until the arrival of adequate medical care.

5. If the victim is not breathing, it will be necessary to apply artificial respiration without delay even though the victim may appear to be lifeless.

The application of artificial respiration should be performed by the use of cardiopulmonary resuscitation (CPR). All electrical maintenance personnel should know the names of employees within the firm who are trained in the practice of CPR, and be familiar with the areas in which they normally may be found. Such information should be prominently posted at various locations throughout the plant, as a part of the plant's normal safety program.

Should you find an employee lying on the floor, lifeless, and it is obvious his condition is due to electroshock, then we give here the proper CPR procedures to follow, as described by the American Heart Association in their pamphlet *Cardiopulmonary Resuscitation (CPR)*. (Courtesy of the American Heart Association, 3550 Wilshire Boulevard, Los Angeles, CA 90010.) Of course, all procedures should be in agreement with any existing plant safety policies.

If you find a collapsed person, determine if the victim is unconscious by gently shaking a shoulder and shouting "Are you all right?" If there is no response, shout for help. If the victim is not lying flat on his or her back, roll the victim over, moving the entire body at one time as a unit. Then open the airway.

To open the victim's airway, lift up the chin gently with one hand while pushing down on the forehead with the other to tilt

Fig. 1-1. Checking victim for breathing.

the head back. Once the airway is open, place your ear close to the victim's mouth:

- Look—at the chest for movement.

- Listen—for sounds of breathing.

- Feel—for breath on your cheek.

If none of these signs is present, victim is not breathing.

If opening the airway does not cause the victim to begin to breathe spontaneously, you must provide rescue breathing.

The best way to provide rescue breathing is by using the mouth-to-mouth technique.

Using the thumb and index finger of the hand that is on the victim's forehead, pinch the victim's nose shut while keeping the

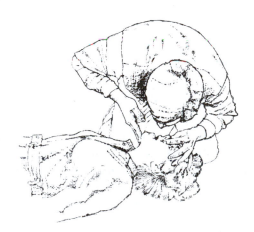

Fig. 1-2. Commencing rescue breathing.

Fig. 1-3. Checking for a heart beat.

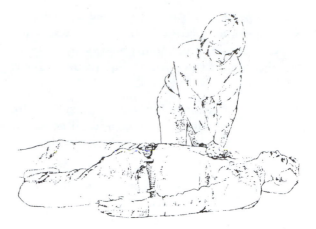

Fig. 1-4. Inducing blood circulation.

heel of the hand in place to maintain head-tilt. Your other hand should remain under the victim's chin lifting up.

Immediately give two full breaths while maintaining an air-tight seal with your mouth on the victim's mouth.

After giving the two full breaths, locate the victim's carotid artery pulse to see if the heart is beating. To find the carotid pulse, take your hand that is supporting the chin and locate the adam's apple (voice box); slide the tips of your fingers down into the groove beside the adam's apple; feel for the pulse.

If you cannot find the pulse, you must provide artificial circulation in addition to rescue breathing. **Activate the Emergency Medical Services (EMS) System: Send someone to call the local emergency number (usually 911).**

EXTERNAL CHEST COMPRESSIONS

Artificial circulation is provided by external chest compressions. In effect, when you apply rhythmic pressure on the lower half of the victim's breastbone, you are forcing the heart to pump blood. To perform external chest compression properly, kneel at the victim's side near the chest. With the middle and index fingers of the hand nearest the legs, locate the notch where the bottom rims of the two halves of

the rib cage meet in the middle of the chest. Place the heel of one hand on the sternum next to the fingers that located the notch. Place your other hand on top of the one that is in position. Be sure to keep your fingers up off the chest wall. You may find it easier to do this if you interlock your fingers.

Bring your shoulders directly over the victim's sternum as you compress downward, keeping your arms straight. Depress the sternum about $1\frac{1}{2}$ to 2 inches for an adult victim. Then relax pressure on the sternum completely. Do not remove your hands from the victim's sternum, but do allow the chest to return to its normal position between compressions. Relaxation and compression should be of equal duration.

If you must provide both rescue breathing and external chest compressions, the proper ratio is 15 chest compressions to 2 breaths. You must compress at the rate of 80 to 100 times per minute.

Ratio of Compressions to Breaths	Rate of Compressions
15 : 2	80–100 times/minute

If you suspect the victim has suffered a neck injury, you must not open the airway in the usual manner. If the victim is injured in a diving or automobile accident, for example, suspect a neck injury and open the airway by using chin-lift without

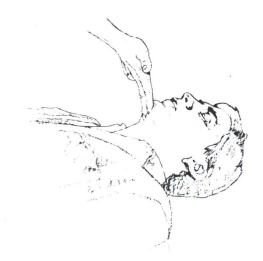

Fig. 1-5. Tilting head when neck is injured.

head-tilt. If the airway remains obstructed, tilt the head slowly and gently until the airway is open.

DO NOT STOP ARTIFICIAL RESPIRATION UNTIL MEDICAL AUTHORITY PRONOUNCES THE VICTIM BEYOND HELP.

M. To Summarize

Take time to be safe when working on electrical circuits and equipment. Carefully study the schematics and wiring diagrams of the entire system, noting what circuits must be deenergized in addition to the main power supply. Remember that electrical equipment frequently has more than one source of power. Be certain that ALL power sources are deenergized before servicing the equipment. Do not service any equipment with the power on unless it is necessary.

It must be borne in mind that deenergizing main supply circuits by opening supply switches will not necessarily "kill" all circuits in a given piece of equipment. A source of danger that has often been neglected or ignored—sometimes with tragic results—is the inputs to electrical equipment from other sources, such as remote control circuits, grounding circuits, etc. Moreover, the rescue of a victim shocked by the power input from a remote source is often hampered because of the time required to determine the source of power and turn it off. Therefore, turn off ALL power inputs before working on equipment.

Remember that the 120-V power supply voltage is not a low, relatively harmless voltage, but is the voltage that has caused more deaths than any other.

DO NOT work with high voltage circuits while alone; have another person (safety observer), who is qualified in first aid for electrical shock, present at all times. The person stationed nearby should also know the circuits and switches controlling the equipment, and should be given instructions to pull the switch immediately if anything unforeseen happens.

Always be aware of the nearness of high voltage lines or circuits. Use rubber gloves where applicable, and stand on approved rubber matting. Remember, not all so-called rubber mats are good insulators.

Items containing metal parts, such as brushes and brooms, should not be used in an area within 4 ft of high voltage circuits or any electric wiring having exposed surfaces.

Inform remote stations as to the circuit on which work is being performed.

Keep clothing, hands, and feet dry if at all possible. When it is necessary to work in wet or damp locations, use a dry platform or wooden stool to sit or stand on, and place a rubber mat or other nonconductive material on top of the wood. Use insulated tools and insulated flashlights of the molded type when required to work on exposed parts.

Do not wear loose or flapping clothing. The use of thin-soled shoes with metal plates or hobnails is prohibited. Safety shoes with nonconducting soles should be worn if available. Flammable articles, such as celluloid cap visors, should not be worn.

When working on electrical apparatus, electricians should first remove all rings, wristwatches, bracelets, ID chains and tags, and similar metal items. Care should be taken that the clothing does not contain exposed zippers, metal buttons, or any type of metal fastener.

DO NOT work on energized circuits unless absolutely necessary. When work is to be performed on deenergized equipment, be sure to take time to lock out (or block out) the switch and tag it. Locks for this purpose should be readily available; if a lock cannot be obtained, remove the fuse and tag it.

Use one hand when turning switches on or off. Keep the doors to switch and fuse boxes closed except when working inside or replacing fuses. Use a fuse puller to remove cartridge fuses after first making certain that the circuit is dead.

All supply switches or cutout switches from which power could possibly be fed should be secured in the OPEN (safety) position and tagged. The tag should read "THIS CIRCUIT WAS ORDERED OPEN FOR REPAIR AND SHALL NOT BE CLOSED EXCEPT BY DIRECT ORDER OF—" (the person making, or directly in charge of, repairs).

Never short out, tamper with, or block open an interlock switch, without giving full consideration to the consequences.

Keep clear of exposed equipment; when it is necessary to work on it, use one hand as much as possible.

Warning signs and suitable guards should be provided to prevent personnel from coming into accidental contact with high voltages.

Avoid reaching into enclosures except when absolutely necessary; when reaching into an enclosure, use rubber blankets to prevent accidental contact with the enclosure.

Do not use bare hands to handle hot parts. Use asbestos gloves if necessary. Use a shorting cable to discharge all high-voltage charges. Before a worker touches a capacitor or any part of a circuit that is known or likely to be connected to a capacitor (whether the circuit is deenergized or disconnected entirely), he should short circuit the terminals to make sure that the capacitor is completely discharged. Grounded shorting cables should be permanently attached to workbenches where electrical equipment using high voltages are regularly serviced.

Make certain that the equipment is properly grounded. Ground all test equipment to the equipment under test.

Turn off the power before connecting alligator clips to any circuit.

When measuring circuits over 300 V, do not hold the test prods.

N. Codes and Standards

One very good indication of the degree of technical and social advance in a country is the extent to which the various industries, such as the trade and professional fields, have adopted systems of regulatory codes and standards. This is necessary to provide a reasonable sense of order, convenience, and interchangeability. In a technically oriented country such as ours, the results would be chaos if there were no codes and standards. Any tradesman who has operated or done work in a country with few or no local codes and standards established will appreciate the convenience they offer.

What is the nature and purpose of codes and standards? Codes, generally speaking, are a set of minimum basic guidelines aimed at protecting health and safety. They are usually, but not always, established by agencies who are conversant in the field or trade involved. Adoption and enforcement are usually left up to regulatory bodies, such as councils, legislatures, etc.

Standards are usually established by agencies consisting of representatives from the trade, industry, or profession directly involved. The purpose is to simplify the design, manufacture, or performance of a product or service as much as possible, consistent with prudence and interchangeability, without unduly restricting competition.

We shall give here a brief listing of those codes and standards with which the plant electrician will most often be concerned.

NATIONAL ELECTRICAL CODE[R] (NEC)[R]

This is a publication produced by the National Fire Protection Association (NFPA) and is considered the "Bible" of the electrical trades. It is available in several forms, and is updated about every three years. A copy should be in every electrical shop or electrician's library. It is concerned primarily with installation of electrical equipment for personal safety and reduction of property hazards.

NATIONAL ELECTRICAL MANUFACTURERS ASSOCIATION (NEMA)

This organization has an on-going program for establishing standards for dimensions, performance data, testing procedures, and related criteria for most of the equipment required in the electrical industry.

UNDERWRITERS LABORATORIES (UL)

This organization consists of several laboratories throughout the country, equipped to perform exhaustive tests on performance and safety of manufactured items, including electrical equipment and materials.

AMERICAN NATIONAL STANDARDS INSTITUTE (ANSI)

ANSI representatives assemble, review, organize, and issue standards submitted by various agencies. Subjects covered are mostly basic components of all phases of engineering, design, construction, manufacturing, and installation.

LOCAL CODES

Practically all municipalities, countries, and states have adopted local codes covering electrical installations in their particular area to meet their own needs. For the most part these codes will adopt portions of the latest edition of the preceding codes, with additions to cover local conditions.

It is the responsibility of the electrical work-ers in the plant to acquaint themselves with the local codes and to keep up to date on all changes. These local codes are usually administered through the Building and Safety department, or similar office.

PERMITS

Most local municipalities require that a permit be obtained before any electrical work is performed in the plant or on the premises. There is usually a charge for this, as the finished work must be inspected and approved before it may be placed into operation. In many cases there may be a substantial fine levied on any work performed without a permit and inspection.

We strongly urge the electrical worker to ascertain the local requirements, and determine his own vulnerability should he be asked to perform electrical work without obtaining a permit. If the work is being done by an outside contractor, then it is usually the contractor's responsibility to obtain the permit and arrange for the final inspection. If the work is being done by the plant staff, then a properly signed work order may absolve the worker from any fines or damages.

Know your legal standing, is the best advice.

Of course, normal repair and maintenance are exempt in most areas from permit requirements, as long as any changes in equipment and wiring are not made.

Note: The various sources for the tables and other data given in this volume are continually updating their published material. Consequently, we caution the reader that the data presented herein should be used as a guide only, and for illustrative purposes. However, in most cases, in the absence of more recent data, the electrical worker will not be far wrong in using this book as a guide.

As a minimum, the electrical worker should also keep on hand a current copy of the NEC® code book. That, with any material published by the manufacturers of equipment in the plant, should be sufficient for the average maintenance problem.

CHAPTER 2
BASIC ELEMENTS

A. What Is Electricity?

Electricity is a concentrated, invisible force of extreme power and magnitude when looked at in the context of the uses found in the average industrial plant. It is true that some applications, notably in the field known as electronics, make use of very small amounts of electricity, some so minute that it is difficult to measure them with the average industrial instruments usually employed for measuring its flow and power. Such applications will not be covered in this course of study. There are many other very worthwhile books written on that subject, which the student may wish to digest after becoming thoroughly versed in the basics covered in this book.

To understand properly the motivating force producing electrical power, we must project ourselves into the basic building block of all material in our universe, the atom.

The atom is the smallest individual particle making up all elements or compounds from which our everyday materials are constructed. Any further breaking down of the atom will produce particles that bear no resemblance to the element of which the atom is a part.

The atom consists of a heavy nucleus containing one or more protons surrounded by revolving particles known as electrons. The heavier the atom, the more electrons there are whirling around the nucleus and the more protons there are in the nucleus. These electrons revolve around the nucleus in fixed orbits, and these orbits are limited in the number of electrons they will hold. See Fig. 2-1 for the simplest of all atoms, the hydrogen atom. There is one proton making up the nucleus, and one electron in its lone orbit revolving around the nucleus.

All other atoms have more than one proton in the nucleus, but the result is the same; the electrons whirl around the nucleus at fantastic speeds. These electrons are held in orbit in part by the electrical attraction caused by one of the basic rules of electricity which we will state here:

Like electrical charged particles repel each other.

Unlike electrical charged particles attract each other.

Also, as you can see from Fig. 2-1, the electrons whirling around in orbit are being acted upon by centrifugal force, which tends to pull the electrons out of orbit and the attraction of the positive (+) charge in the nucleus. It is the balance between these two forces that holds the electrons in orbit.

However, the electrons in the outer orbit may, under certain conditions, be removed from their orbit, and deposited on other atoms, as will be seen by example subsequently.

Referring again to Fig. 2-1, you will note that the proton in the nucleus has been given the

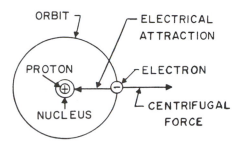

Fig. 2-1. Hydrogen atom.

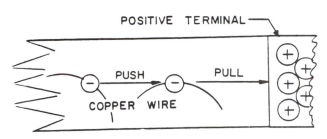

Fig. 2-2. Movement of electrons in a wire.

(+) or positive sign, and the electron has been assigned a (−) or negative sign. This is purely for convenience, and has been universally accepted as the proper designation for these particles. Thus, all electrons are considered to be negatively charged (−) and all protons are considered to be positively (+) charged.

Owing to the powerful attraction between the electrons (−) and the protons (+) wherever one of them migrates in the universe, the opposite charged particles tend to follow. As the protons are many times heavier than the electrons, the flow is always one way; the electrons (−) flow toward the protons (+). This is why you will see, in electrical diagrams, the flow of electricity from the negative (−) to the positive (+) terminal.

It is this powerful attraction between the electrons (−) and the protons (+) that produces the flow of electricity (electrons) in the wires and bus bars of the plant's electrical system and that produces the electrical power and motive forces with which so much of our work is done in our society.

The theory explaining the flow of electricity involves the electrons in the outer orbit of the conductor along which the electricity is expected to flow. We explained in Fig. 2-1 how the hydrogen atom is composed, but the normal conductors of electricity are more complex. However, it is only the electrons in the outer orbits that are considered to transmit electrical energy.

Referring to Fig. 2-2 we show the forces acting upon the electrons in the outer orbit of a conductor. Remembering that unlike charges attract and like charges repel, we see that each electron is acted upon by both a pull and a push, tending to remove the electron from its orbit, and propel it toward the positive terminal in the circuit. Each electron that leaves its orbit leaves an orbit behind hungry for another electron. Thus, electrons are passed from orbit to orbit along the conductor, and each transfer releases electrical energy, which is passed on to the positive terminal. This happens at a speed of 186,000 miles a second.

The best conductors of electricity have only one electron in the outer orbit, such as copper, silver, and gold.

We will not delve any further into the atomic elements or their orbits, but will leave you with this brief review. If you understand the material so far, it is all that will be needed for continuing this study material.

As you may have noted when reading the foregoing discussion on the make-up of electrical energy, the electrician is always working with one of the forces that holds our universe together. Remember that always, and you will learn to have a healthy respect for the power at your disposal.

B. Static Electricity

We stated in the previous discussion that electrons (−) have a very strong attraction for protons (+), and it is this attraction that produces the work which is performed when this force is harnessed properly. This work cannot be performed if the electrons are prohibited from traveling to the protons. In this case, we have a static condition, and there is a build up of electrons waiting for an opportunity to combine with the protons. This is known as "static electricity," in the true sense of the word. As the number of electrons accumulate, they attract an equal number of protons on the opposite side of the barrier between them. This accumulation of electrons is known as an electrostatic force, and it is measured in volts. This term will be explained and defined subsequently.

If this electrostatic force increases sufficiently to overcome the barrier between them, the electrons jump across the barrier and combine with the protons, often in a violent reaction, which can have disastrous effects. It is the duty of the electricians and the electrical engineers in our society properly to channel and direct the flow of these electrons to make them do useful work without harm to the environment, and especially to human beings.

Static electricity is more commonly known by its effect, to the individual who is shocked by touching a metal object under certain conditions, and by the effects that static electricity has upon our industrial processes.

How is this static electricity produced? To answer this we refer you to the previous section, wherein we explained that electrons (−) are charged particles whirling around the outer orbit of the atom. These electrons may under certain conditions be removed from those orbits and deposited in the outer orbits of other atoms. The common method by which this occurs is by friction. Some materials, which we wear, use in our homes, or are used in machinery and commerce, when rubbed together will cause some of these electrons to be displaced from their normal orbits and be deposited upon the surfaces of other materials. When this happens, they attract equal and opposite positive charges, and the two strive to combine, as explained previously.

If these two materials rubbing together are what is known as nonconductors, then the electrons and protons do not readily combine, and if this condition persists, there is a build up of both charges on the two materials, until, as was explained, the electrostatic force is sufficient to cause them to combine, often in a discharge of a spark, or a continuous stream of sparks.

This is what causes the lightning streaks in our skies when a storm is brewing, as lightning is nothing but a gigantic spark, caused by the negatively charged electrons in the clouds building up an equal positive charge on the ground. When the electrostatic force is high enough to overcome the resistance of the air, the electrons break down the air barrier and combine with the protons on the ground, with the release of enormous amounts of electrical energy. This energy appears in the form of heat, sufficient to cause the air to glow with a distinctive crackling noise.

Lightning is nothing more nor less than an enlarged version of a spark between two electrodes, or between two ends of a live wire in the home or in the plant, or between a downed electric high line and the ground.

C. Static Electricity in the Plant

We stated previously that static electricity is formed by friction between two materials. If this friction is a continuous process, as many processes in industrial plants are, there will be a continual release of electrons caused by this friction. If the two materials producing the friction are excellent conductors of electricity, then the electrons will combine with the protons as fast they are stripped from their orbits, and there will be little, if any, build up of an electrostatic force to cause trouble.

However, if one or both of the two rubbing materials are poor conductors of electricity, then the electrons can build up faster than they can combine with the protons, and soon a steady stream of sparks will be observed between the two rubbing materials. In some cases, this static electricity may not appear as sparking, but may show up in other forms, usually troublesome.

A typical example of the formation of static discharges in industry is shown in Fig. 2-3, in which a V-belt drive is running constantly over a cast iron or steel sheave at high speed. There is always some slippage in a V-belt drive, and this slippage causes friction. This friction in

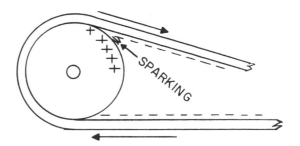

Fig. 2-3. V-belt sparking.

turn strips electrons off the atoms making up the belt and the sheave, leaving positively charged atoms behind. The fast moving belt carries these electrons away from the sheave, and they soon build up until there is a steady stream of sparks flowing between the belt and the sheave, just at the point where the belt leaves the sheave. This sparking may sometimes be observed if the room is dark. These sparks may, of course, cause explosive atmospheres to explode with disastrous effects.

The cure for this is to be sure that nonsparking belts are used, and be sure the machinery is properly wired to ground. This will be covered in more detail.

There are several common characteristics of static-causing materials that will help in understanding its behavior:

1. Friction is usually the cause of static charges being produced. This can be friction between metals, nonmetals, liquids, and between a liquid and other materials, or air to air, air to liquid, or air to metal.

2. The static charges build up on the surfaces of these materials, on the interfaces between them, or on finely divided particles. This is because of the strong attraction between the negatively (−) and positively (+) charged particles, which pulls them as close together as possible in their striving to combine.

3. If one of the surfaces upon which these charges collect is a nonconductor or a poor conductor of electricity, these charges may accumulate and increase in magnitude until there is danger of a spark being produced.

4. If, in their striving to combine, there is a thin film of substance, such as paper, cloth, or plastic between them, this material is almost certain to be pulled toward one of the charged surfaces.

5. If the motion causing the build up of static charges is stopped, the charges will attempt to find a path between them to effect a union. The better the conductor, the quicker the charges will neutralize.

We shall now give you some practical examples of each of the preceding characteristics to illustrate them better.

We have already given you one example of the first item in the list, in the discussion on the V-belt drive. Another example is the formation of lightning in the cloudy sky, where the air currents in motion through moisture-laden clouds produce a large build up of negatively charged particles on the underside of the clouds. This is an illustration of characteristic numbers 1, 2, and 3.

As an example of number 4, many of the readers may have noted that in plants where a high speed film of paper, roofing, or plastic sheet is rolling off a press or a series of forming rolls, that the film tends to hold on to the surface of the roll and appears to be reluctant to break away from it. This is the result of the static charges pulling the sheet down to the roll, unless the roll is applying a coating or liquid to the film, which acts as an adhesive between the roll and the film.

Characteristic number 5 above may best be illustrated by observing a gasoline tank truck when it comes into a loading or unloading station. It has been under motion for some time, and there has been a continual build up of static between the surface of the splashing gasoline and the surface of the tank. The well-known drag chain or cable helps to discharge the excess while the truck is in motion, but this is not always reliable. It is common practice for the attendants to wait for at least 1 min after the truck has stopped at the station, before attempting to attach the grounding connection or the filling or unloading apparatus. This wait is to allow the static charges to neutralize naturally, so there will be no possibility of the attendant providing the path for the static charges to effect a violent union, with the possibility of an explosion.

Now that you have a general idea of the behavior and causes of static electricity, what can you do to prevent trouble in the plant? We shall explore this subject next.

This is where the plant maintenance department can really show its imagination and ingenuity, if for no other reason than they are dealing with something that is invisible, and can only be detected by instruments, and the results that are highly visible, for such is the nature of electrical energy. We will describe only in general terms some of the more common methods you may consider in combatting the effects of static electricity in your plant, when it appears to cause you concern.

The first thing to remember is that, in order for you to eliminate the build up of static charges, they must be drained away as fast as they are formed, and this is best accomplished by providing a path to carry the negatively charged particles into the ground, where they may unite with the positively charged particles. In effect, you must provide a short circuit for the negatively charged particles.

One of the best ways to provide a short circuit to ground is to place a series of finely divided fingers of highly conductive metal, such as copper or aluminum, as close as possible to the point where the negatively charged particles are being accumulated. These conductors must then be connected to the ground, through water pipes, gas pipes, or iron or copper rods driven into the earth. If the machinery being treated is connected to ground already, then all that is necessary is to connect the static collectors to the frame of the machine.

Some of the typical forms of these collectors are fine copper wires, aluminum foil strips, or the common metallic tinsel rope used to decorate Christmas trees.

Where to attach these static collectors? Figures 2-4 through 2-8 will give you a few very practical ideas for the most common problem spots in the plant. But in the final analysis, it is the ingenuity of the plant personnel that really

pays the best dividends when attacking the problem of static electricity.

Let us now take a closer look at Figs. 2-4 through 2-8, for these illustrations cover enough of the common situations found in industrial plants to help you solve many of the problems you will encounter. They will serve to tickle your imagination so that you can visualize what is happening before your eyes when static electricity build-up plays havoc with a process or machine.

Figure 2-4 shows the principle behind the grounding connection on a tank truck being loaded with static-producing liquids, such as gasoline. The grounding leads are *always* connected up *first*, and disconnected *last*, to discharge any accumulation of static charges before the filling lines are hooked up.

In the case of top loading, the arms or hoses should touch the side of the filling port, and the fillng nozzle should extend down to within 4 in. of the bottom of the tank. All precautions should be taken to minimize the amount of splashing at the nozzle outlet, for it is the high-velocity splashing that produces the friction, which in turn produces static electricity.

The trend now is toward bottom filling, with the filling nozzle being of a special design that accomplishes the filling of the tank with a minimum of splashing, and, at the same time, the coupling produces an effective grounding connection with the tank. However, there is still a grounding and electrical supply connection to be hooked up to the truck, *before* making the

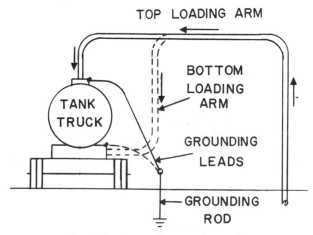

Fig. 2-4. Grounding tank trucks.

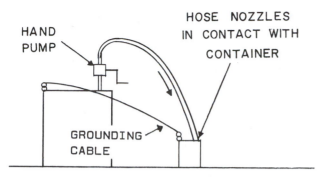

Fig. 2-5. Transferring inflammable liquids.

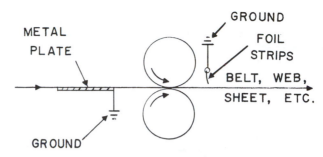

Fig. 2-7. Typical static problem solution.

hook up with the filling nozzle. This electrical plug is designed to serve double duty, supplying electrical power to operate the truck filling controls and to ground the tank at the same time.

Figure 2-5 explains the proper method of filling small portable cans or containers from drums or small tanks with hand pumps. The drawing gives the proper method when the two vessels are sitting on a nonconductive floor, such as concrete, wood, or brick. Here again, the grounding connection should be made *first*, and disconnected *last*.

There is a simpler method, possible only when the two containers are grounded through the floor they are sitting on, such as on steel floor plates, grating, etc. This can be risky if there is any doubt about the floor being conductive, or if either of the containers is not properly grounded. This can happen if either one of them has a heavy layer of paint or dirt on the lower or bottom edge, preventing a good connection through the floor. If in doubt at all, use the method shown in Fig. 2-5.

In any industry where an impression is printed upon paper stock, plastic film, or even sheet metal being passed through high speed rolls, you can be sure that static charges are being produced, and are probably causing a problem some place in the machine or process. Figure 2-6 is a typical example of a printing press arrangement and shows one of the methods of picking up the static charges as they are formed before they have a chance to cause trouble.

Figure 2-7 gives two more ideas for static charge pick-ups. The belt, film, or sheeting is passing at high speed over a flat metal plate, which is grounded. On the leaving side of the rolls, a bank of foil fingers is placed across the width of the stock, close enough to the surface to provide a good static pick-up. This bank of foil fingers is, in turn, grounded.

In Figs. 2-8 and 2-9 are shown two common methods of making static pick-up devices. In the first one, common Christmas tree tinsel tape is used, as long as it is the metallic type, and not plastic. The finely frazzled tinsel does a very good job of collecting the static charges and conducting them along the rope to the grounding connection. Of course, the tinsel

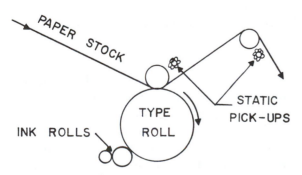

Fig. 2-6. Combatting static in a printing press.

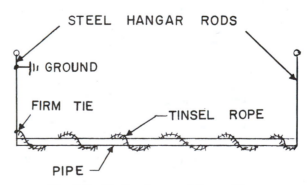

Fig. 2-8. Static charge pickup, #1.

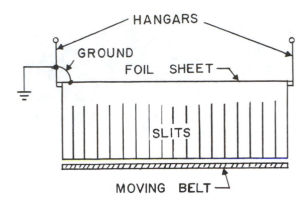

Fig. 2-9. Static charge pickup, #2.

rope has to be securely attached to the wire or steel support, which in turn is grounded.

Another method of providing an effective static pick-up is shown in Fig. 2-9, with aluminum foil being used to collect the charges and conducting them to the grounding connection. In place of the aluminum foil, many copper wire fingers may be used, depending on the possibility of damage to the film being swept by the fingers. It is not always possible to prevent them from touching the film or stock passing by under the fingers.

We have shown you how to combat static electricity, and have alerted you to some of the causes of it. If there is any doubt in your mind as to the presence of static charges existing in any portion of your plant's process, machinery, or atmosphere, there are several types of instruments on the market that will detect the presence of this trouble maker. Some of these instruments will even indicate the amount of the voltage with which you have to contend. These are covered in Chapter 10.

Some of the most common industrial processes which are plagued by this problem are

Printing, coating, or drying of paper, plastics, or textiles.

Pumping, mixing, or transferring of petroleum products or their by-products.

Grinding, transfer, and blending of grain dusts and metal powders, or any dusts which one finds accumulating in plants having very poor housekeeping. Such dusts are very difficult to neutralize.

Any spraying operation, such as paint spraying, air release into the atmosphere, or the free release of steam into the atmosphere; all may produce their own individual static problems.

Materials handling equipment are prime producers of static electricity, wherever there is material, belts, or sheaves in relative motion to each other.

The preceding items may all be summed up in one general statement: you will probably find static electricity in any operation where materials are in relative motion to each other.

This material concerning static electricity just touches the surface of the subject. There are many other sources of data and help available for your use. We suggest that you start by contacting; The American Petroleum Institute, 1220 L St., N.W., Washington D.C. 20005, Publications and Distribution Section and obtain the following publications which they produce:

API RP 2003 Protection Against Ignitions Arising Out of Static, Lightning and Stray Currents.

API Bulletin 1003 Precaustions Against Electrostatic Ignition During Loading of Tank Truck Motor Vehicles.

There will be a nominal cost for these publications, but it will be well worth it. While you are contacting them, it would be well to request an index and price list of all of their publications. You no doubt will find several others that will be of excellent use to you in your work.

D. Magnetism

A substance is said to be a magnet if it has the property of magnetism—that is, if it has the power to attract such substances as iron, steel, nickel, or cobalt, which are known as *magnetic materials.* A steel knitting needle, magnetized, exhibits two points of maximum attraction (one at each end) and no attraction at its center. The points of maximum attraction are called magnetic poles. All magnets have at least two poles. If the needle is suspended by its middle so that it rotates freely in a horizontal plane about its center, the needle comes to rest on an approximately north–south line of direction. The same pole will always point to the north, and the other will always point toward the south. The magnetic pole that points northward is called the *north pole*, and the other is the *south pole*.

A *magnetic field* exists around a simple bar magnet. The field consists of imaginary lines along which a *magnetic force* acts. These lines emanate from the north pole of the magnet, and enter the south pole, returning to the north pole through the magnet itself, thus forming closed loops.

A *magnetic circuit* is a complete path through which magnetic lines of force may be established under the influence of a magnetizing force. Most magnetic circuits are composed largely of magnetic materials in order to contain the magnetic flux. These circuits are similar to the *electric circuit*, which is a complete path through which current is caused to flow under the influence of the electromotive force. (More about this later.)

Magnets may be conveniently divided into three groups.

1. *Natural magnets*, found in the natural state in the form of a mineral called magnetite.

2. *Permanent magnets*, bars of hardened steel (or some form of alloy such as Alnico) that have been permanently magnetized.

3. *Electromagnets*, composed of soft iron cores around which are wound coils of insulated wire. When an electric current flows through the coil, the core becomes magnetized. When the current ceases to flow, the core loses most of its magnetism.

Permanent magnets and electromagnets are sometimes called *artificial magnets* to further distinguish them from natural magnets.

The space surrounding a magnet, in which the magnetic force acts, is called a *magnetic*

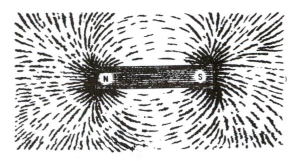

Fig. 2-10. Magnetic field pattern around a magnet.

field. Michael Faraday was the first scientist to visualize the magnetic field as being in a state of stress and consisting of uniformly distributed lines of force. The entire quantity of magnetic lines surrounding a magnet is called *magnetic flux*. Flux in a magnetic circuit corresponds to current in an electric circuit.

The terms *flux* and *flow* of magnetism are frequently used in textbooks. However, magnetism itself is not thought to be a stream of particles in motion but is simply a field of force exerted in space.

A visual representation of the magnetic field around a magnet can be obtained by placing a plate of glass over a magnet and sprinkling iron filings onto the glass. The filings arrange themselves in definite paths between the poles. This arrangement of the filings shows the pattern of the magnetic field around the magnet, as in Fig. 2-10.

The magnetic field surrounding a symmetrically shaped magnet has the following properties:

1. The field is symmetrical unless disturbed by another magnetic substance.

2. The lines of force have direction and are represented as emanating from the north pole and entering the south pole.

LAWS OF ATTRACTION AND REPULSION

If a magnetized needle is suspended near a bar magnet, as in Fig. 2-11, it will be seen that a north pole repels a north pole and a south pole repels a south pole. Opposite poles, however,

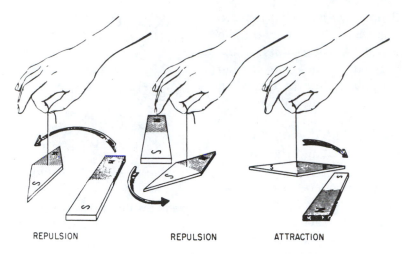

REPULSION   REPULSION   ATTRACTION

Fig. 2-11. Laws of attraction and repulsion.

will attract each other. Thus, the two laws of magnetic attraction and repulsion are:

1. Like magnetic poles repel each other.

2. Unlike magnetic poles attract each other.

The flux patterns between adjacent unlike poles of bar magnets, as indicated by lines, are shown in Fig. 2-12 (A). Similar patterns for adjacent like poles are shown in Fig. 2-12(B). The

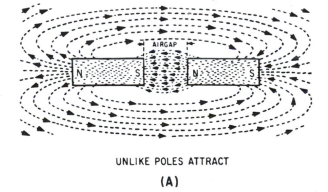

UNLIKE POLES ATTRACT

(A)

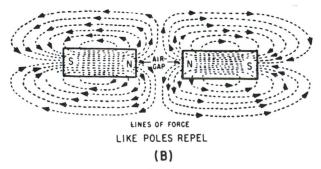

LINES OF FORCE
LIKE POLES REPEL

(B)

Fig. 2-12. Lines of force between unlike and like poles.

lines do not cross at any point, and they act as if they repel each other.

E. Electrical Units

As is necessary in any field of science, engineering, commerce, or trade, it is absolutely essential that a uniform system of terminology and measurement be adopted. This makes it much easier for one person to pass on his knowledge to another, for if everyone who worked at a trade used a different system of terms or measurements, it would be almost impossible for anyone else to gain anything from his experience. Electricity is no different, so we shall now give you the basic system of measurements used in electricity.

The usual procedure in the electrical trade is to compare the flow of electricity with the flow of water in a pipe, since most people are familiar with water flowing in pipes. This is the method we shall adopt, also.

For electricity to flow from one terminal to another, we have already explained that the flow must be from the negative terminal to the positive terminal. Once a path has been provided for the negative charges to flow to the positive charges and combine with them, the rate with which the flow takes place is determined partly by what is known as a "difference in potential" between the two terminals. The more negatively charged particles available at the negative terminal, the higher the difference in potential.

This may be compared to the flow in a water main under pressure. The flow of water between the point of high pressure to the point of lower pressure will depend on the difference between the two points. This leads to the first definition or comparison:

Hydraulic Term	Electrical Term Used	Symbol
Pressure difference	Potential difference	Voltage (E)

Thus, electricians use the term "voltage" to indicate the pressure, or force, available to push the electricity through the conductor from the negative to the positive side of the circuit. The quantity of force is the "volt," and when we speak of the flow of electricity, we use the term, "current," which is caused by the amount of voltage available.

The hydraulic equivalent of current flow is any expression, such as gallons per minute, which defines a rate of flow of water in a pipe. The equivalent terms in electrical jargon is "amperes," or usually abbreviated as "amps." Its symbol is "I."

Now if we multiply the voltage in a circuit by the amperes flowing in the circuit, we get the electrical power available to do work, and this is called "watts," after one of the early pioneers in the field of electricity. Put into equation form, we now have the following:

Electrical power
$$\text{in watts} = \text{Volts} \times \text{Amps} = \text{watts}$$
$$P = E \times I \qquad = \text{watts}$$

While we are on this subject, you should also be aware of the following relationships:

$$1 \text{ kilowatt} = 1000 \text{ watts}$$
(abbreviated as "kW")
$$1 \text{ megawatt} = 1,000,000 \text{ watts} = 1000 \text{ kW}$$
(abbreviated as "MW)

When water flows in a pipe, its flow is resisted by friction between the water and the surface of the pipe. When electricity flows along a conductor or path, it is also opposed by internal friction, and this is termed, appropriately, "resistance." Its symbol is "R," and it is

measured in "ohms," after another early pioneer in the field of electricity.

The relationships between the above terms, quantities and symbols will now be tabulated for your easy reference:

$$E = IR, I = \frac{E}{R}, R = \frac{E}{I}, P = EI, P = I^2R$$

Thus, if any two of the three quantities are known, you can easily calculate the remaining unknown quantity. For example, take a common 120 volt light bulb rated at 60 watts, then $P = 60$ watts and $E = 120$ V, therefore

$$I = \frac{P}{E} = \frac{60}{120} = 0.5 \text{ amps}$$

$$\text{Resistance} = R = \frac{E}{I} = \frac{120}{0.5} = 240 \text{ ohms}$$

So we see that a 60 watt light bulb uses $\frac{1}{2}$ amp of electricity, and it has an internal resistance of 240 ohms.

As a convenient aid to the use of the basic equations that you will be using, we refer you to Fig. 2-13, which is a graphic display of Ohm's law. The entire circle is divided into quadrants, each quadrant containing the three methods of calculating the item shown in the center circle. Thus, current, I, may be determined by

$$I = \frac{E}{R} = \frac{W}{E} = \sqrt{\frac{W}{R}}$$

and likewise with watts (W), resistance (R), and volts (E).

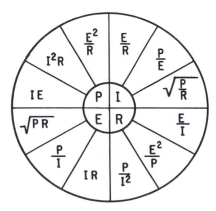

Fig. 2-13. Graphic display of Ohm's law.

There is another term used to denote power, and that is kilovolt-amp, or kVA as it is usually designated, and it is equal to 1000 volt-amps. The equations for determining the kVA of any flow of electricity or rating of an item of electrical equipment is

$$\text{Single-phase kVA} = \frac{E \times I}{1000}$$

$$\text{Three-phase kVA} = \frac{E \times I \times 1.73}{1000}$$

That wasn't too hard, was it? Study this section carefully, since the preceding equations are the basic ones you will be using in your daily work. They should become second nature to you, so that in a very short time you will not have to refer to this text to refresh your memory.

F. Basic Circuits

Now that we have laid the groundwork for the subject, we shall proceed to more practical matters. The first matter to come under our surveilance is a simple, basic circuit.

However, we should here define what is meant by a circuit. An electric circuit is a complete system by which electrical energy is put to useful work, and it consists of the following elements as a minimum:

1. Source of electrical power.

2. Means of conveying the electrical power throughout the system.

3. The load, or useful work, for which the system is designed.

4. Means of controlling the flow of electrical power.

Figure 2-14 illustrates the preceding elements in one of the simplest forms, and we shall explore each element in turn, since you will be working with each of them in one form or another throughout your daily chores in the industrial plant.

You will note that the system in Fig. 2-14 has a source, a supply conductor, a switch, a load, and a return conductor, which is usually, but not always, the same material as the supply conductor.

Starting with the source, we have many choices, but in practical aspects, there are only a very few in common use. The figure uses the symbol for a battery, which may be used for any low-powered system completely self-contained, entirely independent of the plant's power system. There are many styles of batteries available today, and we will discuss some of them subsequently.

In addition, the source may be an electrical generator driven by a diesel engine or similar engine, or by a steam turbine. The source could also be, in this case, a device known as a rectifier. A rectifier converts the alternating current of the plant's electrical system to direct current, similar to the power produced by the storage battery in this circuit.

The terms we are using will be defined and explained, so we ask your indulgence for now.

Some of the lesser known and little used sources of electricity are:

Friction machines

Solar generators

Wind-powered generators

Sound powered generators

Thermoelectric generators

Pressure power generators

Some of these sources have more sophisticated titles, but we will not explore them, since there is a very small possibility that you will see them in your plant.

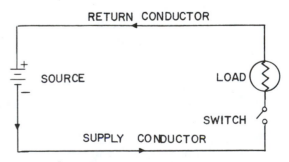

Fig. 2-14. Basic DC circuit.

The supply and return conductors are usually of copper, in solid wire form, or in cable made up of many fine copper wires twisted together in a flexible protective cover. There are other materials sometimes used as conductors of electricity, and we shall cover them all. For now suffice it to say that the conductors shown have been properly chosen and sized for the application.

The flow of electrical power is controlled by a switch ahead of the load. This switch is merely a device for allowing the passage of electricity when it is connected at both ends by the ends of the conductor, or for stopping the flow of electricity when, like an open door, it swings out and disconnects the ends of the conductor. We say that it is open when the path is broken, and it is closed when the path is completed.

There are hundreds of different styles, sizes, and forms of switches available today, and we shall cover some of them in due time. They may be hand operated, timer operated, temperature operated, pressure operated, machinery operated, or electrical solenoid operated. The result is the same, however; they control the flow of electrical power in an electrical circuit.

The load here is indicated as a common light bulb, but it also may take several forms. They all have one thing in common; they all require the supply of electrical power to perform the work for which they were designed. Also, they have other internal electrical characteristics, one of which is a resistance to the flow of electricity.

The load may not be a light bulb, but may be a bell, a solenoid for operating another switch, a small electric motor, or any number of devices requiring electrical power to function.

We are going to ask you to study this circuit in detail with us, since it contains the basic elements that you must be able to recognize in your daily work with electrical equipment in the plant.

First, we point out that this is what is known as a direct current system, since the electricity (or current) flows in only one direction, from the negative pole or terminal of the battery, around the entire circuit to the positive pole or terminal. The symbol shown for the source is that of a battery, with the terminals so marked as (−) and (+). As we stated earlier, the source may be the (−) terminal of a rectifier, or other source of direct current.

We will at this point explain that direct current is usually abbreviated simply by the letters "DC" and that is the term we will use from now on.

There is another basic principle illustrated here, which is that for electricity to flow in a circuit and do useful work, there must be a complete path for it from the (−) terminal, to and through the load, and back to the (+) terminal. In short, the circuit must be "closed." Anything that interrupts this continuous path will cause the flow of electricity to stop, and no work will be performed. However, there may be, at times and under certain conditions, an exception to this rule, which we shall now explore.

In Fig. 2-15, we illustrate what is commonly called a "short circuit." This is simply a condition in which electricity is by-passed around the load, and to the (+) terminal. It then has a path to take a short-cut in its route from the (−) terminal to the (+) terminal. This short circuit path may be accidental, which is the common situation when we call it a "short circuit," or it may be deliberate.

Usually we term it a short circuit when it is accidental, and is not wanted in our system. However, there are times when it is deliberately imposed in the circuit in a controlled

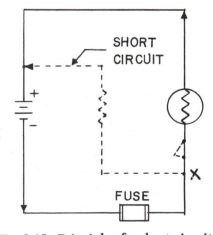

Fig. 2-15. Principle of a short circuit.

manner to moderate or vary the flow of power to the load.

You will note that the short circuit path is shown with a zig-zag symbol for a resistance. If the resistance through the short circuit is exceedingly small relative to the resistance through the load, we may have what is called a "dead short," or a "short to ground," both of which are dangerous in most cases where electric power of any magnitude is involved. This condition is brought about by damaged insulation or covering on the wires, the wire becoming pinched between some grounded metal objects, or the covering burning off due to too much electricity flowing through the wire. This is one of the most common causes of electrical fires in homes in our country.

Note now that this by-pass circuit may be used to control the flow of electrical current to the load, in this manner. When the current reaches the junction marked "X," it can flow in two directions. It can flow through the load, which has a certain resistance to its flow, or it can flow through the by-pass circuit, which also has a resistance to its flow. Naturally, it will split up and part of it will flow through the load, and part will flow through the by-pass. Furthermore, it can be seen readily that the split will depend on the relative resistances between the two paths. For example, if the resistance through the load is twice that of the resistance through the by-pass, it can be seen easily that the flow through the by-pass is going to be twice that through the load. This is as far as we shall take this line of development, since we bring it up only to show one of the basic principles of flow of electricity through a system.

You may have noted that there is another element in this circuit in Fig. 2-15, which we have not mentioned as yet. We have shown the symbol for a fuse placed in the supply conductor. The fuse is there simply to protect the circuit from being damaged from the uncontrolled passage of too much electricity. It has a known current-carrying ability, and when that point is reached, it melts, thus opening the circuit and stopping the flow of current into the system. Without it, when a short circuit develops, the rush of current around the load from the (−) terminal to the (+) terminal could cause the source to fail. If it is a battery, it will discharge completely and go dead. If it is a rectifier, it will either blow an internal fuse or burn up. This uncontrolled flow of current could also cause the wires to overheat, damaging the covering, and possibly starting a fire.

As we stated earlier, the principles involved in this section will be explored in more detail in this book, in one form or another. If you have followed the discussion so far, you will have no trouble in recognizing these principles as they occur. The remainder of this book is almost exclusively devoted to variations of what has been covered up to this point.

Figure 2-16 carries this discussion one step further. This illustration is here to acquaint you with the fact that electricity does not require copper wire to complete its path. Note that there are many other materials which under the right conditions will serve as a conductor, and allow the circuit to be completed. Under some applications, the return path to the positive terminal is deliberately routed through, for instance, railroad rails, water pipes, or the ground. A good example of using

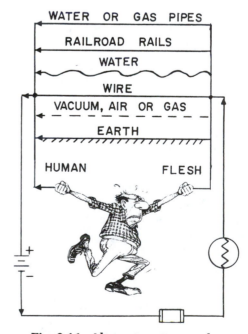

Fig. 2-16. Alternate return paths.

rails for the return conductor is the miniature electric train, and its full size running mate, the electric train or streetcar.

Another purpose of this illustration is to make you aware of the various possibilities which exist to produce a short circuit. Any one of the paths shown may short out the circuit, and produce trouble. For this to occur, it is not necessary that the inlet connection to the short circuit be ahead of the load, for when connected as shown in Fig. 2-16, you can still get a very dangerous jolt if you are the human flesh shown in the picture! In the case of the DC circuit shown here, the load will reduce the jolt somewhat, but don't rely on it! If this were a typical alternating current (AC) circuit, it would make no difference whether the inlet to the short circuit was ahead of or downstream from the load, for reasons which we shall explain later.

Of course, not all of the paths shown have the same ability to conduct an electric current. The particular conditions existing at the time of the short circuit connection will determine the effectiveness of the short circuit. For instance, the ground's ability to conduct an electric current will depend greatly upon its moisture content, as does also that of air or gas. The current-carrying ability of water will vary with its mineral content, such as dissolved salts. The matter of human flesh providing the short circuit path is covered in Chapters 1 and 11.

We have covered some of the various materials through which an electric current may be conducted, although in a very brief manner. We shall now take up the more common materials, notably copper and aluminum, and go much deeper into their use, since they are the ones you will be using most of the time in your plant.

C H A P T E R 3
CONDUCTORS

A. Electrical Pathways

In the last chapter we mentioned that electricity must have a path along which it travels, and we called this path CONDUCTOR. As you might expect, these paths have considerable differences in their ability to conduct an electric current. This was touched upon during the discussions on basic circuits, and we shall expand on this subject now.

Table 3-1 will give you a general idea of the variations in this current carrying ability for a number of the more common materials. The numbers are merely orders of magnitude, and do not represent any finite values. They show the ranking for the different materials, that is all.

What Table 3-1 tells us is that copper has 2030 times the ability to conduct electricity than carbon. To compare one metal with another, use the following example as your guide.

To compare copper with aluminum in determining relative conductance (RC) of aluminum to copper, thus:

$$\frac{RC(copper)}{RC(alum.)} = \frac{2030}{1277} = 1.59.$$

Which means that copper has 1.59 times the current carrying capacity of aluminum.

You can do the same with any material listed in the table. For example, compare platinum with aluminum:

$$\frac{RC(plat.)}{RC(alum.)} = \frac{350}{1277} = .275.$$

Which means that platinum has only 27.5% of the current carrying capacity of aluminum.

B. A Common Problem

To help in understanding what we are attempting to explain in this book, we shall at this time set up a simple, hypothetical problem, one which is often encountered in a plant undergoing alterations or expansion. We shall then proceed to help you solve this problem, step by step. Please bear in mind that it will be somewhat simplified, and there may be other factors to keep in mind should you be faced with a similar problem in your plant.

First, we shall list the approximate procedure to be followed for the project, as a guide only. It may be altered to suit your conditions.

1. Make a sketch of the proposed change, including routing, distances, major obstacles, equipment required, and any special notes, such as temperatures, etc.

2. Check the **NEC**® book, and any other applicable codes for compliance. List all pages and sections governing the final selection of the materials and equipment.

Table 3-1. Electrical Characteristics of Various Common Materials

Metal	Relative Conductance	Relative Resistance
Silver	2207	1.0
Copper	2030	1.09
Gold	1471	1.50
Aluminum	1277	1.73
Tungsten	634	3.48
Zinc	561	3.93
Brass	461	4.79
Platinum	350	6.31
Iron	304	7.26
Nickel	263	8.39
Tin	248	8.90
Steel	235	9.39
Lead	159	13.88
Mercury	37	59.65
Nichrome	34	64.91
Carbon	1	2207

3. Make a list of all equipment and materials required, including quantity, catalog numbers, source, price, etc.

4. Obtain all necessary permits and authorization from in-house management or local governing bodies, such as Building and Safety departments, etc.

5. Place orders for all materials not on hand, obtaining the approximate delivery dates. Tag and reserve all materials available in plant inventory.

6. Schedule the work, including dates, times, manpower required, special conditions to be established, etc.

7. Perform the work as close as possible to the original plan. Record any deviations necessary not foreseen in the planning stage, both for plant records and for clearance by the inspector.

8. Check out all circuits, using proper instruments and procedures, for continuity, shorts, and other abnormalities, before putting into service. Make all necessary corrections.

9. Arrange for final inspection, and go through the system with the inspector, if he requests it.

10. Make any changes required to meet inspection.

11. When cleared by the inspector, activate the system and perform all running tests, checking results against any manufacturer's data available.

12. Make all necessary entries into the plant records. This includes final sketch of system, running data, and any file material mentioned in Chapter 18.

It is desired to install a new piece of equipment in the plant, having in its operation a new 50 horsepower electric motor, standard torque, 440 volt, 3 phase, 60 cycle service required.

The motor will be wired from an existing motor control center (MCC), in which there is space available for the new switchgear required.

The routing for the wiring and conduit covers a total distance of 225 ft. from the MCC to the motor, and the area through which it passes is of a fairly normal temperature, seldom over 125° F. From page 310 in the Appendix we find that this motor will probably draw about 63 amps of current.

It is required that the switchgear be determined for this problem, the size and type of conductor wire, the size and type of conduit, and the size and style of switchgear at the motor location; as well as all the various connections, intermediate boxes, and miscellaneous other items necessary to meet all applicable codes.

As we progress through this book, the various portions of this problem will be attacked, so that when you are through, you will have a good idea of what is involved.

C. Circular Mil

To compare the resistance and size of one conductor with that of another, a standard or unit size of conductor must be established. A convenient unit of wire measurement, as far as the diameter of a piece of wire is concerned, is the mil (0.001 of an inch); and a convenient unit of wire length is the foot. The standard unit of size in most cases is the MIL-FOOT; that is, a wire will have unit size if it has a diameter of 1 mil and a length of 1 foot.

The circular mil is the standard unit of wire cross-sectional area used in American and English wire tables. Because the diameters of round conductors, or wires, used to conduct electricity may be only a small fraction of an inch, it is convenient to express these diameters in mils, to avoid the use of decimals. For example, the diameter of a wire is expressed as 25 mils instead of 0.025 inch. A circular mil is the area of a circle having a diameter of 1 mil, as shown in Fig. 3-1. The area in circular mils of a round conductor is obtained by squaring the diameter measured in mils. Thus, a wire having a diameter of 25 mils has an area of 25^2 or 625 circular mils.

In comparing square and round conductors, it should be noted that the circular mil is a smaller unit of area than the square mil, and therefore there are more circular mils than square mils in any given area. The comparison is shown in Fig. 3-1. The area of a circular mil is equal to 0.7854 of a square mil. Therefore, to determine the circular-mil area when the square-mil area is given, divide the area in square mils by 0.7854. Conversely, to determine the square-mil area when the circular-mil

area is given, multiply the area in circular mils by 0.7854.

For example, a No. 12 wire has a diameter of 80.81 mils. What is (1) its area in circular mils and (2) its area in square mils?
Solution:

(1) $A = D^2 = 80.81^2 = 6,530$ circular mils
(2) $A = 0.785 \times 6,530 = 5,126$ square mils.

A rectangular conductor is 1.5 inches wide and 0.25 inch thick. (1) What is its area in square mils? (2) What size of round conductor in circular mils is necessary to carry the same current as the rectangular bar?
Solution:

(1) $1.5'' = 1.5 \times 1,000 = 1,500$ mils
$0.25'' = 0.25 \times 1,000 = 250$ mils
$A = 1,500 \times 250 = 375,000$ sq. mils.

(2) To carry the same current, the cross-sectional area of the rectangular bar and the cross-sectional area of the round conductor must be equal. There are more circular mils than square mils in this area as, therefore,

$$A = \frac{375,000}{0.7854} = 477,464 \text{ circular mils.}$$

A wire in its usual form is a slender rod or filament of drawn metal. In large sizes, wire becomes difficult to handle, and its flexibility is increased by stranding. The strands are usually single wires twisted together in sufficient numbers to make up the necessary cross-sectional area of the cable. The total area in circular mils is determined by multiplying the area of one strand in circular mils by the number of strands in the cable.

A circular-mil foot, as shown in Fig. 3-2, is actually a unit of volume. It is a unit conductor 1 foot in length and having a cross-sectional area of 1 circular mil. Because it is considered a unit conductor, the circular-mil-foot is useful in making comparisons between wires that are made of different metals. For example, a basis of comparison of the RESISTIVITY (to be treated later) of various substances may be made by determining the resistance of a circular-mil-foot of each of the substances.

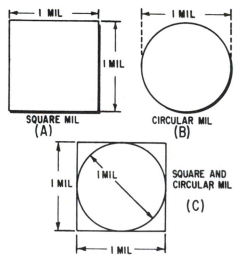

Fig. 3-1. Basic wire measurements. (A) Square mil. (B) Circular mil. (C) Square and circular mil.

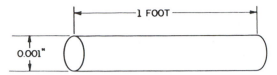

Fig. 3-2. Circular mil-foot.

D. Specific Resistance (SR)

Specific resistance (SR), or resistivity, is the resistance in ohms offered by one circular-mil-foot of a substance to the flow of electric current. Resistivity is the reciprocal of conductivity. A substance that has a high resistivity will have a low conductivity, and vice versa.

Thus, the specific resistance of a substance is the resistance in ohms of a volume of the substance 1 foot long and 1 circular mil in cross-sectional area. The temperature at which the resistance measurement is made is also specified. If the kind of metal of which a conductor is made is known, the specific resistance of the metal may be obtained from a table. The specific resistances of some common substances are given in Table 3-2.

Table 3-2. Specific Resistance

Substance	Specific resistance at 20°C Circular-mil-foot (ohms)
Silver	9.8
Copper (drawn)	10.37
Gold	14.7
Aluminum	17.02
Tungsten	33.2
Brass	42.1
Steel (soft)	95.8
Nichrome	660.0

The resistance of a conductor of uniform cross section varies directly as the product of the length and the specific resistance of the conductor, and inversely as the cross-sectional area of the conductor. Therefore, the resistance of a conductor may be calculated if the length, cross-sectional area, and specific resistance of the substance are known. Expressed as an equation, the resistance R, in ohms, of a conductor is

$$R = SR \times \frac{L}{A}$$

where (SR) is the specific resistance in ohms per circular-mil-foot, L the length in feet, and A the cross-sectional area in circular mils.

For example, what is the resistance of 1,000 feet of copper wire having a cross-sectional area of 10,400 circular mils (No. 10 wire), the wire temperature being 20° C?

Solution:

The specific resistance, from Table 3-2, is 10.37. Substituting the known values in the preceding equation, the resistance, R, is determined as

$$R = SR \times \frac{L}{A} = 10.37 \times \frac{1,000}{10,400} = 1 \text{ ohm,}$$
$$\text{approximately.}$$

If R, SR, and A are known, the length may be determined by a simple mathematical transposition. This is of value in many applications. For example, when it is desired to locate a short to ground in a telephone line, special test equipment is used that operates on the principle that the resistance of a line varies directly with its length. Thus, the distance between the test point and the fault can be computed accordingly.

Conductance (G) is the reciprocal of resistance. When R is in ohms, the conductance is expressed in mhos. Where resistance is opposition to flow, conductance is the ease with which the current flows. Conductance in mhos is equivalent to the number of amperes flowing in a conductor per volt of applied emf. Expressed in terms of the specific resistance, length, and cross section of a conductor

$$G = \frac{A}{(SR)L}.$$

When A is in circular mils, SR is in ohms per circular-mil-foot, L is in feet, and G is in mhos. The relative conductance of several substances is given in Table 3-1. Table 3-1 also gives the Relative Resistance of the same substances. Here, again, the figures are purely relative to each other, and have no definite units or values assigned to them. They are for your use in the

Table 3-3. Standard Annealed Solid Copper Wire Characteristics

| Size AWG/MCM | Area Cir. Mils | Conductors | | | | DC Resistance at 75°C (167°F) | | |
| | | Stranding | | Overall | | Copper | | Aluminum |
		Quantity	Diam. In.	Diam. In.	Area In.²	Uncoated ohm/MFT	Coated ohm/MFT	ohm/MFT
18	1620	1	—	0.040	0.001	7.77	8.08	12.8
18	1620	7	0.015	0.046	0.002	7.95	8.45	13.1
16	2580	1	—	0.051	0.002	4.89	5.08	8.05
16	2580	7	0.019	0.058	0.003	4.99	5.29	8.21
14	4110	1	—	0.064	0.003	3.07	3.19	5.06
14	4110	7	0.024	0.073	0.004	3.14	3.26	5.17
12	6530	1	—	0.081	0.005	1.93	2.01	3.18
12	6530	7	0.030	0.092	0.006	1.98	2.05	3.25
10	10380	1	—	0.102	0.008	1.21	1.26	2.00
10	10380	7	0.038	0.116	0.011	1.24	1.29	2.04
8	16510	1	—	0.128	0.013	0.764	0.786	1.26
8	16510	7	0.049	0.146	0.017	0.778	0.809	1.28
6	26240	7	0.061	0.184	0.027	0.491	0.510	0.808
4	41740	7	0.077	0.232	0.042	0.308	0.321	0.508
3	52620	7	0.087	0.260	0.053	0.245	0.254	0.403
2	66360	7	0.097	0.292	0.067	0.194	0.201	0.319
1	83690	19	0.066	0.332	0.087	0.154	0.160	0.253
1/0	105600	19	0.074	0.373	0.109	0.122	0.127	0.201
2/0	133100	19	0.084	0.419	0.138	0.0967	0.101	0.159
3/0	167800	19	0.094	0.470	0.173	0.0766	0.0797	0.126
4/0	211600	19	0.106	0.528	0.219	0.0608	0.0626	0.100
250	—	37	0.082	0.575	0.260	0.0515	0.0535	0.0847
300	—	37	0.090	0.630	0.312	0.0429	0.0446	0.0707
350	—	37	0.097	0.681	0.364	0.0367	0.0382	0.0605
400	—	37	0.104	0.728	0.416	0.0321	0.0331	0.0529
500	—	37	0.116	0.813	0.519	0.0258	0.0265	0.0424
600	—	61	0.992	0.893	0.626	0.0214	0.0223	0.0353
700	—	61	0.107	0.964	0.730	0.0184	0.0189	0.0303
750	—	61	0.111	0.998	0.782	0.0171	0.0176	0.0282
800	—	61	0.114	1.03	0.834	0.0161	0.0166	0.0265
900	—	61	0.122	1.09	0.940	0.0143	0.0147	0.0235
1000	—	61	0.128	1.15	1.04	0.0129	0.0132	0.0212
1250	—	91	0.117	1.29	1.30	0.0103	0.0106	0.0169
1500	—	91	0.128	1.41	1.57	0.00858	0.00883	0.0141
1750	—	127	0.117	1.52	1.83	0.00735	0.00756	0.0121
2000	—	127	0.126	1.63	2.09	0.00643	0.00662	0.0106

These resistance values are valid ONLY for the parameters as given. Using conductors having coated strands, different stranding type, and especially, other temperatures, change the resistance.

Formula for temperature change: $R_2 = R_1 [1+\alpha(T_2-75)]$ where: $\alpha_{cu} = 0.00323$, $\alpha_{AL} = 0.00330$.

Conductors with compact and compressed stranding have about 9 percent and 3 percent, respectively, smaller bare conductor diameters than those shown. See Table 5A for actual compact cable dimensions.

The IACS conductivities used: bare copper = 100%, aluminum = 61%.

Class B stranding is listed as well as solid for some sizes. Its overall diameter and area is that of its circumscribing circle.

(FPN): The construction information is per NEMA WC8-1976 (Rev 5-1980). The resistance is calculated per National Bureau of Standards Handbook 100, dated 1966, and Handbook 109, dated 1972.

same manner as the Relative Conductance values.

E. Wire Measure

Wires are manufactured in sizes numbered according to a table known as the American Wire Gage (AWG). As may be seen in Table 3-3, the wire diameters become smaller as the gage numbers become larger. Larger and smaller sizes are manufactured but are not commonly used.

A wire gage is shown in Fig. 3-3. It will measure wires ranging in size from number 0 to number 36. The wire whose size is to be measured is inserted in the smallest slot that will just accommodate the bare wire. The gage number corresponding to that slot indicates the wire size. The slot has parallel sides and should not be confused with the semicircular opening at the end of the slot. The opening simply permits the free movement of the wire all the way through the slot. The measurement is taken across the parallel sides of the entrance slots.

F. Wire and Cable

A WIRE is a slender rod or filament of drawn metal. If a wire is covered with insulation, it is properly called an insulated wire. Although the term "wire" properly refers to the metal, it is generally understood to include the insulation.

A CONDUCTOR is a wire or combination of wires not insulated from one another, suitable for carrying an electric current.

A STRANDED CONDUCTOR is a conductor composed of a group of wires or of any combination of groups of wires. The wires in a stranded conductor are usually twisted together.

A CABLE is either a stranded conductor (single-conductor cable) or a combination of conductors insulated from one another (single-conductor cable). The term cable is a general one, and in practice it is usually applied only to the larger sizes of conductors. A small cable is more often called a stranded wire or cord. Cables may be bare or insulated. The insulated cables may be sheathed (covered) with lead or protective armor.

Figure 3-4 shows some of the different types of wire and cables used in industry.

Conductors are stranded mainly to increase their flexibility. The overall flexibility may be increased by further stranding of the individual strands.

Figure 3-5 shows a typical 37-strand cable. It also shows how the total circular-mil-cross-sectional area of a stranded cable is determined.

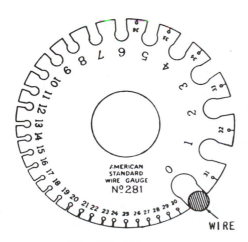

Fig. 3-3. Wire gage.

G. Wire Sizing

Several factors must be considered in selecting the size of wire to be used for transmitting and distributing electric power.

One factor is the allowable power loss (I^2R loss) in the line. This loss represents electrical energy converted into heat. The use of large conductors will reduce the resistance and therefore the I^2R loss. However, large conductors are more expensive initially than small ones; they are heavier and require more substantial supports.

A second factor is the permissible voltage drop (IR drop) in the line. If the source maintains a constant voltage at the input to the line, any variation in the load on the line will cause

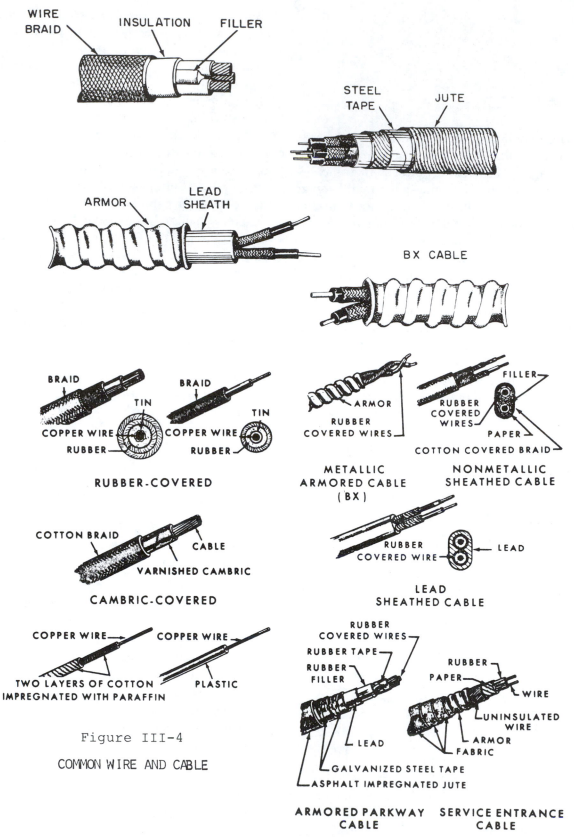

Figure III-4

COMMON WIRE AND CABLE

Fig. 3-4. Common wire and cable.

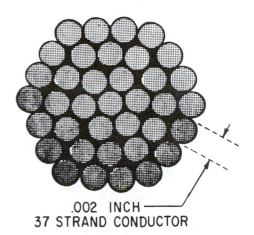

.002 INCH
37 STRAND CONDUCTOR

DIAMETER OF EACH STRAND = .002 INCH
DIAMETER OF EACH STRAND, MILS = 2 MILS
CIRCULAR MIL AREA OF
EACH STRAND = D^2 = 4 CM
TOTAL CM AREA OF
CONDUCTOR = 4 x 37 = 148 CM

Fig. 3-5.

a variation in line current, and a consequent variation in the IR drop in the line. A wide variation in the IR drop in the line causes poor voltage regulation at the load. The obvious remedy is to reduce either I or R. A reduction in load current lowers the amount of power being transmitted, whereas a reduction in line resistance increases the size and weight of conductors required. A compromise is generally reached whereby the voltage variation at the load is within tolerable limits and the weight of line conductors is not excessive.

A third factor is the current-carrying ability of the line. When current is drawn through the line, heat is generated. The temperature of the line will rise until the heat radiated, or otherwise dissipated, is equal to the heat generated by the passage of current through the line. If the conductor is insulated, the heat generated in the conductor is not so readily removed as it would be if the conductor were not insulated. Thus, to protect the insulation from too much heat, the current through the conductor must be maintained below a certain value. Rubber insulation will begin to deteriorate at relatively low temperatures. Varnished cloth insulation retains its insulating properties at higher temperatures; and other insulation—for example, asbestos or silicon—is effective at still higher temperatures.

Electrical conductors may be installed in locations where ambient (surrounding) temperature is relatively high; in which case the heat generated by external sources constitutes an appreciable part of the total conductor heating. Allowance must be made for the influence of external heating on the allowable conductor current, and each case has its own specific limitations. The maximum allowable operating temperature of insulated conductors is specified in tables and varies with the type of conductor insulation being used.

Tables have been prepared giving the safe current ratings for various sizes and types of conductors covered with various types of insulation. For example, the allowable current-carrying capacities of copper conductors at various ambient temperatures and with various types of insulation are given on page 305 in the Appendix. The required conduit sizes are also given.

The best way to explain the line sizing method is by the example, in Section B, which we shall now do.

In the table on page 310 in the Appendix, we found that a 50 hp motor will require a running current of about 63 amps, for a 440 volt supply. Now in the table on page 305, at the bottom of the page, last note, we are instructed to size the wire for 125% of the current draw. This calculates out at 78.75 amps (63 × 1.25 = 78.75), but we shall assume 85 amps.

In the same table, under the first wire sizing column, opposite 85 amps, we read that #4 wire is required for type TW and THWN wire. See Section H for a description of the various common types of insulated wire. Directly beside it, in the next column, we find that a 1″ conduit is required to handle the three wires for the three-phase motor supply. The choice of conduit material is narrowed down to only two—aluminum or steel—and we would recommend the steel as being the safest, easiest, and the most damage resistant. The fittings for it are readily available and easy to install.

We would not use plastic for exposed conduit, as it is too easily damaged and subject to fire damage. It is used for underground service to a great extent, where it can be easily protected. Also, plastic conduit cannot be used as an electric ground, whereas metal conduit is often used for this purpose. When plastic conduit is used, a separate grounding wire must be run, covered in green insulation, along with the other wiring in the conduit.

We must now check for voltage loss and power loss to see if the tabulated selection is correct for our application.

Table 3–3 tells us that at 167 degrees F, the size 3 wire has a resistance to the flow of current equal to .254 ohms per 1,000 feet. However, our wiring is 225 feet long, so to find the total resistance we would use the following formula:

$$\text{Total Resistance} = \frac{R/1000' \times \text{Length}}{1000}$$

$$\text{Total Resistance} = \frac{0.254 \times 225}{1000}$$

$$= 0.057 \text{ ohm for 225 feet.}$$

The next step is to determine what the voltage drop will be in our run of wiring, and to do this, we would use the following formula:

$$\text{Voltage Drop} = I \times R = 63 \text{ amps} \times 0.057$$
$$= 3.6 \text{ volts}$$

which simply means that at the end of our length of size 3 wire, we would have about 3.6 volts less than the voltage available at the motor control center, or about 436 volts. This drop is well within the maximum of 3% drop permitted.

We have selected the wire size for the problem posed in Section B. It wasn't hard, was it?

COPPER VERSUS ALUMINUM CONDUCTORS

Although silver is the best conductor, its cost limits its use to special circuits where a substance with high conductivity is needed.

The two most generally used conductors are copper and aluminum. Each has characteristics that make its use advantageous under certain circumstances. Likewise, each has certain disadvantages.

Copper has a higher conductivity; it is more ductile (can be drawn out), has relatively high tensile strength, and can be easily soldered. It is more expensive and heavier than aluminum.

Although aluminum has only 60 percent of the conductivity of copper, its lightness makes possible long spans, and its relatively large diameter for a given conductivity reduces corona—that is, the leakage of electricity from the wire when it has a high voltage. The leakage is greater when smaller diameter wire is used than when larger diameter wire is used. However, aluminum conductors are not easily soldered, and aluminum's relatively large size for a given conductance does not permit the economical use of an insulation covering.

The use of aluminum for service wiring has gained a bad reputation, due to the tendency of the terminal connections to loosen after prolonged periods of time. The resultant arcing has been suspected as the source of many fires of otherwise undetermined origin. For that reason, we strongly urge that the local inspector be contacted before using it in the plant.

H. Conductor Insulation

To be useful and safe, electric current must be forced to flow only where it is needed. It must be "channeled" from the power source to a useful load. In general, current-carrying conductors must not be allowed to come in contact with one another, their support hardware, or personnel working near them. To accomplish this, conductors are coated or wrapped with various materials. These materials have such a high resistance that they are, for all practical purposes, nonconductors. They are generally referred to as insulators or insulating material.

Because of the expense of insulation and its stiffening effect, together with the great variety of physical and electrical conditions under which the conductors are operated, only the necessary minimum of insulation is applied for any particular type of cable designed to do a specific job. Therefore, there is a wide variety of insulated conductors available to meet the requirements of any job. There are so many different types and materials of insulation available today that it would normally be very difficult to keep track of them by name and material alone. Therefore, a system of code letters has been adopted to make the task easier. Such a code list is given on page 306, in the Appendix, and you will find it to be very complete. The codes by which the various types are known are listed under the column headed "Type letter." The other columns give a rather complete picture of the material and its use.

Referring back to Section B, we shall now choose the type of insulation for this same problem of feeding electricity to a hypothetical 50 hp motor.

The table on page 306 contains practically all of the selections which are available, and we should start at the beginning and read down the columns until we can make a reasonable selection. Most of the selections shown in the table will not be found in the average industrial plant. The usual plant operation will require types T, TW, THHN, THW, or THWN. There may be some isolated instances where one of the other more exotic types will be needed, and some factory wired equipment may use one of them. For the most part, however, the industrial plant electrical worker will only be handling one or more type T styles given above.

First, rubber deteriorates in time, so we recommend all of the rubber insulation listings be bypassed, down to the thermoplastic covers, which are much more durable and long lasting. In fact, we could probably choose type T if the job was considered to be a short payout application, or a temporary one.

If we expect the installation to be as permanent as the remainder of the plant, then we suggest you choose at least the TW or type THWN.

In pulling conductor through conduit, there are available several brands of grease, especially formulated for this service. It saves energy in performing the task, and helps considerably in preventing tearing or gouging the conductor insulation resulting from sharp edges or turns in the conduit.

We have illustrated the general method of choosing and sizing the wire required for a typical plant problem. There may be other factors to be considered in solving your problem, which will require a careful study of the **NEC**® handbook. We refer you to page 354 for the index of articles in the **NEC**® handbook. It is very important that the footnotes to all tables therein be read and followed very carefully.

C H A P T E R 4
WIRING DIAGRAMS

A. Purpose and Use of Symbols

Like all trades and professions, electricians have a language of their own, with symbols, abbreviations, and diagrams by which they communicate with each other. By this means, the design formulated by one mind can easily be passed on to another electrician, and the end result will usually, but not always, perform as the originator intended it. This language system has been formalized by the American National Standards Institute, into a well-recognized standard, which we present in Table 4–1. We do not expect you to memorize every symbol on these lists, but in time you should become familiar with most of them, at least those pertinent to your plant electrical system.

Some of the wiring diagrams, for that is what we are discussing here, are extremely complicated, and tend to discourage the novice in pursuing this trade. We shall be covering only those simpler systems which make up the average electrical system in the small to medium size industrial plant. So stay with us, and you will learn the jargon of the trade in easy steps.

B. Single Line Diagrams

The first type of electrical diagram, and the simplest, is the single line representation of a system, containing a logical flow path of the electrical power, from start to finish. In this method, one single line is used to represent any number of current-carrying lines running in parallel from the source to the load. In the case of DC systems, the diagrams are shown as single line diagrams, as each line represents the conductor of current along the flow path, from the source, to the load, and back to the source to complete the circuit. You have already been exposed to this type of diagram in Chapter 2, and probably were not aware of it. The diagrams are a logical method of explaining how each particular circuit carries out its function. That is the purpose of the whole system, and if you keep that thought in mind, the rest of this textbook will be much easier to understand.

Note that in the case of DC circuits, a single line diagram includes the return path for the current flow, as well as the supply path to the load. In the case of alternating current (AC) circuits, the single line diagram represents both supply and return, due to the nature of alternating current. This will be explained in Chapter 7.

Figure 4-1 shows a single line diagram of part of the distribution system for a typical industrial plant, consisting of the front end of the system only. It is actually the start of the single line diagram for a typical industrial plant system, as covered in Chapter 16.

Figure 4-1 needs very little explanation. The single line flow path is labelled as shown to in

Table 4-1. Electrical Diagram Symbols—I

Name	Symbol	Name	Symbol
Battery		Crossing of conductors—not connected	
Capacitor, fixed		Crossing of connected conductors	
Circuit breakers		Joining of conductors—not crossing	
Air circuit breaker		*Contacts (electrical) Normally closed contact (NC)	
Three-pole power circuit breaker (Single throw) (with terminals)		Normally open contact (NO)	
Thermal trip air circuit breaker		*NO* contact with time closing (TC) feature	
Coils		Winding symbols Three phase wye (ungrounded)	
Non-magnetic core—fixed		Three phase wye (grounded)	
Magnetic core—fixed		Three phase delta Note: Winding symbols may be shown in circles for all motor and generator symbols.	
Magnetic core—adjustable tap or slide wire			
Operating coil			
Blowout coil		Rectifier, dry or electrolytic, full wave	
Blowout coil with terminals			
Series field		Relays Overcurrent or overvoltage relay with 1 *NO* Contact	
Shunt field		Thermal overload relay having 2 series heating elements and 1 *NC* contact	
Commutating field			
Connections (wiring) Electric conductor—control			
Electric conductor—power		Resistors Resistor, fixed, with leads	
Junction of conductors		Resistor, fixed, with terminals	
Wiring terminal		Resistor, adjustable tap or slide wire	
Ground		Resistor, adjustable by fixed leads	

*Note: NO (Normally Open) and NC (Normally Closed) designate the position of the contacts when the main device is in the de-energized or nonoperated position.

Table 4-1. Electrical Diagram Symbols—II

Name	Symbol	Name	Symbol
Resistor, adjustable by fixed terminals		Contactor, single-pole, electrically operated, with blowout coil	
Instrument or relay shunt		Note: Fundamental symbols for contacts, coils, mechanical connections, etc., are the basis of contactor symbols	
Switches		Fuse	
Knife switch, single-pole (SP)			
Knife switch, double-pole single-throw (DPST)		**Indicating lights**	
		Indicating lamp with leads	
Knife switch, triple-pole single-throw (TPST)		Indicating lamp with terminals	
		Instruments	
Knife switch, single-pole double-throw (SPDT)		Ammeter, with terminals	OR
Knife switch, double-pole double-throw (DPDT)		Voltmeter, with terminals	OR
Knife switch, triple-pole double-throw (TPDT)		Wattmeter, with terminals	OR
		Machines (rotating)	
Field-discharge switch with resistor		Machine or rotating armature	
		Squirrel-cage induction motor	
Pushbutton normally open (NO)		Wound-rotor induction motor or generator	
Pushbutton normally closed (NC)			
Pushbutton open and closed (spring-return)		Synchronous motor, generator, or condenser	
Normally closed limit switch contact	LS	D-C compound motor or generator	
Normally open limit switch contact	LS	Note: Commutating, series, and shunt fields may be indicated by 1, 2, and 3 zigzags, respectively. Series and shunt coils may be indicated by heavy and light lines or 1 and 2 zigzags, respectively.	
Thermal element (Fuse)			
Transformers			
I Phase two-winding transformer			
Autotransformer single-phase		*NC contact with time opening (TO) feature	TO

*Note: NO (Normally Open) and NC (Normally Closed) designate the position of the contacts when the main device is in the de-energized or nonoperated position.

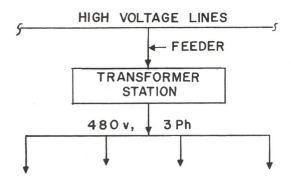

HIGH VOLTAGE LINES

FEEDER

TRANSFORMER STATION

480v, 3 Ph

Fig. 4-1. Single line diagram.

dicate the various line services being used, and with equipment labels as required to clarify the purpose of the diagram. There are few rules concerning the degree of labelling and the amount of detail which must be shown. That is left up to the designer of the system to decide, based upon what he feels is necessary to accomplish his desired end.

In Fig. 4-1 each line could be labelled with the number and size of the conductors, the code letters for them, and the size of conduit in which the conductors are installed. The square boxes could be defined in more detail, and the internal equipment could be shown in continuity from the inlet to the outlet. We refer you to Figs. 16-1 and 16-2 for examples of two different degrees of detailing the single line diagram.

C. Elementary Diagrams

These are sometimes known as "schematics," as they are in closer detail than the single line diagram, and are used to explain the actual workings of the individual pieces of equipment, or the "scheme" of the system. In this method we are concentrating on the part that each item of electrical gear in the system plays in the overall system displayed by the single line diagram.

The main thing about the elementary diagram to remember is that it is designed to permit tracing the electrical circuits through from start to finish, and is an excellent method of troubleshooting the equipment. All elements are shown in their correct relation to each

other, permitting the electrician to logically trace the actions of each item on others in the train. The individual items are not, therefore, necessarily shown in their physical location inside the equipment. They are electrically in their correct location, but not always physically.

Figure 4-2 is an elementary diagram of a simple magnetic across-the-line starter for a three phase motor, of a type which is described in Chapter 17. As you can see, the wiring has been laid out beside the three leads to the motor, and the layout does not tell you anything about the physical layout of the individual parts of the system.

The control circuit is powered by the voltage across lines L1 and L2 by the lead taken from point 1, through the control device, point 2, through the magnetic coil, point 3, through the thermal overload, point 4, and back to point 5, on lead L2. This completes the circuit which powers the control system to actuate the starter. The action of these parts in turning the motor on and off is as follows.

The two wire control device, either a manual switch, timer, pressure, temperature, or level switch, closes the normally open contacts, point 2. This energizes the magnetic coil, point 3, which in turn closes the contacts marked "M" on the three motor leads, thus starting the motor. If the contacts in the control device, point 2, should open, then the magnetic coil

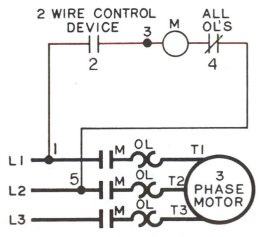

Fig. 4-2. Elementary diagram of starter. (Courtesy of Square D Co.)

would release the contacts on the three motor leads, marked "M," thus stopping the motor.

There are any number of combinations of auxiliary devices and wiring arrangements which may be used in the control side to actuate the magnetic coil, and we shall cover a few of them in later chapters.

Referring back to Fig. 4-2, the terminal 3 is for the convenience of wiring some of the auxiliary control devices into the circuit.

The thermal overload contact, point 4 in Fig. 4-2, is to protect the motor, and there may actually be more than one in the control device circuits.

You may have noticed that there is nothing in the elementary diagram which indicates any direct connection between the magnetic coil and the three motor lead contacts marked "M," except the "M" beside the contacts and the coil. This is typical of elementary diagrams, and is

Table 4-2. Common Markings and Symbols Used on Electrical Circuit Diagrams

Device	Contractor designation	Relay designation	Other equipment designation	Device	Contractor designation	Relay designation	Other equipment designation
Accelerating	A	AR		Master switch			MS
Ammeter switch			AS	Motor circuit			MCS
Autotransformer			AT	switch			
Brake	B	BR		Motor field			MF
Capacitor			C	Overload		OL	
Circuit breaker			CB	Overspeed		OSR	
Closing coil		CCR	CC	Overspeed			OSS
Control		CR		switch			
Control switch			CS	Plugging	P	PR	
Counter EMF		CEMF		Plugging		PF	
Current limit		CLR		forward			
Current			CT	Plugging reverse		PR	
transformer				Potential			PT
Down	D			transformer			
Dynamic	DB	DBR		Power factor		PFR	
braking				Power factor			PF
Emergency			ES	meter			
switch				Pushbutton			PB
Exciter field	EF	EFR		Rectifier			REC
Field	F	FR		Resistor			RES
Field	FA	FAR		Reverse, run,	R		
accelerating				raise			
Field discharge	FD	FDR		Sequence		SPR	
Field economy	FE	FER		protective			
Field loss	FL	FLR		Slow down		SR	
(failure)				Squirrel-cage		SCR	
Field weakening	FW	FWR		protective			
Float, flow			FS	Start	S		
switch				Switch			SW
Forward	F	FR		Time closing			TC
Full field	FF	FFR		Time opening			TO
Ground detector			GD	Time relay		TR	
High speed	HS	HSR		Transfer relay		TRR	
Hoist	H	HR		Trip coil			TC
Jam, jog	J	JR		Undervoltage	UV	UVR	
Kickoff	KO	KOR		Up	U		
Limit switch			LS	Voltage		VRG	
Lowering	L	LR		regulator			
Low speed	LS	LSR		Voltmeter			VS
Main breaker			MB	switch			

Table 4-3. Power-Terminal Markings

	Direct current	Alternating current
Brake	B1, B2, B3	B1, B2, B3
Brush on commutator (armature)	A1, A2	A1, A2, A3
Brush on slipring (rotor)		M1, M2, M3
Field (series)	S1, S2	
Field (shunt)	F1, F2	F1, F2
Line	L1, L2	L1, L2, L3
Resistance (armature)	R1, R2, R3	R1, R2, R3
Resistance (shunt field)	V1, V2, V3	
Stator		T1, T2, T3
Transformer (high voltage)		H1, H2, H3
Transformer (low voltage)		X1, X2, X3

true to some extent in many wiring diagrams, also, for symbols of that type which augment those found in Table 4-1. It is simply not feasible to include all possibilities in making up a system of standard symbols to be used in the many types of electrical circuits and equipment in the industry. Often there will be a table of extra symbols on design drawings which will indicate special applications used in that particular design.

Tables 4-2 and 4-3 include additional symbols which will aid in making and reading wiring diagrams.

D. Full Wiring Diagrams

We come now to that type of diagram which most nearly approaches the appearance of the actual hardware, which is the full wiring diagram. Figure 4-3 is an excellent example, as it is a development of Fig. 4-2. If you were to open the cover of this same magnetic starter, you would see something similar to Fig. 4-3, as the internal layout would look much like the upper half, including the incoming terminals at the top and the outgoing terminals at the bottom, with all other components in about the same

position, except the motor and its leads, which are all external to the starter box.

We suggest you follow the wiring and the connections to the various items inside the starter box in Fig. 4-3, and assure yourself that it is, indeed, the same as the elementary diagram in Fig. 4-2. There are some missing links, due to the fact that some of the wiring is not visible. For instance, the link between L1 and terminal 2, point 2, is part of the starter mechanism. By the electrician making the necessary control and power line connections to the terminals marked by a small open o, the starter is ready to perform its work.

Note that the large open circle representing the magnetic coil is located in the diagram close to the three contactors which it operates. Note, also, that the power supply for the control circuit is taken off the incoming power terminals. We will show later in this book that this does not have to happen that way, as the control circuit power may come from a source outside the starter mechanism. The magnetic coil may be operated from any convenient power source as long as the coil has the correct current characteristics.

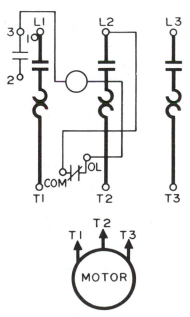

Fig. 4-3. Wiring diagram. (Courtesy of Square D Co.)

E. Ladder Diagrams

If we were to take the control circuit line diagram, or schematic diagram of Fig. 4-2, and stretch it out so that point 1 is on the left of a straight horizontal line, with point 5 on the right, as in Fig. 4-4, we would then have one segment of what is known as a Ladder Diagram. You will note that the elements are all there as in the elementary diagram, and in the same order from L1 to L2. In this case, the diagram only appears to resemble one rung of a ladder, but in more complicated control diagrams, the general appearance of a ladder will be obvious.

Ladder diagrams are used only on 2-wire control circuits, not for the power circuits of the driven or controlled equipment. They are constructed for one control circuit or loop for each diagram, and they all start at the top and progress downward, rung by rung, with the terminals often numbered or labelled at the junction of the control line with the supply voltage line.

Figure 4-5 illustrates the principle and the logic behind each line of a ladder diagram. Comparing this figure with Fig. 4-4, we can see that item 2 on Fig. 4-4 is in the Signals section of Fig. 4-5, item 3 is in the Decisions section, and item 4 is in the Action section.

The signal is always in the form of a switch, and this switch may be manually operated, automatic, or mechanically operated from some activity or function being monitored.

When the Signal has been sent to the Decision section, the circuitry of the Decision section determines what action has to be taken. The action taken may be as simple as starting a motor, or it may bring into action a long chain of events, such as motors and other con-

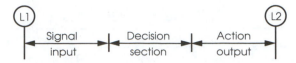

Fig. 4-5. Control circuit functions.

trols on a conveyor chain, boiler control panel, or similar system.

In further analysis of the ladder diagram, the following basics should be kept in mind.

Control circuits are to be read from left to right, and downward from the L1 terminal.

The signal takes no action, but simply starts the Decision function.

There is only one load (action) on each horizontal line of the diagram.

All loads (action) are connected directly into the L2 line on the right of the diagram.

All loads requiring power must be connected in parallel, and never in series with other loads in the diagram.

Protective devices not requiring power for operation may be connected in the Action section in series with each other or with loads.

All elements of the ladder diagram, and any control diagram, are shown in the deenergized condition.

Some ladder diagrams become very complicated, and only the most experienced electrician will be able to analyze such a maze of lines, elements, and logic. Therefore, we recommend that the electrical maintenance worker adopt the procedure shown in Fig. 4-6. Notice

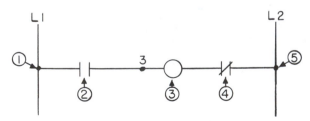

Fig. 4-4. Basic element of a ladder diagram.

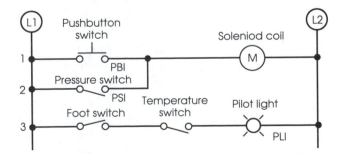

Fig. 4-6. Line numbering for easy analysis.

that on the left vertical line marked L1, starting at the top and reading down, all horizontal lines are numbered, 1, 2, and 3. If the diagram being analyzed does not contain these numbers, then we suggest they be placed on the diagram as an aid in determining the overall logic of the control system. Each numbered line is capable of providing a complete circuit from L1 to L2, if all switches and overload devices are closed.

Notice that in the case shown in Fig. 4-6, there is no definite "Decision" section, the switches operate the loads directly. The Decision is made by the operating elements connected to the four switches in the three circuits—two by the person running the system, and two by pressure and temperature sensors. This may be the case only where the "Signal" section is capable of providing the full power to perform the "Action."

It is important to remember that the numbers designating lines 1, 2, and 3 do not identify terminals or junctions. They are not to be confused with terminal or junction numbers on any associated wiring diagram.

Now that the basics have been discussed giving the essentials of the ladder diagram, we show a composite diagram in Fig. 4-7. Notice that there are three overload relays in series with the load coil in line 1, Action section. Usually, these overload relays are indicated by one symbol, with the notation "All" above it, in place of showing all three in series. This is true only if all OL's in series are identical in function and purpose.

Figure 4-8 contains duplicate symbols for identical elements in a ladder diagram, which at first would appear to be confusing. For instance, CR1 appears in lines 1, 2, 3, and 4, and the impression at first glance would be that there are four identical control relays, all connected in some way. In fact, there is only one control relay in line 1, with four sets of contacts.

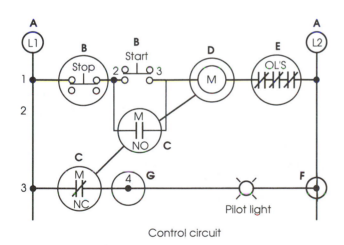

Control circuit

A Terminal screw connection

B Momentary contact push buttons
   Depressing button against spring pressure
   opens upper contacts and closes lower
   contacts. Releasing button returns contacts
   to the normal condition shown.

C Auxiliary contacts-Operate when parent start
   switch B powers coil M. In this case, normally
   open (NO) contacts close and normally closed
   (NC) open when coil M is energized.

D Operating coil of contactor

E Overload relay contacts

F See Table 4-1

G Reference point-Often used on starter,
   corresponds with the number shown in
   the push button station wiring diagram

Fig. 4-7. Typical ladder diagram explained.

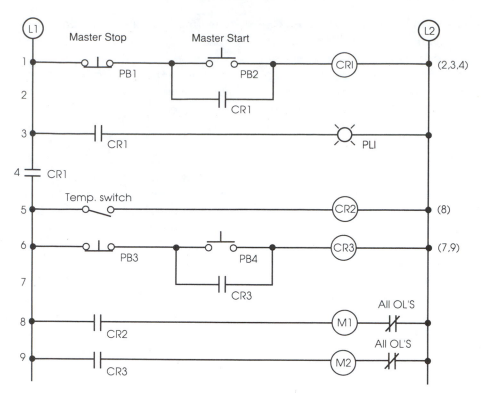

Fig. 4-8. Duplicate elements in a control diagram.

If Control Relay #1 (CR1), shown in line 1 with one set of contacts, were to be drawn with four sets of contacts, the resulting control diagram would be very complicated and confusing. It is easy to see that a much simpler method is to show the same CR in four different locations, as done in the ladder diagram in Fig. 4-8.

At the top of power leg L2, at the end of horizontal line 2, there are the figures (2, 3, 4). These numbers refer to the lines in which CR1 contacts appear. Notice that line 2 contains a CR1, which is used as a holding contact, line 3 contains a CR1 which operates a pilot light, and line 4 contains a CR1 which activates the control functions in lines 5 through 9. The CR1 in line 1 performs no other function than to activate the rest of the control system.

Once the control system has been activated, the Master Start PB2 can be released, and CR1 in line 2 will maintain contact, powering the entire control circuit, until the Master Stop

PB1 is pushed, opening the power to CR1, and thus deactivating the entire system.

Once the CR1 on line 4 is closed, FL1 controls CR2, which in turn controls M1 and PL2. Lines 6 through 9 are similar to lines 1, 2, and 3.

As you can now see, relays, contactors, and motor starters often have more than one set of contacts, each of which may appear in several different locations in the control diagram.

As normal procedure is to show all elements on the ladder diagram in the Normally Open (NO) condition, then it is necessary to show all Normally Closed (NC) elements in the NC position, or label the element as NC, so there can be no mistaking the action to be expected from it.

Many switches found in the industrial plant have more than one set of contacts which operate with one specific action by the control, and these sets of contacts may be either NO, NC, or multiples of each, all operating simultaneously. These different actions are usually found in different lines on the same ladder diagram, and some means must be used to indicate that they operate from the same switch action.

One method is to draw a dashed line between the two switch actions which are part of one switching device. This works very well if they are close enough together so as not to clutter the diagram. Should the two switch actions be some distance apart, then referral numbers placed beside each switch, referring to the line on which the mating switch is located, will do rather nicely in most cases. This is indicated by the small numbers in parentheses along the right side of Fig. 4-8. For instance, (2, 3, 4) indicates that CR1s in lines 2, 3, and 4 are all operated by CR1 in line 1. CR1 thus has four sets of contacts.

Any logical, simple method of nomenclature which will assist in analyzing a diagram is perfectly all right to use, as long as others who may later have to analyze the diagram are made aware of the symbols devised.

That ends the discussion on ladder diagrams. Some experience and practice in the analysis of them, along with some ingenuity and logical thinking, will assist the electrical maintenance worker in analyzing and trouble shooting control circuits in the plant.

F. Harness Diagrams

Replacement wiring for panels and other electrical equipment of a complicated nature are often furnished in a package, ready to be inserted into the equipment in place of the old package, or bundle, and the terminals hooked up as were the originals. The terminals, brought out from the bundle at the proper locations, will have tags giving the terminal numbers or letters, for identification and for attaching to the proper mating terminals on the panel or other piece of equipment. Figure 4-9 illustrates this arrangement. All that is necessary then is to remove the old bundle, and place the new bundle in the space vacated, then hook up the terminals according to the original wiring and terminal plan.

This method may also be used to change the

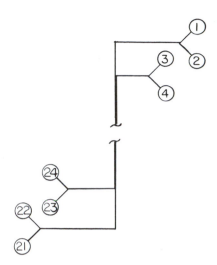

Fig. 4-9. Harness diagram.

wiring sequence of existing equipment, as the replacement will be tagged on the terminals to show the revised hook-ups and thus achieve the altered function desired of the equipment.

Examples of this method of wiring replacement will be found in the automobile industry, aircraft, military and space equipment, electronics, and high-tech industries. The time savings are very impressive, as the alternative is to replace each wire individually, one at a time, which could take days, instead of hours, with the package concept.

G. Building Wiring Diagrams

Now that you have mastered the basics of wiring diagrams for the simpler electrical elements, it is now time to incorporate them into the plant master wiring diagrams, as relating to the electrical distribution services. This we shall do by means of the following three diagrams in Figs. 4-10, 4-11, and 4-12.

Figure 4-10 is a typical one-line diagram showing the electrical distribution system for a technical trades instruction building. As there are some symbols here that are not described elsewhere, we shall cover them by the following legend.

1—Incoming service, dual voltage, 4160/2400 volts, 3 phase, 4 wire system in a 600 amp bus

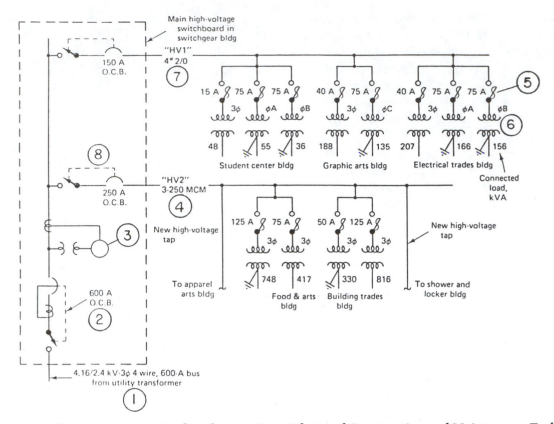

Fig. 4-10. One-line diagram: service distribution. From *Electrical Construction and Maintenance Techniques*, by Gray/Bailey, © 1986. Reprinted by permission of Prentice-Hall, Inc., Englewood Cliffs, NJ.

2—600 amp oil-filled circuit breaker, overload tripping

3—Watt-Hour meter

4—High Voltage circuit #2, consisting of three wires of 250 thousand circular mils

5—Oil fused cutouts

6—Step-down transformers for three phase 240V and single phase at 120/240 volts

7—High voltage circuit #1, consisting of 4 size 00 wires

8—Oil circuit breakers.

Notice that this wiring diagram is a combination of existing services and new additions, labeled "New High Voltage Tap." Its purpose would be as an orientation tool, showing how the new services tie into the existing scheme, and also to serve as a record drawing, "as built," for the files.

Figure 4-11 is a detailed one-line diagram of Building "D" for the same project as Fig. 4-10.

The legend for the symbols is as follows:

1—High voltage feeder from HV2, Fig. 4-10

2—HV feeder consists of 3 size 0000 wires, rated 5Kv through a $3\frac{1}{2}''$ conduit

3—Step-down transformer and circuit breaker station. Transformer is filled with askarel type oil, wired on the primary side in delta configuration, reducing to 120/208 volts on the secondary side, in Y configuration

4—Similar to #3, except secondary side of transformer is 240 volt, in 3 phase delta configuration

5—Normal supply to the elevator motor

6—Feeder to item #7, consisting of 3 size 000 wires in a $2\frac{1}{4}''$ conduit

7—Power panel "A" for building "D," second floor

8—Motor control panel "A" for building "D," second floor

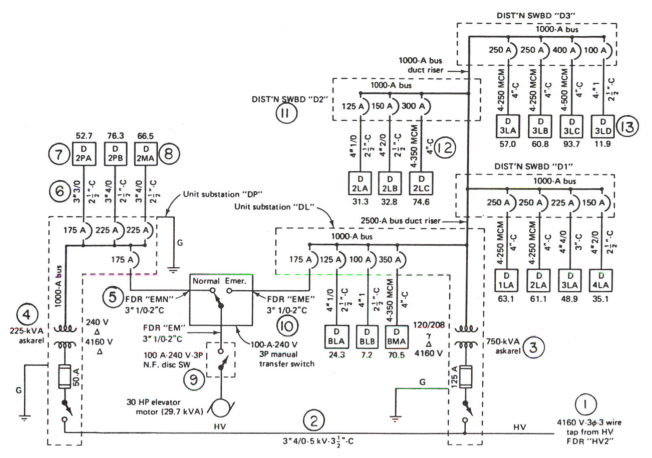

Fig. 4-11. One-line diagram: building power distribution. From *Electrical Construction and Maintenance Techniques*, by Gray/Bailey, © 1986. Reprinted by permission of Prentice-Hall, Inc., Englewood Cliffs, NJ.

9—Non-fused disconnect switch for the elevator

10—Emergency feeder from Substation "DL," to the elevator, requiring manual switchover in an emergency on loss of power from the normal supply

11—Distribution switchboard handling the second floor of building "D," lighting circuits

12—Four wires of 350 thousand circular mils in a 4" conduit, supplying lighting panel "C" on second floor of building "D," 74.6 amps connected load

13—Same as item #12, except lighting panel "D" is on the third floor of building "D," being served from Distribution Switchboard "D3" on the third floor, total connected load 11.9 amps.

The last one is a Riser Diagram, Fig. 4-12, typically for the auxiliary systems which normally come under the electrical contract. These systems are:

Public address speaker system

Public address microphone system

Clock, master and slave, system

Public telephone system

Closed circuit television system

Fire alarm system.

The purpose of the Riser Diagram is to show the relative locations of the various electrical panels and equipment. The locations are shown horizontally by floors, and vertically by column row numbers, as shown on Fig. 4-12.

Some of the elements shown in Figs. 4-10, 11, and 12 are new at this stage in the book, but

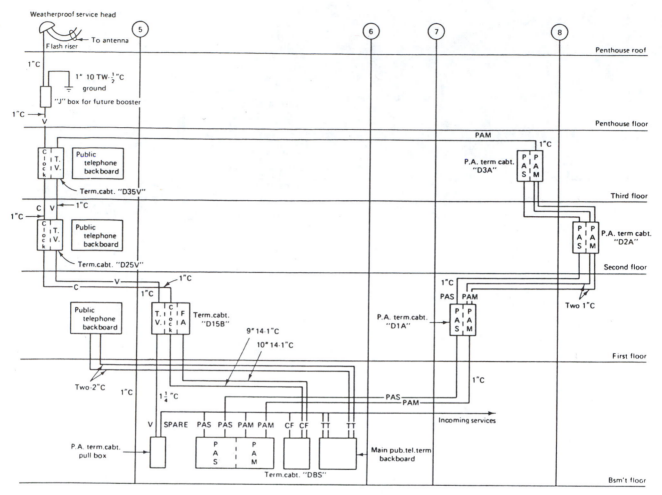

Fig. 4-12. Riser diagram. From *Electrical Construction and Maintenance Techniques*, by Gray/Bailey, © 1986. Reprinted by permission of Prentice-Hall, Inc., Englewood Cliffs, NJ.

most of them will be covered in detail later. You may wish to come back and review these three diagrams after you have gained a more thorough knowledge of the items comprising the various electrical circuits in the average industrial plant.

H. Conductor Color Coding

As most of the electrical wiring in the plant is concealed throughout its entire length, except at the terminals, it is essential that some method be used to identify each wire, and the service it performs, at both terminals. If all wiring in a system were the same size and color, with nothing to tell the electrician which wire or wires he should be working with, a consider-

able amount of time would be required simply to sound out the individual wires each time with instruments. Also, there would be a good chance of errors creeping into the final effort.

The **National Electrical Code**® has established standard color coding methods for the various services which are considered to be the most dangerous if a mix-up occurs in the wiring. These standards apply only to the grounding system, as this is the most important function from a safety angle.

Referring to Fig. 4-13, a typical grounding system is shown for a three phase load supply from a panel or cabinet. There are two general grounding systems involved; the electrical service grounding system consisting of the wiring, and the equipment grounding system, consisting of the distribution and control apparatus,

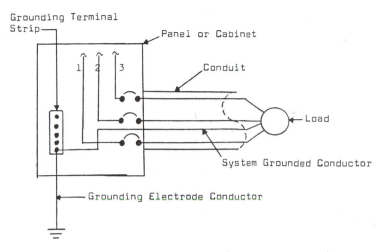

Fig. 4-13. Typical grounding diagram.

such as transformers, motor control centers, distribution centers, conduit, junction boxes, disconnect switches, magnetic starters, motor housings, etc.

The standard color scheme established by the **National Electrical Code®** has taken into consideration the possible difficulty of obtaining the necessary color and length of wire for contractors to complete a job on time, as the specified wire may not always be available when needed.

Referring to Fig. 4-13 again, the Code specifies that the System Grounded Conductor shall be white or gray in size #6 or smaller. In larger sizes, any color may be used, providing that it is re-identified white at the terminal ends, visible to the electrician servicing the system.

Equipment Grounding Conductors may be green or green with yellow stripes. If the conduit and distribution system is continuous with no electrical breaks, then no additional grounding wires are required. However, if the conduit is not continuous, or if it is plastic, then separate grounding wiring is required, with the colors given above, or the wire may be bare.

For ungrounded hot conductors, the **NEC®** recommends, for safety and for load balancing, that all legs be identified with any chosen color, used consistently throughout the system, providing the chosen color is not the same as used for grounding. Black is most generally used for hot legs.

In any group of stationary lugs, studs, or screws designed to be used as electrical terminals, the electrician should look for the silver colored terminal, as that is the grounded terminal.

The subjects of grounding and color coding will be mentioned in several other places throughout the remainder of this book.

This ends the chapter on wiring diagrams. We wish to warn the reader that he will find many variations in the symbols and procedures of wiring diagrams from those shown here. Most of them will reflect the individual or specific requirement of the particular piece of equipment being represented. They will all have sufficient similarities to permit the trained electrical mechanic to work out the basic logic and desires of the person producing the diagrams. Look for the table or legend of symbols on the wiring diagram.

C H A P T E R 5
DIRECT CURRENT CIRCUITS

A. Switches

We stated in Chapter 2 that there must be a complete path, known as a "closed circuit," for the electrical current, or amperage, to follow from the negative terminal of a DC power source, through the load, and to the positive terminal of the power source, if we expect to get work out of our DC power.

Always keep in mind, regardless of what type of source of power you are handling, whether electrical, hydraulic, heat, atomic, or pressure, that it is essential that the power be completely under our control at all times. This means it must be supplied with devices which permit you to safely shut it off, turn it on, limit its strength, or control its direction of flow. If any one of these are not provided, either by inherent design in the system or by the addition of proper devices, there is always danger of that power source getting out of control and wreaking untold damage. And as we stated earlier, the one person almost certain to be directly hurt is you.

This section is about methods used to shut off the flow of an electrical current or to turn it on, either at your demand, or upon the automatic action of a system designed to act in someone's behalf.

Turning to Fig. 2-15, you will note that a simple switch is shown, and its action or purpose is described briefly. If this circuit was for a doorbell, which is a common one, then the switch would be one which is spring loaded to hold the contact button open, or away from the electrical contact points. Normally, the circuit is open. When the doorbell button is pushed, the contacts complete the circuit, which permits current to flow through the switch and to the doorbell. When the button is released, the button breaks the circuit, due to the spring action inside the button, and the bell stops ringing.

Now take a look at Fig. 5-1, and see what can be done with a bank of simple switches. The switches shown here, and in the example of Fig. 5-2, are not spring loaded, but are knife type, or make-and-break switches. The voltages handled are low enough to permit the use of these switches directly in the flow of power. They may be operated by hand or by small auxiliary devices, with a minimum of sparking across the contact points when the switch is opened or closed.

If you study Fig. 5-1 carefully, you will see that all of the switches must be closed before the path from the negative terminal to the load is completed. This is known as a series switch circuit, as all of the switches are wired in series, one with the others, all on the common conductor or path.

This type of arrangement is often used when wiring safety cut-out switches, or "lockout" switches as protective devices in a circuit. In

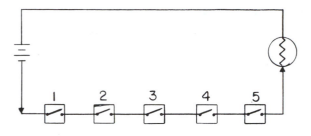

Fig. 5-1. Switches in series.

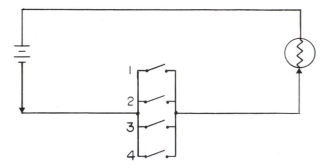

Fig. 5-2. Switches in parallel.

Fig. 5-1, the first switch, NO. 1, may be the main or "demand" switch which is closed to start the system in operation. But in order for the system to start its work, all of the other switches, which in our example here may be safety cut-out switches, must be closed before the system will start. For example, this circuit could be on a watercooled air compressor, and the cut-out switches could be operated by any of the following conditions:

low cooling water flow or pressure

high lubricating oil temperature

low lubricating oil pressure

high discharge air temperature.

Thus, if you closed the main or "start" button and nothing happened, then you would have to check the position of all of the other switches in the circuit to see which one was being held open, indicating that the proper conditions did not exist for the system to operate in a correct or safe manner. In practice, when one of the safety cut-out switches operates to shut the system down, a light may flash on an alarm panel, telling the operator which condition is not satisfactory.

Many circuits, however, do not have this feature so that it is necessary for the maintenance mechanic to check out every lock-out to find the one holding the circuit open.

There may be any number of these safety cut-out or lock-out switches in a series circuit. Keep in mind that the more switches in a series circuit, the safer the system will be, but the more prone the system is to shut-downs and maintenance calls.

Turning now to Fig. 5-2, we see a different arrangement. This is what is known as a "parallel switch circuit" as the switches here are all wired into the circuit in parallel, forming parallel paths for the current to take through the circuit.

You will note that in this arrangement, closing any one of the switches will complete the flow path for the current to take. Thus, only one switch needs to be closed in order for the system to operate. Obviously, this is normally a "demand" type of control as the important consideration in this circuit is that the system be started up on demand from different locations, reasons, or conditions. It may be to meet safety requirements of another system, or to permit starting a system up from several different locations throughout a plant.

There is another very common application of the parallel switch arrangement shown in Fig. 5-2, as used in military and space exploration equipment. Figure 5-2 is known as a "redundant" circuit, as any one of the switches will operate to provide the action desired. This is a distinct advantage when it is absolutely essential that the circuit be closed upon demand, for if all redundant switches are operated by the same activating element, all but one of the switches can fail, and the circuit will still serve its purpose. This increases the reliability of the system immensely. The same principle holds true for multiple switches in series when the circuit absolutely must be opened.

When using this parallel arrangement to permit starting equipment from various locations in the plant, keep in mind that this may not be the safe thing to do. It is important, if this arrangement is desired, that the following precautions be kept in mind:

> The "on" or "off" condition of the load should be either visible from, or indicated at, all "start" locations.

> There should be signs posted at the load, warning personnel that the device may start at any moment.

> There should be a master monitoring panel under the supervision of someone in authority in the plant, informing him of the condition of the load, and the various switches in the systems, so he will know where the "start" signal originated for the system.

As might be expected, Fig. 5-3 is a combination of the series and the parallel switch circuits, and it combines the features of both. Here we have the ability to start the equipment up from several locations throughout the plant, providing that all safety cut-outs are closed, indicating that it is safe to start the system in operation. This is a more normal type of control system than either of the previous ones, as it combines all of the safety features of the series circuit with the flexible demand features of the parallel circuit. The same warnings and conditions apply to this combination as applies to both the series and parallel circuits, which can make for some interesting and complicated circuits in such equipment as package conveyors, long assembly lines, or any continuous flow automatic machinery.

The switch arrangements above are all based upon simple DC circuits. The same principles are valid in the case of alternating current, which we shall cover in a later chapter.

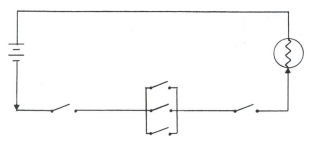

Fig. 5-3. Series-parallel switch combination.

B. Multiple Batteries in Direct Current Circuits

So far, we have assumed that all of the DC circuits have been powered by one power source, and that one was shown in the diagrams as a battery. In this section we shall continue to use batteries (or dry cells) as our source of DC power, but we shall hook them into the circuit in multiples. And, as we shall show, there are three general ways in which they can be connected into a DC circuit.

In many cases, a battery-powered device may require more electrical energy than one dry cell can provide. The device may require either a higher voltage or more current, and in some cases both. Under such conditions, it is necessary to combine, or interconnect, a sufficient number of cells to meet the higher requirements. Cells connected in series provide a higher voltage, while cells connected in parallel provide a higher current capacity. To provide adequate power when both voltage and current requirements are greater than the capacity of one cell, a combination series–parallel network of cells must be interconnected.

SERIES CONNECTED CELLS

Assume that a load requires a power supply with a potential of 6 volts and a current capacity of $\frac{1}{8}$ ampere. Since a single cell normally supplies a potential of only 1.5 volts, more than one cell is obviously needed. To obtain the higher potential, the cells are connected in series as shown in Fig. 5-4.

In a series hookup, the negative electrode of the first cell is connected to the positive electrode of the second cell, the negative electrode of the second to the positive of the third, etc.

The positive electrode of the first cell and negative electrode of the last cell then serve as the power takeoff terminals of the system. In this way, the potential is boosted 1.5 volts by each cell in the series line. There are four cells, so the output terminal voltage is 1.5 × 4 = 6 volts. When connected to the load, $\frac{1}{8}$ ampere flows through the load and each cell of the battery. This is within the capacity of each cell.

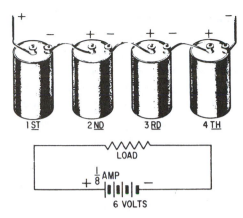

Fig. 5-4. Batteries in series.

Therefore, only four series-connected cells are needed to supply this particular load.

Please note that our discussion and the illustrations are based on the ordinary $1\frac{1}{2}$ volt battery, but the principle is equally true for all other types of storage batteries. This includes all batteries having multiple cells within one case also.

PARALLEL CONNECTED CELLS

In the case of parallel-connected cells, assume an electrical load requires only 1.5 volts, but will draw $\frac{1}{2}$ ampere of current. (Assume that a cell will supply only $\frac{1}{8}$ ampere.) To meet this requirement, the cells are connected in parallel, as shown in Fig. 5-5 (top). In a parallel connection, all positive cell electrodes are connected to one line, and all negative electrodes are connected to the other. No more than one cell is connected between the lines at any one

point; so the potential between the lines is the same as that of one cell, or 1.5 volts. However, each cell may contribute its maximum allowable current of $\frac{1}{8}$ ampere to the line. There are four cells, so the total line current is $\frac{1}{8} \times 4 = \frac{1}{2}$ ampere. Hence, four cells in parallel have enough capacity to supply a load requiring $\frac{1}{2}$ ampere at 1.5 volts.

SERIES-PARALLEL CONNECTED CELLS

Figure 5-6 depicts a battery network supplying power to a load requiring both a voltage and current greater than one cell can provide. To provide the required 4.5 volts, groups of three 1.5-volt cells are connected in series. To provide the required $\frac{1}{2}$ ampere of current, four series groups are connected in parallel, each supplying $\frac{1}{8}$ ampere of current.

The following material in this chapter delves much deeper into electrical circuits than is really necessary for the basic understanding of the use and maintenance of plant equipment. We give it primarily to tickle the imagination of those readers and students who may have a desire to go deeper into the subject.

One thing the reader will note, if he will take the effort to study the following sections on various parts of the Direct Current circuits, is that electrical engineering, and the design of electrical circuits, relies very heavily upon mathematics. Therefore, if mathematics is not your strong point, you will be forgiven if you skip over the material lightly. It will, however, help you in your understanding of some of the problems involved in design of electrical equipment and circuits, We leave the choice up to your discretion and conscience.

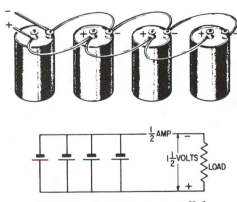

Fig. 5-5. Batteries in parallel.

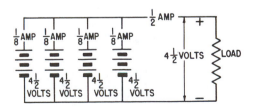

Fig. 5-6. Series-parallel battery combination.

C. Series Circuit Characteristics

As previously mentioned, an electric circuit is a complete path through which electrons can flow from the negative terminal of the voltage source, through the connecting wires or conductors, through the load or loads, and back to the positive terminal of the voltage source. A circuit is thus made up of a voltage source, the necessary connecting conductors, and the effective load.

If the circuit is arranged so that the electrons have only ONE possible path, the circuit is called a SERIES CIRCUIT. Therefore, a series circuit is defined as a circuit that contains only one path for current flow. Figure 5-7 shows a series circuit having several lamps.

Referring to Fig. 5-7, the current in a series circuit, in completing its electrical path, must flow through each lamp inserted into the circuit. Thus, each additional lamp offers added resistance. In a series circuit, THE TOTAL CIRCUIT RESISTANCE (R_T) IS EQUAL TO THE SUM OF THE INDIVIDUAL RESISTANCES.

As an equation:

$$R_T = R_1 + R_2 + R_3 \ldots R_n.$$

NOTE: The subscript n denotes any number of additional resistances that might be in the equation.

Example #1: Three resistors of 10 ohms, 15 ohms, and 30 ohms are connected in series across a battery whose emf (electromotive force) is 110 volts. What is the total resistance?

$$\begin{aligned}
\text{Given:}\quad R_1 &= 10 \text{ ohms} \\
R_2 &= 15 \text{ ohms} \\
R_3 &= 30 \text{ ohms} \\
R_T &= ?
\end{aligned}$$

$$\begin{aligned}
\text{Solution:}\quad R_T &= R_1 + R_2 + R_3 \\
R_T &= 10 + 15 + 30 \\
R_T &= 55 \text{ ohms.}
\end{aligned}$$

Since there is but one path for current in a series circuit, the same current must flow through each part of the circuit. To determine the current throughout a series circuit, only the current through one of the parts needs to be known.

The fact that the same current flows through each part of a series circuit can be verified by inserting ammeters into the circuit at various points as shown in Fig. 5-8. If this were done, each meter would be found to indicate the same value of current.

At this point, to clarify the following examples and illustrations, we will explain that in electrical jargon, loads have resistances of defi-

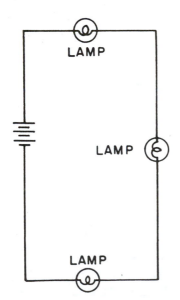

Fig. 5-7. Basic series circuit.

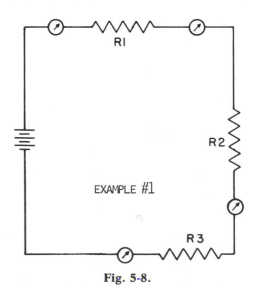

Fig. 5-8.

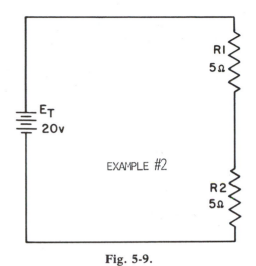

Fig. 5-9.

nite size listed in ohms, and are shown here by the symbol R_1, R_2, etc. Thus, in Fig. 5-9, example #2, the loads consist of resistances of 5 ohms each.

As stated previously, the voltage drop across the load in the basic circuit is the total voltage across the circuit and is equal to the applied voltage. The total voltage across a series circuit is also equal to the applied voltage, but consists of the sum of the individual voltage drops. In any series circuit, the SUM of the individual voltage drops must equal the source voltage. This statement can be proven by an examination of the circuit shown in Fig. 5-9. In this circuit, a source potential (E_T) of 20 volts is impressed across a series circuit consisting of two 5 ohm loads. The total resistance of the circuit is equal to the sum of the two individual resistances, or 10 ohms. Using Ohm's Law, the circuit current may be calculated as follows:

$$I = \frac{E_T}{R_T}$$

$$I = \frac{20}{10}$$

$$I = 2 \text{ amperes.}$$

Knowing the resistances of the loads to be 5 ohms each, and the current through the loads to be 2 amperes, the voltage drops across the loads can be calculated. The voltage (E_1) across R_1 is therefore:

$E_1 = IR_1$
$E_1 = 2 \text{ amperes} \times 5 \text{ ohms}$
$E_1 = 10 \text{ volts.}$

Since R_2 is the same resistance as R_1 and carries the same current, the voltage drop across R_2 is also equal to 10 volts. Adding these two 10 volt drops together gives a total drop of 20 volts, exactly equal to the applied voltage. For a series circuit then:

$$E_T = E_1 + E_2 + E_3 \ldots E_n.$$

Example #3: A series circuit consists of three loads having resistances of 20 ohms, 30 ohms, and 50 ohms, respectively. Find the applied voltage if the current through the 30 ohm resistor is 2 amperes.

To solve the problem, a circuit diagram is first drawn and labelled as shown in Fig. 5-10.

Given: $R_1 = 20$ ohms
$R_2 = 30$ ohms
$R_3 = 50$ ohms
$I = 2$ amperes.

Solution: Since the circuit involved is a series circuit, the same 2 amperes of current flows through each load. Using Ohm's law, the voltage drops across each of the three loads can be calculated and are:

$E_1 = 40$ volts
$E_2 = 60$ volts
$E_3 = 100$ volts.

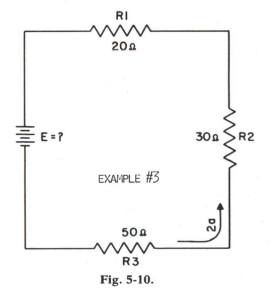

Fig. 5-10.

Once the individual drops are known, they can be added to find the total or applied voltage:

$$E_T = E_1 + E_2 + E_3$$
$$E_T = 40v + 60v + 100v$$
$$E_T = 200 \text{ volts.}$$

NOTE: In using Ohm's Law, the quantities used in the equation MUST be taken from the SAME part of the circuit. In the above example, the voltage across R_2 was computed using the current through R_2 and the resistance of R_2.

The important factors governing the operation of a series circuit are listed below. These factors have been set up as a group of rules so that they may be easily studied, and they must be understood before more advanced circuit theory is undertaken.

1. The same current flows through each part of a series circuit.

2. The total resistance of a series circuit is equal to the sum of the individual resistances.

3. The total voltage across a series circuit is equal to the sum of the individual voltage drops.

4. The voltage drop across a load in a series circuit is proportional to the resistance of the load.

5. The total power dissipated in a series circuit is equal to the sum of the individual power dissipations.

D. Parallel Circuit Characteristics

In stepping-stone fashion, the discussion of series DC circuits will now be followed by a consideration of the characteristics of parallel DC circuits. It will be shown how the principles applied to series circuits can be used to determine the reactions of such quantities as voltage, current, and resistance in parallel circuits.

Problems involving the determination of re-sistance, voltage, current, and power in a parallel circuit are solved as simply as in a series circuit. The procedure is the same—(1) draw a circuit diagram, (2) state the values given and the values to be found, (3) state the applicable equations, and (4) substitute the given values and solve for the unknown.

A parallel circuit is defined as one having more than one current path connected to a common voltage source. Parallel circuits, therefore, must contain two or more load resistances which are not connected in series. An example of a basic parallel circuit is shown in Fig. 5-11.

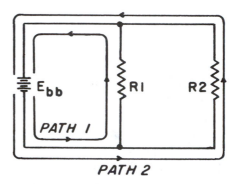

Fig. 5-11. Basic parallel circuit.

Commencing at the voltage source (E_{bb}) and tracing counterclockwise around the circuit, two complete and separate paths can be identified in which current can flow. One path is traced from the source through resistance R_1 and back to the source; the other is traced from the source through resistance R_2 and back to the source.

You have seen that the source voltage in a series circuit divides proportionately across each load in the circuit. In a parallel circuit, the same voltage is present across all the loads of a parallel group. This voltage is equal to the applied voltage (E_{bb}). The foregoing statement can be expressed in equation form as

$$E_{bb} = E_{R1} = E_{R2} = E_{Rn}.$$

Voltage measurements taken across the loads of a parallel circuit, as illustrated by Fig. 5-12, verify the above equation. Each voltmeter indicates the same amount of voltage. Notice that the voltage across each load is the same as the applied voltage.

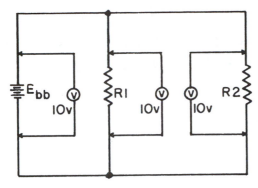

Fig. 5-12. Voltages in parallel circuits.

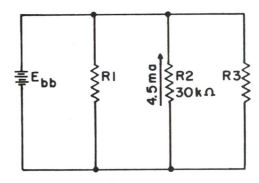

Fig. 5-13. Example #4.

Example #4: Assume that the current through a resistor of a parallel circuit is known to be 4.5 milliamperes (ma) and the value of the resistor is 30,000 ohms.* Determine the voltage across the resistor. The circuit is shown in Fig. 5-13.

Given:

R_2 = 30,000 ohms
I_{R_2} = 4.5 milliamperes = .0045 amperes.

Find:

$$E_{R_2} = ?$$
$$E_{bb} = ?$$

Solution: Select the proper equation from Chapter 2.

$$E = IR.$$

Substitute known values:

$E_{R_2} = I_{R_2} \times R_2$
E_{R_2} = .0045 amperes $\times$ 30,000 ohms
E_{R_2} = 4.5 $\times$ 30
Resultant: E_{R_2} = 135v
Therefore: E_{bb} = 135v.

Having determined the voltage across one load (R_2) in a parallel circuit, the value of the source voltage (E_{bb}) and the voltages across any other loads that may be connected in parallel with it are known.

The current in a circuit is inversely propor-

tional to the circuit resistance. This fact, obtained from Ohm's law, establishes the relationship upon which the following discussion is developed.

A single current flows in a series circuit. Its value is determined in part by the total resistance of the circuit. However, the source current in a parallel circuit divides among the available paths in relation to the value of the resistors in the circuit. Ohm's law remains unchanged. For a given voltage, current varies inversely with resistance.

The behavior of current in parallel circuits will be shown by a series of illustrations using example circuits with different values of resistance for a given value of applied voltage.

Example #5: Part (A) of Fig. 5-14 shows a basic series circuit. Here the total current must pass through the single resistor. The amount of current is determined as

$$I_{R_1} = \frac{E_{bb}}{R_1} = \frac{50}{10} = 5 \text{ amperes}$$
$$I_{R_2} = \frac{E_{bb}}{R_2} = \frac{50}{10} = 5 \text{ amperes.}$$

However, it is apparent that if 5 amperes of current flows through each of the two loads, there must be a total current of 10 amperes drawn from the source. The distribution of current in the simple parallel circuit shown in Fig. 5-14 (B) is as follows.

---

*Extremely large or extremely small numbers are often expressed in powers of 10, to make them easier to handle in mathematics. When you see a number as 4.5×10^{-3}, you simply move the decimal point 3 places to the left of the number, thus: .0045. Likewise, if the power of 10 is a positive number, as in 30×10^3, the decimal point is moved 3 places to the right of the number, which in this case results

in 30,000. As a further example:

$$6.8 \times 10^5 = 680,000$$
$$9.4 \times 10^{-4} = .00094.$$

(See page 363 in the Appendix for methods of handling calculations using this system.)

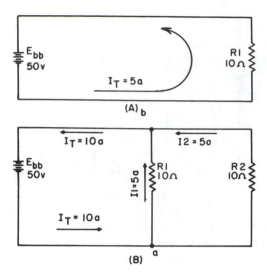

Fig. 5-14. Example #5.

The total current of 10 amperes leaves the negative terminal of the battery and flows to point a. Since point a is a connecting point for the two loads, it is called a junction. At junction a, the total current divides into two smaller currents of 5 amperes each. These two currents flow through their respective loads and rejoin at junction b. The total current then flows from junction b back to the positive terminal of the source. Thus, the source supplies a total current of 10 amperes and each of the two equal loads carries one-half the total current.

Each individual current path in the circuit of Fig. 5-14 (B) is referred to as a branch. Each branch will carry a current that is a portion of the total current. Two or more branches form a network.

From the foregoing observations, the characteristics of current in a parallel circuit can be expressed in terms of the following general equation:

$$I_t = I_1 + I_2 \ldots I_n.$$

Power computations in a parallel circuit are essentially the same as those used for the series circuit. Since power dissipation in loads consists of a heat loss, power dissipations are additive regardless of how the loads are connected in the circuit. The total power dissipated is equal to the sum of the powers dissipated by the individual loads. Like the series circuit, the total power consumed by the parallel circuit is

$$P_t = P_1 + P_2 \ldots P_n.$$

Example #6: Find the total power consumed by the circuit in Fig. 5-15. Solution:

$$P_{R1} = E_{bb} \times I_{R1}$$
$$P_{R1} = 50 \times 5$$
$$P_{R1} = 250 \text{ w}$$

$$P_{R2} = E_{bb} \times I_{R2}$$
$$P_{R2} = 50 \times 2$$
$$P_{R2} = 100 \text{ w}$$

$$P_{R3} = E_{bb} \times I_{R3}$$
$$P_{R3} = 50 \times 1$$
$$P_{R3} = 50 \text{ w}$$

$$P_t = P_1 + P_2 + P_3$$
$$P_t = 250 + 100 + 50$$
$$P_t = 400 \text{ w.}$$

Note that the power consumed in the branch circuits is determined in the same manner as the power consumed by individual loads in a series circuit. The total power (P) is then obtained by summing up the powers consumed in the branch loads.

Since, in the example shown in Fig. 5-15, the total current is known, the total power could be determined by the following method:

$$P_t = E_{bb} \times I_t$$
$$P_t = 50v \times 8a$$
$$P_t = 400 \text{ w.}$$

Rules for solving parallel DC circuits are summarized as follows.

1. The same voltage exists across each branch of a parallel circuit and is equal to the source voltage.

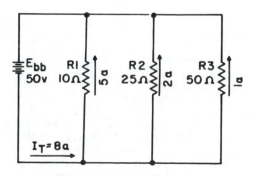

Fig. 5-15. Example #6.

2. The current through a branch of a parallel network is inversely proportional to the amount of resistance of the branch.

3. The total current of a parallel circuit is equal to the sum of the currents of the individual branches of the circuit.

4. The total power consumed in a parallel circuit is equal to the sum of the power consumption of the individual loads.

In the preceding discussions, series and parallel DC circuits have been considered separately. However, the technician will seldom encounter a circuit that consists solely of either type of circuit. Most circuits consist of both series and parallel elements. A circuit of this type will be referred to as a combination circuit. The solution of a combination circuit is simply a matter of application of the laws and rules discussed up to this point.

The foregoing material comprises only the basic principles involved, and are by no means the total forms or types of circuits which may be found in direct current circuits. We give them here merely as an illustration of what is possible with mathematics in the hands of an electrical engineer or technician.

E. Combination Circuits

So far, we have described the cases of either series or parallel circuits alone, and described the method of solving them for current, volt-age, and resistance. Now we shall give the general method for solving circuits in which parallel circuits are in combination with series circuits. The method is more involved, but not beyond the range of anyone who has the basic knowledge of ordinary mathematics, and the patience, skill at logic, and the initiative to stay with it until the solution is found.

The circuit is shown in Fig. 5-16, and the procedure for solving it is as follows.

1. Draw the circuit diagram.

2. Place upon it all of the known values and the values to be found.

3. List all of the equations needed for solution.

4. Calculate the resulting series-parallel circuits with the method given in Sections C and D.

So go at it, and here are the steps, the stages, and the solution.

1. Find the resistance of branch (a).

2. Find the resistance of branch (b).

3. Find the total circuit resistance.

4. Find the total circuit current.

5. Find the voltages E_{R1}, E_a, and E_b.

6. Find the current for branch (a) and (b).

7. Find the voltages E_{R2} and E_{R5}.

8. Find the currents I_1, I_2, I_3, and I_4.

9. Find the voltages E_{R3}, E_{R6}, and E_{R7}.

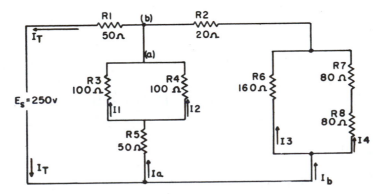

Fig. 5-16. Typical series-parallel circuit.

10. Find the power for R_8, branches (a) and (b), and R_1.

11. Find the total circuit power.

Solutions:

1. The resistance of branch (a) R_a is

$$R_a = \frac{R_3 \times R_4}{R_3 + R_4} + R_5$$

$$R_a = \frac{100 \times 100}{100 + 100} + 50$$

$$R_a = 50 + 50 = 100 \text{ ohms.}$$

2. The resistance of branch (b) R_b is

$$R_b = R_2 + \frac{(R_7 + R_8) R_6}{R_6 + R_7 + R_8}$$

$$R_b = 20 + \frac{(80 + 80) 160}{80 + 80 + 160}$$

$$R_b = 20 + 80 = 100 \text{ ohms.}$$

3. The total circuit resistance R_T is

$$R_T = \frac{R_a \times R_b}{R_a + R_b} + R_1$$

$$R_T = \frac{100 \times 100}{100 + 100} + 50$$

$$R_T = 50 + 50 = 100 \text{ ohms.}$$

4. The total circuit current is $I_T = \dfrac{E}{R_T}$

$$I_T = \frac{E}{R_T} = \frac{250}{100} = 2.5 \text{ amperes.}$$

5. The voltage drop of R_1 is

$$E_{R1} = I_1 R_1 = 2.5 \times 50 = 125 \text{ volts.}$$

The voltage for (a) is

$$E_a = E_S - E_{R1} = 250 - 125 = 125 \text{ volts.}$$

The voltage for (b) is

$$E_b = E_S - E_{R1} = 250 - 125 = 125 \text{ volts.}$$

6. The current for branch (a) is

$$I_a = \frac{E_a}{R_a} = \frac{125}{100} = 1.25 \text{ amperes.}$$

The current for branch (b) is

$$I_b = \frac{E_b}{R_b} = \frac{125}{100} = 1.25 \text{ amperes.}$$

7. The voltage drop across R_2 is

$$E_{R2} = I_b R_2 = 1.25 \times 20 = 25 \text{ volts.}$$

The voltage drop across R_5 is

$$E_{R5} = I_a R_5 = 1.25 \times 50 = 62.5 \text{ volts.}$$

8. The current I_1 is

$$I_1 = \frac{E_{R3}}{R_3} = \frac{62.5}{100} = 0.625 \text{ ampere.}$$

The current I_2 is

$$I_2 = I_a - I_1 = 1.25 - 0.625 = 0.625 \text{ ampere.}$$

The current I_3 is

$$I_3 = \frac{E_{R6}}{R_6} = \frac{100}{160} = 0.625 \text{ ampere.}$$

The current for I_4 is

$$I_4 = I_b - I_3 = 1.25 - 0.625 = 0.625 \text{ ampere.}$$

9. The voltage drop across R_3 is

$$E_{R3} = I_1 R_3 = 0.625 \times 100 = 62.5 \text{ volts.}$$

The voltage drop across R_6 is

$$E_{R6} = I_3 R_6 = 0.625 \times 160 = 100 \text{ volts.}$$

The voltage drop across R_8 is

$$E_{R8} = I_4 R_8 = 0.625 \times 80 = 50 \text{ volts.}$$

10. The power consumed by R_8 is

$$P_{R8} = I_4 E_{R8} = 0.625 \times 50 = 31.25 \text{ watts.}$$

The power consumed by branch (a) is

$$P_a = I_a E_a = 1.25 \times 125 = 156.25 \text{ watts.}$$

The power consumed by branch (b) is

$$P_b = I_b E_b = 1.25 \times 125 = 156.25 \text{ watts.}$$

The power consumed by R_1 is

$$P_{R1} = I_{R1} E_{R1} = 2.5 \times 125 = 312.5 \text{ watts.}$$

11. The total power consumed by the circuit is

$$P_T = P_{R1} + P_a + P_b = 312.5 + 156.25 + 156.25$$
$$= 625 \text{ watts}$$

or

$$P_T = EI_T = 250 \times 2.3 = 625 \text{ watts.}$$

C H A P T E R 6
STORAGE BATTERIES

A. Basic Principles

In this chapter we shall cover the subject of storage batteries, their principle characteristics, the various designs, and some of the applications. First, we discuss the various types which you may have occasion to use, and we shall save the most common one—the lead and acid storage battery—until the last, covering it in great detail, as it is the one you will probably come into contact with most frequently. It is also one of the most dangerous ones to handle, operate, and service. For that reason, we shall devote considerable coverage to the problems involved, in an effort to enable the plant maintenance man to give these power sources the respect and care they need and deserve.

Batteries are widely used as sources of direct-current electrical energy in automobiles, boats, aircrafts, ships, portable electric/electronic equipment, and lighting equipment. In some instances, they are used as the only source of power; while in others, they are used as a secondary or standby power source.

A battery consists of a number of cells assembled in a common container and connected together to function as a source of electrical power.

A cell is a device that transforms chemical energy into electrical energy. The simplest cell, known as either a galvanic or voltaic cell, is shown in Fig. 6-1. It consists of a piece of carbon (C) and a piece of zinc (Zn) suspended in a jar that contains a solution of water (H_2O) and sulfuric acid (H_2SO_4).

The electrodes are the conductors by which the current leaves or returns to the electrolyte. In the simple cell, they are carbon and zinc strips that are placed in the electrolyte; while in the dry cell, Fig. 6-2, they are the carbon rod in the center and the zinc container in which the cell is assembled.

The electrolyte is the solution that acts upon the electrodes which are placed in it. The electrolyte may be a salt, an acid, or an alkaline solution. In the simple galvanic cell and in the automobile storage battery, the electrolyte is in a liquid form; while in the dry cell, the electrolyte is a paste.

A primary cell is one in which the chemical action eats away one of the electrodes, usually the negative. When this happens, the electrode must be replaced or the cell must be discarded. In the galvanic type cell, the zinc electrode and the liquid solution are usually replaced when this happens. In the case of the dry cell, it is usually cheaper to buy a new cell. Some primary cells have been developed to the state where they can be recharged.

A secondary cell is one in which the electrodes and the electrolyte are altered by the chemical action that takes place when the cell delivers current. These cells may be restored to their original condition by forcing an electric current through them in the opposite direction to that of discharge. The automobile storage

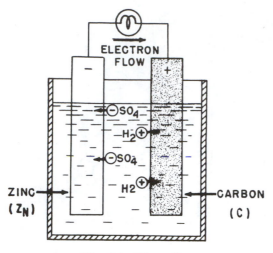

Fig. 6-1. Wet cell.

battery is a common example of the secondary cell.

Referring now to Fig. 6-1, if a conductor is connected externally to the electrodes of a cell, electrons will flow under the influence of a difference in potential across the electrodes from the zinc (negative) through the external conductor (load) to the carbon (positive), returning within the solution to the zinc.

When current flows through a cell, the zinc is gradually dissolved in the solution and the acid is neutralized. The chemical action that occurs in the cell, Fig. 6-1, while the current is flowing causes hydrogen bubbles to form on the surface of the positive carbon in great numbers until the entire surface is surrounded. This action is called polarization. Some of these bubbles rise to the surface of the solution and

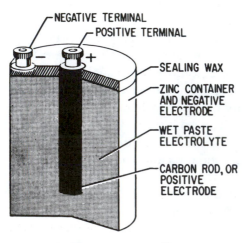

Fig. 6-2. Dry cell.

escape into the air. This hydrogen may be dangerous if it is allowed to accumulate in a poorly ventilated room.

The hydrogen tends to set up an electromotive force in the opposite direction to that of the cell, thus increasing the effective internal resistance, reducing the output current, and lowering the terminal voltage.

When the external circuit is opened, the current ceases to flow, and theoretically all chemical action within the cell stops. However, commercial zinc contains many impurities, such as iron, carbon, lead, and arsenic. These impurities form many small cells within the zinc electrode in which current flows between the zinc and its inpurities. Thus, the zinc is oxidized even though the cell itself is an open circuit. This wasting away of the zinc on open circuit is called local action, which explains why some batteries deteriorate while connected in an unused circuit.

B. Types of Batteries

The development of new and different types of batteries in the past decades has been so rapid that it is virtually impossible to have a complete knowledge of all of the various types currently being developed or now in use. A few recent developments are the silver-zinc, nickel-zinc, nickel-cadmium, magnesium-magnesium perchlorate, mercury, thermal, and water-activated batteries.

DRY (PRIMARY) CELL

The dry cell, Fig. 6-2, is so called because its electrolyte is not in a liquid state. Actually, the electrolyte is a moist paste. If it should become dry, it would no longer be able to transform chemical energy to electrical energy. The name "dry cell," therefore, is not strictly correct in a technical sense.

Binding posts are attached to the electrodes so that wires may be conveniently connected to the cell.

Since the zinc container is one of the electrodes, it must be protected with some insulating material. Therefore, it is common practice for the manufacturer to enclose the cells in cardboard containers.

One of the popular sizes in general use is the standard, or No. 6, dry cell. It is approximately $2\frac{1}{2}$ inches in diameter and 6 inches in length. The voltage is about $1\frac{1}{2}$ volts when new but decreases as the cell ages. When the open-circuit voltage falls below 0.75 to 1.2 volts (depending upon the circuit requirements), the cell is usually discarded. The amount of current that the cell can deliver, and still give satisfactory service, depends upon the length of time that the current flows. For instance, if a No. 6 cell is to be used in a portable radio, it is likely to supply current constantly for several hours. Under these conditions, the current should not exceed $\frac{1}{8}$ ampere, the rated constant-current capacity of a No. 6 cell. If the same cell is required to supply current only occasionally, for only short periods of time, it could supply currents of several amperes without undue injury to the cell. As the time duration of each discharge decreases, the interval of time between discharges increases, the allowable amount of current available for each discharge becomes higher, up to the amount that the cell will deliver on short circuit.

The short-circuit current test is another means of evaluating the condition of a dry cell. A new cell, when short-circuited through an ammeter, should supply not less than 25 amperes. A cell that has been in service should supply at least 10 amperes if it is to remain in service.

A popular size of dry cell, the size D, is $1\frac{3}{8}$ inches in diameter and $2\frac{3}{4}$ inches in length. It is also known as the unit cell. The size D cell voltage is 1.5 volts when new.

A cell that is not being used (sitting on the shelf) will gradually deteriorate because of slow internal chemical action (local action) and changes in moisture content. However, this deterioration is usually very slow if cells are properly stored. Highgrade cells of the larger sizes should have a shelf life of a year or more. Smaller size cells have a proportionately shorter shelf life, ranging down to a few months

for the very small sizes. If unused cells are stored in a cool place, their shelf life will be greatly increased; therefore, to minimize deterioration, they should be stored in refrigerated spaces (10° F to 35° F) that are not dehumidified (dry).

MERCURY CELLS

With the advent of the national space program and the development of small transceivers and miniaturized equipment, a power source of miniaturized size was needed. Such equipment requires a small battery which is capable of delivering maximum electrical capacity per unit volume while operating in varying temperatures and at a constant discharge voltage. The mercury battery, which is one of the smallest batteries, meets these requirements.

Present mercury batteries are manufactured in three basic structures, as shown in Fig. 6-3.

NOTE: Mercury batteries have been known to explode with considerable force when shorted. Caution should be exercised to ensure that the battery is not accidentally shorted.

SECONDARY (WET) CELLS

Secondary cells function on the same basic chemical principles as primary cells. They differ mainly in that they may be recharged, whereas the primary cell is not normally recharged. As mentioned earlier, some primary cells have been developed to the state where they may be recharged. Some of the materials of a primary cell are consumed in the process of changing chemical energy to electrical energy. In the secondary cell, the materials are merely transferred from one electrode to the other as the cell discharges. Discharged secondary cells may be restored (charged) to their original state by forcing an electric current from some other source through the cell in the opposite direction to that of discharge.

NICKEL-CADMIUM BATTERIES

The nickel-cadmium batteries are far superior to the lead-acid type. Some are physically and electrically interchangeable with the lead-acid type, while some are sealed units which use standard plug and receptacle connections which are used on other electrical components. These batteries generally require less maintenance than lead-acid batteries throughout their service life in regard to the adding of electrolyte or water.

The nickel-cadmium and lead-acid batteries have capacities that are comparable at normal

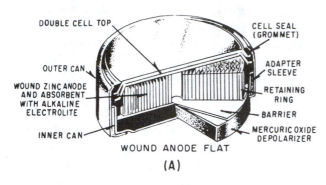

WOUND ANODE FLAT
(A)

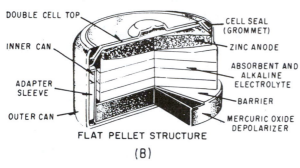

FLAT PELLET STRUCTURE
(B)

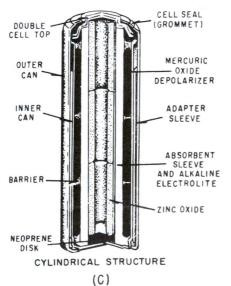

CYLINDRICAL STRUCTURE
(C)

Fig. 6-3. Mercury batteries. (A) Wound anode flat. (B) Flat pellet structure. (C) Cylindrical structure.

discharge rate, but at high discharge rates the nickel-cadmium battery can:

1. be charged in a short time;

2. deliver a large amount of power;

3. stay idle in any state of charge for an indefinite time and keep a full charge when stored for a long time;

4. be charged and discharged any number of times without any appreciable damage; and

5. the individual cells may be replaced if a cell wears out; the rest of the cells do not have to be replaced.

Due to their superior capabilities, nickel-cadmium batteries are being used extensively in many military applications that require a battery with a high discharge rate. A prime example is the aircraft storage battery.

Some lead-acid batteries are equipped with the same quick-disconnect receptacle and plug used on nickel-cadmium batteries. In distinguishing a lead-acid battery, the nameplate of each battery should be checked, since the physical appearance could be the same. See Fig. 6-4.

The electrolyte used in a nickel-cadmium battery is a 30-percent-by-weight solution of potassium hydroxide in distilled water. Chemically speaking, this is just about the exact opposite to the diluted sulfuric acid used in the lead battery. As with lead batteries, there are limitations on the concentration of electrolyte solution that can be used in nickel-cadmium cells. The specific gravity of the solution should not be outside the range of 1.240 to 1.300 at 70° F. It is not possible to determine the charge state of a nickel-cadmium battery by checking the electrolyte with a hydrometer; neither can the charge be determined by a voltage test because of the inherent characteristic that the voltage remains constant during 90 percent of the discharge cycle.

No external vent is required since gassing of this type battery is practically negligible. As a safety precaution, however, relief valves have been installed in the fill hole cap of each cell in

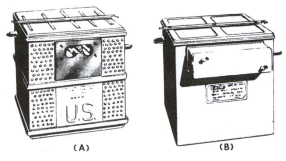

Fig. 6-4. Typical lead-acid battery.

order to release any excess gas that is formed when the battery is charged improperly.

At the present time, a simple method of determining the exact state of charge has not been developed. In an aircraft battery, the only practical method is to measure the open circuit battery voltage. If the open circuit voltage exceeds 1.4 volts per cell and the battery voltage is 26.0 volts (normally 28.0 volts) or more, it can be assumed that the battery is fully charged. If the cell voltage is less than 1.0 volt under load, or the battery voltage measures 25 volts or less, it can be assumed that the battery has expended over 90 percent of its capacity and is in need of charging.

In a battery shop, either of two methods may be used for determining the state of charge of a nickel-cadmium battery; namely, the constant potential method or the discharge method. The constant potential method consists of connecting a constant potential of 28.5 ± 0.3 volts across the battery and observing the charging current. If the current falls to 3 amperes or less within 5 minutes, the battery is charged. The discharge method consists of placing a 15-ampere load across the battery for 5 minutes. If the voltage does not drop below 22 volts during the discharge period, the battery may be returned to service after being recharged.

The available ampere-hour capacity cannot be accurately determined. Therefore, it is recommended that any battery whose charge is unknown or subject to doubt be discharged to or beyond the manufacturer's set end point of 1.0 or 1.1 volts, and then be recharged in accordance with the appropriate instructions. This process will prevent possible damage to the cells from overcharge.

Nickel-cadmium batteries should preferably be charged at an ambient temperature of 70° F to 80° F. Never allow a battery on charge to exceed 100°F as this may cause overcharging and gassing. In the battery shop, a thermometer should be placed between the central cells in such a manner that the bulb of the thermometer is located below the top of the cell. Whenever the temperature of the battery is 100° F or higher, the battery should not be charged.

The rate of charging of a nickel-cadmium battery is dependent upon two factors, the first being the charging voltage, and the second being the temperature of the battery. In hot weather where the air temperature approaches 90° F or higher, the battery can be adequately charged at 27 volts. In mild air temperatures ranging from 35° F to 85° F, the battery can be satisfactorily charged at 27.5 volts. In cold, sub-freezing weather, the battery requires a charging voltage of 28.5 volts.

The nickel-cadmium battery was designed and constructed to operate without gassing of the cells. The charging voltage should be maintained below the gassing voltage (approximately 29.4 volts at 80° F) so that the life of the battery is prolonged. Therefore, on constant potential charging in the battery shop, the voltage should be set at 28 volts or less. Under no circumstances should this voltage exceed 28.5 volts.

If the battery has never been placed in service, follow the manufacturer's instruction accompanying the battery for the initial charge. If possible, the battery should be charged by the constant potential method.

For constant potential charging, maintain the battery at 28 volts for 4 hours, or until the current drops below 3 amperes. Do not allow battery temperature to exceed 100° F.

For constant current charging, start the charge at 10 to 15 amperes and continue until the voltage reaches 28.5 volts—then reduce the current to 4 amperes and continue charging until the battery voltage reaches 28.5 volts, or until the battery temperature exceeds 100° F and the voltage begins to decline.

Never add electrolyte unless the battery is fully charged. Allow the fully charged battery to stand for a period of 3 or 4 hours before distilled water is added to bring the electrolyte to the proper level. A hydrometer or syringe can be used for introduction of the distilled water—just enough to cover the top of the plates. The battery solution is then recycled to stir the water and prevent it from freezing during cold-weather operation.

The electrolyte used in nickel-cadmium batteries is potassium hydroxide (KOH). This is a highly corrosive alkaline solution, and should be handled with the same degree of caution as sulfuric acid (H_2SO_4). Personnel should always wear rubber gloves, a rubber apron, and protective goggles when handling and servicing these batteries. If the electrolyte is spilled on the skin or clothing, the exposed area should be rinsed immediately with water or, if available, vinegar, lemon juice, or boric acid solution. If the face or eyes are affected, treat as above and report immediately for medical examination and treatment.

The battery shop used for nickel-cadmium batteries should be separately isolated from the lead-acid battery shop.

SILVER-ZINC BATTERIES

The silver-zinc battery was developed for one major and one secondary purpose. The major purpose was to secure a large quantity of electrical power for emergency operations. The secondary purpose was to permit a design weight savings in new batteries. A lightweight, silver-zinc battery provides as much electrical capacity as a much larger lead-acid or nickel-cadmium battery.

Operational silver-zinc batteries have a nominal operating voltage of 24 volts, obtained with sixteen 1.5 volt cells. Cell electrolytic levels should be monitored and adjusted periodically. The other required operations that might be considered maintenance are the normal recharging of the battery and keeping the top surfaces of the cells reasonably clean.

Because of its extremely low internal resistance, the silver-zinc battery is capable of discharge rates of up to 30 times its ampere-hour rating. Silver-zinc batteries are capable of producing as much as six times more energy per unit of weight and volume than other types. Silver-zinc cells have been built with capacities ranging from tenths of ampere-hours to thousands of ampere-hours.

Good voltage regulation is provided by the relatively constant voltage discharge characteristic of the silver-zinc battery. Terminal voltage is essentially constant throughout most of the discharge when discharged at rates higher than a 2- or 3-hour rate.

Silver-zinc batteries have a maximum service cycle life which is less than that of other types, but their life expectancy compares favorably with that of other types of batteries that are designed for maximum capacity per unit of space and weight such as nickel-cadmium batteries.

As with other types of alkaline cells, and unlike lead-acid cells, the electrolyte does not take part in the chemical transformations and therefore its specific gravity does not change with the state of charge of the cell. As long as the plates are covered, the electrical capacity of the battery is independent of the amount of electrolyte present.

In general, silver-zinc batteries require maintenance which is similar in many respects to that which is required of the lead-acid batteries. Testing the open circuit voltage of the battery is the method by which its state of charge is determined.

A silver-zinc battery tester or a voltmeter which reads accurately to 0.1 volt should be used to test the open circuit voltage of the battery. If the reading is below 25.6 volts, remove the battery cover and inspect the top of the battery for corrosion or damaged cells. If any damage is evident, remove and replace the battery.

The silver-zinc battery is usually shipped in a dry condition. Only the special electrolyte furnished in the filling kit provided with each new battery should be used. The filling kit contains detailed instructions for filling and should be followed in detail. (NOTE: Batteries that will not be used within 30 days should be stored in the dry state.)

Silver-zinc batteries are sensitive to excessive voltage during charging and may be damaged if the voltage exceeds 2.05 volts per cell. Precaution must therefore be taken to ensure that the charging equipment is adjusted accurately to cut off the current at 28.7 volts.

Where charging is not monitored automatically or periodically, a voltage cut off system must be used which will interrupt the charging current when the voltage rises to 28.7 volts.

If possible, charging should be performed at an ambient temperature of 60° to 90° F, and the battery temperature during charging should not exceed 150°F as measured at the intercell connections.

While silver-zinc batteries do not generate any harmful gasses during normal charge and discharge operations, they do generate both oxygen and hydrogen gasses during excessive overcharging. All vent caps and sponge-rubber plugs must be removed from the vent hole during charging operations. If electrolyte is forced from the vent holes or if excessive gassing is evident, it is an indication of overheating, and the charging should be interrupted for 8 hours to allow the battery to cool. After charging, the batteries should be allowed to stand idle at least 3 hours.

The level of the electrolyte of each cell of the battery should be checked after charging, and the level adjusted by either removing any excess electrolyte or by adding distilled water if low.

The safety precautions relating to silver-zinc batteries are the same as those which have been presented for the nickel-cadmium batteries.

SILVER-CADMIUM BATTERY

One of the recent developments in storage batteries is the silver-cadmium battery. Generally, the most important requirements for evaluating and designing a battery are for high energy density, good voltage regulation, long shelf life, repeatable number of cycles, and long service life expectancy. The silver-cadmium battery is designed to offer the overall maximum performance in all of these expectations.

The silver-cadmium battery has more than twice the wet shelf life of the silver-zinc battery. The long shelf life plus the good voltage regulation makes the silver-cadmium battery a highly desirable addition to the family of electric storage batteries. Limitations include lower cell voltage than other rechargeable batteries and high initial cost.

C. Lead-Acid Storage Batteries*

Storage batteries do not actually store electrical energy. They, instead, accept the electrical energy delivered to them during charging periods and convert it into chemical energy which is slowly accumulated as the charge progresses. A battery in use is said to be on discharge. During discharge, the chemical energy in the battery is converted into usable electrical energy.

A lead-acid storage battery consists of cells with positive and negative plates, called "electrodes," which are physically separated from each other and immersed in an electrolyte, such as sulphuric acid solution. The active materials of the electrodes are Lead Peroxide for the positive plates and Sponge Lead for the negative, as shown in Figs. 6-5 and 6-6.

In a fully charged cell, the electrolyte has a specific gravity that varies from 1.200 to 1.325 depending on the type of service required of the cell. When fully charged, each cell has a voltage of approximately 2 volts on open circuit. However, a cell may have a voltage of from 2.12 to 2.70 volts while being charged.

A storage battery cell develops a voltage potential when any two dissimilar metals are immersed in a suitable electrolyte. The two metals used in lead-acid cells result in a voltage potential of 2 volts per cell, and this potential does not vary regardless of cell size. However, cell size does vary the ability to deliver this

---

*Material in this section was supplied by GNB Industrial Battery Co. Although the descriptions cover GNB products, the discussion is applicable to other similar products being used in today's industrial plants.

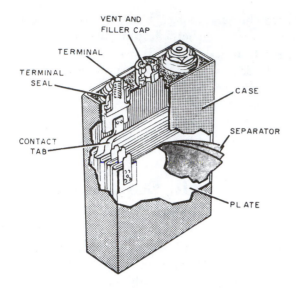

Fig. 6-5. Lead-acid battery cell.

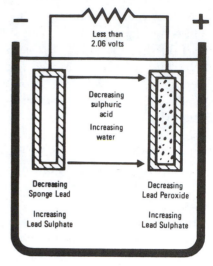

Fig. 6-7. Discharging cell. (Courtesy GNB, Inc.)

voltage under varying loads and for varying periods of time.

During cell discharge, Lead Peroxide and Sponge Lead combine with sulphuric acid to form Lead Sulphate on both plates, as shown in Fig. 6-7. This action decreases cell voltage as the two electrodes approach being of the same chemical composition (Lead Sulphate). As the sulphuric acid is removed from the electrolyte solution, the specific gravity of the electrolyte decreases and approaches the specific gravity of water. This condition is shown in Fig. 6-8.

Specific gravity is the weight of electrolyte as compared to an equal amount of water.

CHARGING THE CELL

When a charging current is applied to a discharged cell, as shown in Fig. 6-9, the Lead Sulphate is broken up, the active materials are restored to their respective plates, and the electrolyte again becomes a sulphuric acid solution. Cell voltage rises as the two elements become increasingly different in composition,

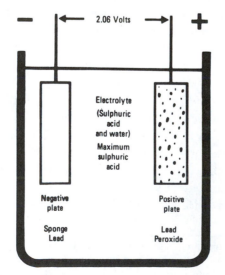

Fig. 6-6. Charged cell. (Courtesy GNB, Inc.)

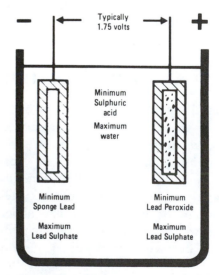

Fig. 6-8. Discharged cell. (Courtesy GNB, Inc.)

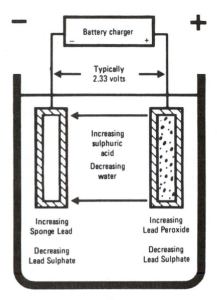

Fig. 6-9. Charging cell. (Courtesy GNB, Inc.)

and the specific gravity of the electrolyte increases as more and more acid is formed.

When a cell is fully charged, the positive electrode is Lead Peroxide, the negative electrode is Sponge Lead, and the electrolyte is a sulphuric acid solution.

BATTERY RATINGS

A single lead-acid cell does not have sufficient power to handle most requirements. However, connecting a number of cells together in series results in a battery capable of supplying higher power demands.

The number of cells is determined by the required nominal operating voltage of the equipment. Since each cell has a nominal voltage of 2 volts, a 24-volt industrial truck will require a 12-cell battery (12 cells × 2 volts/cell = 24 volts).

The ampere hour rating or capacity of a battery is the same as the rating of any one of its cells. This rating is the number of amperes (current) the cell will deliver for a fixed period of time. Example: 120 amperes for six hours = 720 ampere hours (six hour rate). Increasing or decreasing the number of cells in the battery does not change this rating. It can only be changed by increasing or decreasing the size or number of plates in a cell.

Battery capacity is also expressed in kilowatt hours (KWH), which is the product of amperes × time × average volts per cell. Example: 120 amperes × 6 hours × 1.93 average volts per cell = 1389.6 watt hours divided by 1000 = 1.3896 KWH. For a 12-cell battery, the capacity would be 1.3896 × 12 = 16.67 KWH.

The Kilowatt Hour rating can be varied by increasing or decreasing the size of the cells or the number in the battery.

Positive plate capacity is the amperes delivered for a fixed period of time (usually 6 hours) for a particular size positive plate. A GNB 120C type positive plate has the capability of delivering 20 amperes for 6 hours or 120 ampere hours (20 × 6 = 120 AH) to a final voltage of 1.70.

This ampere hour rating or capacity can be varied by increasing or decreasing the number of positive plates in the cell. In the previous examples, the battery is a 12 cell, 120C-13 plate unit. To determine the number of positive plates in each cell, subtract 1 from the total number of plates in the cell and divide by 2. Example: 13-1 = 12 divided by 2 = 6 positive plates × 120 ampere hours each = 720 AH.

The above ratings are based on a temperature of 77° F (25° C) and with a fully charged specific gravity of 1.280 to 1.290.

RECEIVING A NEW BATTERY

When receiving a new battery, it is extremely important to examine the exterior of the packing. Examine for wet spots on sides and bottom, which may indicate leaking jars broken in shipment. If any damage has occurred, take immediate and proper claim measures with the carrier.

If any jars are damaged and the electrolyte has leaked out, make immediate repairs and replace broken jars at once. If a replacement jar is not immediately available, withdraw the elements from the damaged jar and place the elements in a glass, porcelain, or rubber vessel containing water suitable for battery use. Sufficient water must be added to completely cover the plates and separators. Damage or complete destruction of the cell may result if these procedures are not followed.

NOTE: Use distilled water or tap water which has been approved for use in batteries.

Special attention should be given to cells that require new jars. The cells should be refilled with electrolyte of the same specific gravity as measured in the balance of cells, and a charge should be applied at a low finishing rate until the specific gravity of the electrolyte ceases to rise. If the specific gravity after charging is lower than that of a normal, fully charged cell, a small amount of the electrolyte should be withdrawn and replaced with electrolyte of 1.400 specific gravity. The battery should then be given an additional charge for $\frac{1}{2}$ to 1 hour to thoroughly mix the liquid. Repeat this process until the normal specific gravity is obtained.

NOTE: **All repairs of this nature should be recorded and the supplier notified.**

PLACING BATTERY IN SERVICE

Upon receipt of the battery, give it a freshening charge for from 3 to 6 hours, until the specific gravity indicates no further rise. The charge should be given only with a direct-current charger, and with the positive terminal of the charger connected to the positive terminal of the battery and the negative terminal of the battery connected to the negative terminal of the charger.

CAUTION: **Permanent damage to the battery may result if the battery is connected incorrectly.**

Cell temperature during charge should not exceed 100° F. All points of contact between charger and battery should be clean to ensure good conductivity through terminal connections. If connections are copper, apply a coat of petroleum jelly or no-oxide grease to prevent corrosion.

If the battery is installed in a vehicle, properly fasten it in place by hold-downs on the battery or bars of the vehicle to reduce vibration and jarring to the minimum. If the battery is to be installed in a metal compartment, make sure the compartment is thoroughly dry and free of moisture prior to installation. If the battery is to be installed in a locomotive, block the battery in position allowing a $\frac{1}{8}$ inch space between the block and battery tray. Do not wedge the battery in position. All connections between the battery and the vehicle must be flexible. All vent caps must be in place while battery is in service. Failure to keep caps in position will result in loss of electrolyte (and therefore capacity) and will cause corrosion to the outside of the battery and the surrounding areas.

OPERATION

Do not overcharge the battery by using too high a rate of charge. Overcharging causes excessive gassing, high battery temperatures, shorter battery life, and a waste of power.

Batteries should be of the proper ampere-hour rating for the specific type of work intended. Extending or exceeding the workload overdischarges the battery and possibly shortens its life.

Limit the discharge level of the cells to the specific gravity values shown in Fig. 6-10 for the type cell involved. To obtain maximum life and serviceability, it is recommended that an 80% discharge not be exceeded. Only when occasional demands require it should cells be discharged to the 100% level. If the service requirements only partially discharge the battery, it is not necessary to recharge after each partial use. If the specific gravity is 1.240 or above, defer recharging until the recommended discharge limit is reached. However, recharge if the next duty cycle is expected to exceed the remaining battery capacity. Periodic hydrometer checks will assist in determining the need for recharging.

NOTE: **The battery must always be recharged immediately following a complete discharge.**

Inspect the battery once each week to ensure that connections are tight. Remove dust or dirt accumulations from the battery top. At least once each month, neutralize any acid on the battery covers and terminals with a baking soda solution (1 pound of soda per gallon of wa-

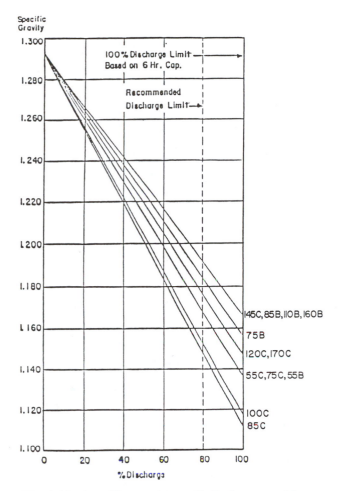

Fig. 6-10. Specific gravity vs. % discharge at 6 hour rate. (Courtesy GNB, Inc.)

ter) and then wash the battery with clean water and dry with compressed air. Keep terminals and metal parts free of corrosion.

Check electrolyte levels daily and replace water lost due to evaporation and electrolysis occurring during the charging process. Never allow electrolyte levels to drop below separator protectors. Do not overfill cells. Fill only to bottom of vent well, as overfilling causes loss of acid, thus reducing battery capacity, and also results in corrosion of the steel tray.

To ensure that water is thoroughly mixed with the electrolyte, additions should be made while the battery is on charge. Only suitable water should be used for batteries, as certain impurities are harmful and will reduce battery life. Water sources in certain geographic areas are not suitable at any time, and in other areas

are only satisfactory during certain seasons of the year.

The recommendation for battery replacement water is shown in this listing below indicating the maximum allowable impurities in parts per million:

TOTAL PARTS	125 PPM
FIXED SOLIDS	75 PPM
ORGANIC AND VOLATILE MATTER	50 PPM
CHLORIDES AS CL	25 PPM
NITRATES AS NO_3	10 PPM
NITRATES AS NO_2	5 PPM
AMMONIA AS NH_3	5 PPM
IRON	4 PPM
MANGANESE	.007 PPM

HYDROMETERS AND THERMOMETERS

The syringe type hydrometers (Fig. 6-11) have four parts—a rubber nozzle, glass barrel, float, and rubber anti-roll device. Inspect all parts carefully for breaks or cracks. Wash the instruments to prevent any foreign material

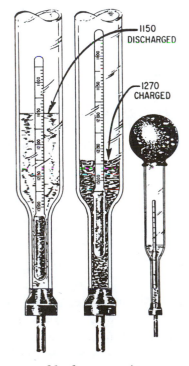

1150 DISCHARGED

1270 CHARGED

Fig. 6-11. Typical hydrometer. (Courtesy GNB, Inc.)

from getting into the battery electrolyte during use.

To assemble a new hydrometer, insert the float into the barrel, weighted end first, then install the rubber insert. Finally, install the rubber bulb and rubber tip. All rubber parts should be lubricated first with water.

CAUTION: Wear rubber gloves and face protection when assembling and using the hydrometer.

Reading Hydrometer: Place the rubber nozzle of the hydrometer into the vent opening of the battery and draw enough liquid into the barrel to permit the float to ride free. The float must not touch the side, top, or bottom of the barrel. Hold vertical and at eye level. If the hydrometer has to be removed from the vent, pinch the nozzle tightly or place gloved finger against the opening to prevent dripping of electrolyte. Read the float scale at the electrolyte level, disregarding the curvature of the liquid.

Temperature Correction: Correction of the specific gravity for temperature is plus or minus three points (one point = .001) of gravity for each 10° F change in electrolyte temperature. Using 77° F as a base, subtract three points for each 10° F below, or add three points for each 10° F above the 77° base value. This correction must be considered a part of normal specific gravity readings. As the temperature of the electrolyte rises, the electrolyte expands and therefore is not as dense. The float then rides low, giving readings lower than normal. Refer to the following example as an aid in more fully understanding temperature correction.

Hydrometer Reading (Example)	Thermometer Reading	Correction	True Specific Gravity
1.250	87°F	+3	1.253
1.210	80°F	+1	1.211
1.180	64°F	−4	1.176

Reading Thermometer: One well-known thermometer, Fig. 6-12, has a scale which ensures quick, accurate readings. Adjacent to the normal temperature scale is a column of red figures with a plus or minus sign before the figures for the amount of correction required for

Fig. 6-12. Battery thermometer. (Courtesy GNB, Inc.)

obtaining True Specific Gravity. For example, if the hydrometer reading is 1.260 and the red scale on the thermometer indicates a plus 8, the true specific gravity is 1.268.

How to Determine Battery Condition: There are three primary methods used to determine the condition of a motive power battery.

   a. Maintain accurate records.

   b. Perform test discharges.

   c. Conduct internal inspections.

Accurate record keeping can provide a complete day-to-day case history of the battery. Daily battery record forms prepared by the user to suit local conditions should include the following information:

   1. Date

   2. Battery number

   3. Identification of truck using battery

   4. Specific gravity of battery when put on charge

   5. Temperature (pilot cell) going on charge

   6. Time put on charge

7. Indication if water added

8. Time taken off charge

9. Specific gravity and temperature

10. Date when put back in truck.

Figure 6-13 shows a suggested Daily Battery Record form. This will help to pinpoint the degree of discharge, including equipment in-

volved, and will also show if the battery went into service fully charged. It will provide the necessary information to calculate the number of charge–discharge cycles. If the battery temperature and specific gravities show a deviation from the normal values coming off charge, possible charger problems can be detected and the unit responsible identified.

Individual cell specific gravity readings, to-

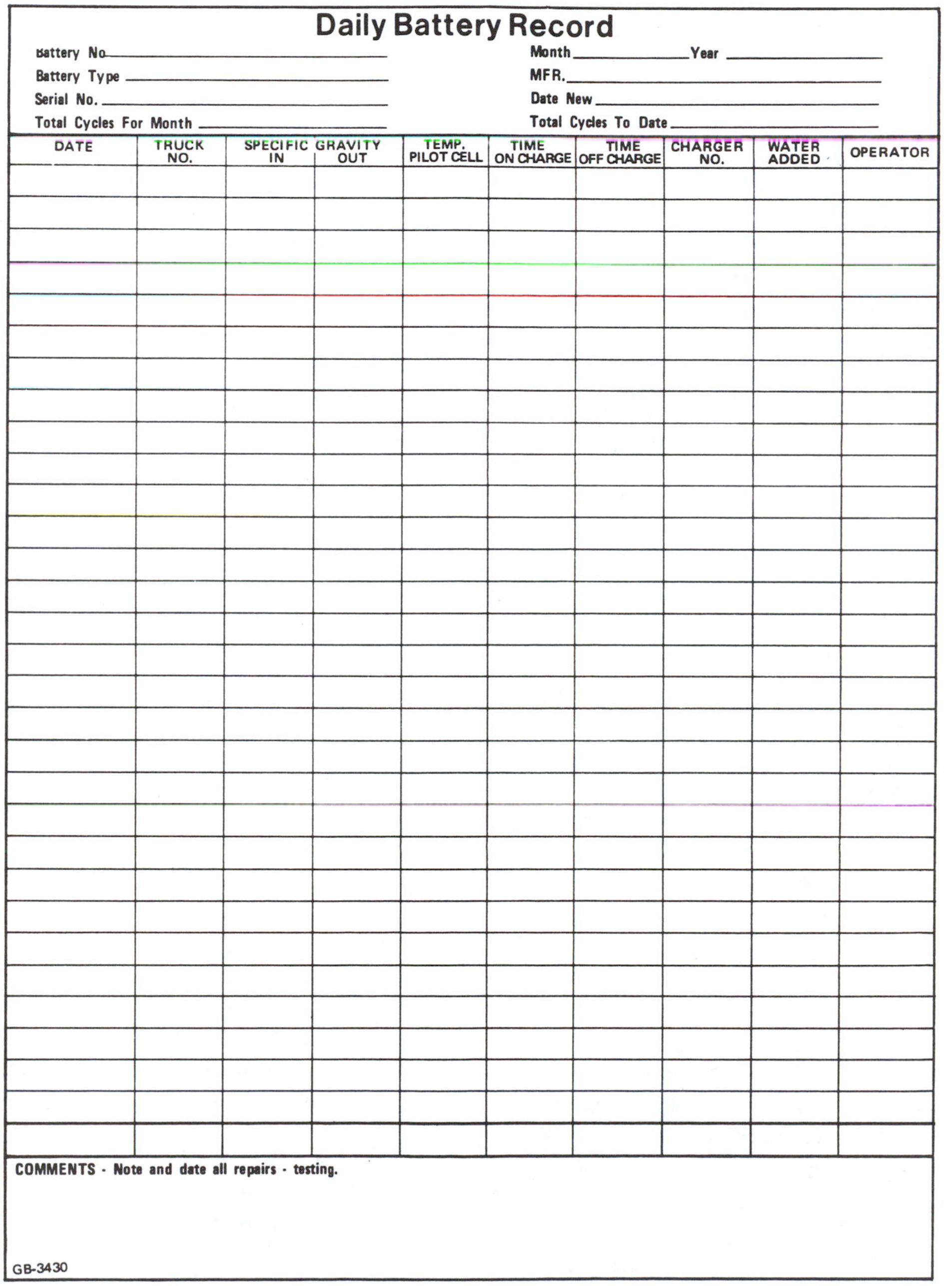

Daily Battery Record

Battery No_____   Month_____Year_____
Battery Type_____   MFR._____
Serial No._____   Date New_____
Total Cycles For Month_____   Total Cycles To Date_____

DATE	TRUCK NO.	SPECIFIC IN	GRAVITY OUT	TEMP. PILOT CELL	TIME ON CHARGE	TIME OFF CHARGE	CHARGER NO.	WATER ADDED	OPERATOR

COMMENTS - Note and date all repairs - testing.

GB-3430

Fig. 6-13. Daily battery record form. (Courtesy of GNB, Inc.)

gether with on-charge cell voltage readings, are valuable in determining if cells are healthy or have internal problems. A supplementary method of determining this is cadmium electrode testing which is performed with the battery on charge, near the end of the charging period.

A new cadmium electrode must be "aged" for at least 72 hours by soaking it in 1.280 specific gravity sulphuric acid. If, after aging, the electrode is allowed to dry out, re-soak it for 30 minutes to ensure reliable readings.

To obtain the positive cadmium reading of a cell, connect the cadmium electrode to the negative lead of a 0–3 volt range voltmeter. Connect the positive voltmeter probe to the positive cell terminal. Immerse the cadmium electrode in the cell electrolyte and continually agitate the electrode to minimize formation of gas bubbles so a reliable reading can be obtained.

Take and record the positive cadmium readings for all cells in the battery while on charge. Also take and record cell voltages and specific gravities of all cells and temperatures on at least three cells (outside–middle–outside). Record data on a form similar to the Battery Inspection Report (Fig. 6-14).

The following table shows typical positive cadmium readings of a six cell battery in a fully charged condition, with cell temperature at 77° F (25° C) as a typical situation.

Cell Number	On-Charge Cell Voltage	Positive Cadmium Voltage
1	2.58	2.44
2	2.56	2.43
3	2.39	2.43
4	2.58	2.42
5	2.39	2.24
6	2.57	2.44

Cell number 5 has a low on-charge voltage of 2.39. In addition, the positive cadmium reading is also low, indicating probable problems with the positive plates, thus requiring an internal cell inspection. Cell number 3 also has a below normal on-charge voltage, however, the positive cadmium reading is normal. This may indicate negative plate trouble. Cadmium read-

ings therefore are useful in identifying whether positive or negative plates are involved in the low cell voltage condition. Refer to the Trouble-Shooting Chart in Section D for probable causes.

Test Discharge: A test discharge determines if a battery can deliver its rated capacity. The test is conducted by discharging a fully charged battery at a constant ampere rate until the battery voltage drops to the accepted discharge termination value of 1.70 volts per cell. By noting the time lapse between when the battery is put on discharge and the time the termination value is reached, it will be indicated whether the battery is delivering rated capacity.

Motive power batteries are generally discharged at a cataloged six-hour rate to standardize and simplify discharging (one-sixth of the ampere-hour capacity at the six-hour rate). Before making the test, the battery is given an equalizing charge, and the fully charged specific gravity is adjusted to normal. The test proceeds as follows.

(a) Record time at which the discharge test is started.

(b) During the test, individual cell voltages and overall battery voltages are recorded at intervals. The first interval should be 15 minutes after starting the test and then at each hour, from starting time, until voltage of any one cell reaches 0.05 volts above the termination value (1.70 volts).

(c) After conditions described in Step (b) have been reached, take voltage readings at 15 minute intervals.

(d) Record the time when each cell voltage goes below termination value.

(e) Stop the test discharge when the majority of the cells reach termination voltage and before any single cell goes into reversal.

(f) Calculate capacity delivery. For example, if test terminated at 5.6 hours, the capacity percentage equals:

$$\frac{5.6 \text{ hours}}{6 \text{ hours}} \times 100 = 93\%$$

Battery Inspection Report Date _____

Battery No. _____ Dept. Used In _____
Battery Type _____ MFR. _____
Serial No. _____ Date New _____
Reading on Charge at _____ Amperes Charger No. _____

CELL NO.	CELL VOLTS	POSITIVE CADMIUM	SPECIFIC GRAVITY	TEMP.	CELL NO.	CELL VOLTS	POSITIVE CADMIUM	SPECIFIC GRAVITY	TEMP.
1					33				
2					34				
3					35				
4					36				
5					37				
6					38				
7					39				
8					40				
9					41				
10					42				
11					43				
12					44				
13					45				
14					46				
15					47				
16					48				
17					49				
18					50				
19					51				
20					52				
21					53				
22					54				
23					55				
24					56				
25					57				
26					58				
27					59				
28					60				
29					61				
30					62				
31					63				
32					64				

COMMENTS

GB-3431

READINGS TAKEN BY

Fig. 6-14. Battery inspection report form. (Courtesy of GNB, Inc.)

(g) Record the specific gravity of each cell immediately after terminating the test discharge. The readings will determine whether the battery is uniform or if any one or more cells are low in capacity. If the battery is uniform and delivers 80% or more of its rated capacity, the battery can be returned to service.

(h) Further guidance for conducting tests can be obtained from NEMA Standards Publication No. IB2-1987 entitled "Determination of Capacity of Lead Acid Industrial Storage Batteries for Motive Power Service." Copies of this can be purchased at a nominal cost. Order from:

NEMA Standards Sales Office
2101 L Street, NW
Washington, DC 20037

Internal Inspection: If the test discharge previously explained indicates that the battery was not capable of delivering more than 80% of rated capacity and all cells are uniform, an internal inspection should be made of one of the cells.

The positive plates, which wear first, should be examined. If it is discovered that the positive plates are falling apart or that the grids have many frame fractures, a replacement battery is needed. If the positive plates are in good mechanical condition and the cells contain little sediment, the battery may be sulphated.

Causes of Sulphated Batteries: A sulphated battery is one that has been left standing in a discharged condition or has been undercharged to the point where abnormal Lead Sulphate has formed on the plates. If this occurs, see Section H in the Appendix for proper cure.

D. Trouble-Shooting Chart, Lead-Acid Batteries

a. Battery Overheats on Discharge

Probable Cause	Possible Remedy
1. Overdischarge	Limit discharge to 1.110 specific gravity. Put more batteries into service. Set up more frequent charging schedule.
2. Excessive load	Determine cause of overload and correct. Put more batteries into service if equipment requirements exceed battery capacity.
3. Not fully charged prior to work assignment, resulting in overdischarge	Needs more frequent and complete charging.
4. Electrolyte levels low	Add water as required. Do not assign battery to work if levels are below top of plates.

Probable Cause	Possible Remedy
5. When overheating is confined to a few cells, nearby operating equipment may be source	Install heat insulating material between equipment and cells with air circulating space between.
6. Operating in high ambient temperature	Provide cool location for charging and good ventilation.

b. Low Electrolyte Level

Probable Cause	Possible Remedy
1. Broken or cracked jar	Replace jar.
2. Water additions neglected or cell missed in previous water addition	Better maintenance supervision. Add water as required to all cells.
3. Overcharging	Adjust charging equipment.
a. Voltage relay set for more cells	a. Connect relay for proper number of cells.
b. Timer set for too many hours	b. Reduce time.
c. Voltage charge rate relay operating	c. Reduce voltage value at which voltage relay operates.

c. Specific Gravity Between Cells Not Uniform

Probable Cause	Possible Remedy
1. Overfilled with water	Do not fill above high level; give equalizing charge and adjust acid.
2. Operating cell with cracked jar	Replace jar and adjust acid.
3. Acid not adjusted properly after jar change	Adjust acid specific gravity.
4. Vent caps removed during operation	Keep vent caps securely in place. Give equalizing charge and adjust acid.
5. Electrolyte leaking through sealing compound	Reseal and adjust acid.
6. Operating battery with broken cell cover	Replace cover and adjust acid.
7. Neutralizing agent in cell	Keep vent caps in place at all times except when adding water. Keep battery clean.
8. Unequal cell voltages	Refer to Item "d."

d. Unequal Cell Voltages

Probable Cause	Possible Remedy
1. Overdischarge; also more than .020 point spread in specific gravity average	Give an equalizing charge and do not discharge below 1.110.
2. Lack of equalizing	Give an equalizing charge periodically.
3. Internal shunt	Make internal inspection of low voltage cell and correct cause. Check for split separator and moss short.
4. Dirty Battery Top	Neutralize and clean top of battery.
5. Cells operated with low electrolyte level	Add water as required. Give equalizing charge.
6. Low specific gravity of fully charged cell	Adjust acid after equalizing charge.
7. Sediment space filled	Replace battery.
8. Half tap on cells for lower voltage circuit	Remove tap and connect load to battery terminals through resistance.
9. External source heating certain cells	Install heat insulating material between heat source and battery.
10. Impurities in cells	Add only distilled or approved water to electrolyte.
11. Variation in charge rate	Take readings when charge rate is constant.

e. Battery Will Not Work Full Shift

Probable Cause	Possible Remedy
1. Uneven cell voltages	Give an equalizing charge.
2. Low electrolyte levels	Refer to Item "b."
3. Battery not charged before work cycle	Check charging schedules. Do not assign discharged battery to work.
4. One or more jars leaking electrolyte	Replace broken jars.
5. Incorrect battery (number of cells) assigned to equipment	Install battery with correct voltage and capacity to equipment.
6. Fully charged specific gravity below normal	Adjust specific gravity to normal.
7. Impurities in electrolyte	Add only distilled or approved water to electrolyte.
8. Operator "riding the brake"	Discourage practice.
9. Using reverse in place of braking	Discourage practice.
10. Inexperienced operator	Instruct him on power conservation.
11. Load excessive	Use larger battery or reduce load.
12. Wheels, axles, and bearings need grease	Plan lubrication schedule.
13. Tires underinflated	Plan periodic check of air pressure.
14. Brakes dragging	Adjust brakes properly.
15. Dirty tracks (if locomotive)	Clean excessive dirt from tracks.
16. Deeply grooved wheels	Replace with new wheels.
17. Deep ruts in road bed if locomotive	Fill in ruts if car is dragging.
18. Series field in motor shorted or grounded	Replace field and remove grounds.
19. Armature needs repair	Replace or repair armature.
20. Ground on equipment	Find grounds and insulate.
21. Excessive grades	Use larger battery or revise battery charge schedule.
22. Capacity or equipment assigned to job is inadequate	Reassign equipment.
23. When batteries are in two halves and discharged half is paired with a charged half	Better supervision. Keep batteries properly paired.
24. Uneven number of cells in two halves where split batteries are used in parallel start-series-run control circuits	Have equal number of cells for each half of battery.

E. Safety Rules for Lift Truck Batteries

The GNB Industrial Battery Company has published the following set of safety rules for handling lift truck batteries, which are normally of the lead-acid type. We reprint them here, with the expectations that today's electrical maintenance man will read and heed!

An industrial battery is no more dangerous than a bathtub or a ladder, but many persons have been injured by improper careless use of both. Observe these simple, commonsense suggestions when working around storage batteries to help prevent injuries to personnel and damage to batteries and equipment.

OBSERVE THESE SAFETY RULES

1. Wear rubber apron, gloves, boots, and goggles when handling, checking, filling, charging, or repairing batteries.

    Why—To protect yourself against accidental spillage of electrolyte—a mixture of sulfuric acid and water.

2. Keep open flames away from storage batteries—do not check electrolyte level with a cigarette lighter or match. USE A FLASHLIGHT OR PROVIDE ADEQUATE PERMANENT LIGHTS. Do not smoke or create sparks.

    Why—Storage batteries contain a dilute solution of sulfuric acid with water. The space between the underside of the cover and the top of the electrolyte in the cell usually contains a hydrogen–oxygen mixture which is explosive when ignited.

3. Have water readily available for use when electrolyte has been accidentally splashed on the skin or clothing of a worker. Extreme care should be taken in flushing electrolyte from the eye. Use plain water only and obtain medical attention immediately.

    Why—Volumes of water applied quickly, continuously, and copiously will prevent serious injury to the skin. Quick medical attention is necessary to assure proper care and treatment.

4. Apply strong neutralizer such as baking soda when acid is spilled on the floor, and clean up promptly.

    Why—Baking soda will neutralize the acid and make it safe to clean or flush from the floor.

5. Take proper care in melting the sealing compound when preparing to seal a battery. Avoid puncturing the hard surface of a partially melted compound with a screwdriver or other sharp tool. DO NOT ALLOW COMPOUND TO IGNITE BY TOO RAPID HEATING.

    Why—The hot liquid may squirt up and burn hands, face, or body of a workman. If it should catch fire, it creates the hazard of an open flame. Burning consumes oil in the compound and impairs its useful characteristics.

6. Remove vent caps from nearby cells and blow out each cell to remove gas when preparing to reassemble a repaired cell into the battery. Cover the vent holes with layers of damp cloth before using lead burning equipment on the intercell connectors in order to integrate the cell into the battery circuit. Also, use safety glasses to protect the eyes. Dark safety glasses should be worn if a carbon arc burner is used.

    Why—Gas is removed from each cell to prevent a possible hydrogen–oxygen concentration from exploding when a flame is lit.

7. Be sure to shut off the power when changing or repairing plugs or receptacles that are connected to the charging equipment. This will prevent a short circuit and arcing.

    Why—A short circuit may injure a workman, and arcing may cause explosion and fire.

8. When repairing a damaged or dirty plug or receptacle, *carefully* remove only one lead at a time and wrap it in electrician's tape. If possible, open the battery circuit first.

    Why—The leads which terminated in the receptacle are live or "hot" with the total voltage of the battery existing across the terminals. If the terminal lugs are accidentally touched together, a short circuit or arcing will occur with its attendant danger.

9. When mixing acid to prepare electrolyte, always pour acid slowly in the water and

never pour water into acid. Always store acid in plastic or glass containers.

Why—If water should be added to acid, it will not readily mix and will splash the acid due to the great difference in the specific gravity of the two liquids. The effect of splashed acid is dangerous.

10. Always lift batteries with mechanical equipment such as a hoist, crane, or lift truck. Move batteries horizontally with power trucks, conveyers, or rollers. Make sure that hoist hooks, spreader bars, and other tools are of ample strength and properly installed. Cover top of battery with rubber mat or other insulating material to prevent external short circuits from chains or cables falling on top of battery.

Why—Batteries are a heavy concentrated load and might easily cause painful strains or injury to handlers' feet or hands. Batteries, too, may be seriously damaged, or electrolyte spilled, if the battery is dropped.

11. Make sure that charging plugs and receptacles are properly locked and all other connections tight, secure, and free from friction.

Why—A loose connection may mean sparking or even arcing with attendant danger of gas explosion.

12. Disconnect the battery from the truck when doing maintenance and repair work on the motor or electrical system.

Why—The live current may cause arcing or short circuit with its attendant damage to equipment or injury to the workers.

13. The battery room should be restricted to authorized personnel only who are qualified, by training and experience, in handling batteries.

Why—Untrained or inexperienced persons may unknowingly break normal rules for proper handling causing injury to themselves and damage to batteries and equipment.

14. Never lay metal tools such as wrenches or other material on top of an open battery.

Why—Sparking and short circuiting will occur and the battery will quickly be discharged, or may cause explosion.

15. Enclose all bare wires and bus bars in the battery room by wire guards, guard rail, or other means of isolation from general plant traffic.

Why—Any open high current transmission equipment is a possible hazard to workmen and plant.

16. Check battery cells for cracks and leaks. Repair compound-seal cells if possible. Otherwise replace. Do not attempt repairs to heat-sealed cells. Replace them with new cells.

Why—Electrolyte will spill on floor or equipment causing corrosion of the steel tray and related equipment. Continuous flow of leaking electrolyte can cause a ground path that reduces battery life and capacity.

17. Familiarize yourself as completely as possible with batteries and the proper rules for their charging, handling, and maintenance.

Why—Full knowledge of batteries and the dangers of improper handling will pay dividends in the elimination of injuries and damage.

NOTE: Your own company, industry, and government safety regulations should be reviewed to help reduce accidents and damage to equipment.

C H A P T E R 7
INTRODUCTION TO ALTERNATING CURRENT

A. General Discussion

So far in this textbook we have covered the basics of Direct Current, or DC, and have learned that it has very closely defined limitations, which were recognized during the latter half of the 19th century when, in an attempt to expand the use of DC electricity, it was discovered that there were a number of stumbling blocks that at the time were almost unsurmountable.

It was not until the development of alternating current (AC) that the use of electricity took huge jumps, which led to the gigantic networks of public electrical systems in our society today.

Like so many other facets of our life, the principle of AC is very simple. The proper control and utilization of it is an entirely different matter, however. In this discussion, we shall only cover the briefest principles of it, to sufficiently enable the reader to understand what it is, how it behaves in our electrical systems, and how it is used and controlled. There are many more advanced books on the subject, which the reader may wish to study to become more familiar with the mathematics and engineering phases of it.

In a direct-current system, the supply voltage must be generated at the level required by the load. To operate a 240-volt lamp, for example, the generator must deliver 240 volts. A 120-volt lamp could not be operated from this generator by any convenient means.

Another disadvantage of direct-current systems is the large amount of power lost due to the resistance of the transmission wires used to carry current from the generating station to the consumer. This loss could be greatly reduced by operating the transmission line at very high voltage and low current. This is not a practical solution in a DC system, however, since the load would also have to operate at high voltage. As a result of the difficulties encountered with direct current, practically all modern power distribution systems use ALTERNATING CURRENT (AC). In an alternating-current system, the current flows first in one direction, then reverses and flows in the opposite direction.

Unlike DC voltage, AC voltage can be stepped up or down by a device called a TRANSFORMER. This permits the transmission lines to be operated at high voltage and low current for maximum efficiency. Then, at the consumer end, the voltage is stepped down to whatever value the load requires by using a transformer. Due to its inherent advantages and versatility, alternating current has replaced direct current in all but a few commercial power distribution systems.

Many other types of current and voltage exist in addition to direct current and voltage. If a graph is constructed showing the magnitude of DC voltage across the terminals of a battery

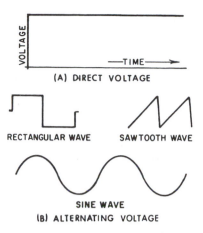

Fig. 7-1. Voltage waveforms. (A) Direct voltage. (B) Alternating voltage.

with respect to time it would appear as in Fig. 7-1 (A). The voltage is shown to have a constant amplitude. Some voltages go through periodic changes in amplitude like those shown in Fig. 7-1 (B).

The pattern which results when these changes in amplitude with respect to time are plotted on graph paper is known as a WAVE-FORM. Of those illustrated, the sine wave will be dealt with here. "Sine wave" is a term from trigonometry. As used in this discussion, it describes the manner in which the AC voltage makes a smooth, uniform transition from point to point throughout the cycle, as shown in Fig. 7-1 (B), which is typical, and the one to be encountered almost exclusively.

B. Generation of Alternating Current

An alternating-current generator converts mechanical energy into electrical energy. It does this by utilizing the principle of electromagnetic induction. In the study of magnetism, it was shown that a current-carrying conductor produces a magnetic field around itself. It is also true that a changing magnetic field may produce an emf in a conductor. If a conductor lies in a magnetic field, and if either field or conductor moves, an emf is induced in the con-

ductor. This effect is called electromagnetic induction.

Figure 7-2 shows a suspended loop of wire (conductor) being rotated (moved) in a counterclockwise direction through the magnetic field between the poles of a permanent magnet. For ease of explanation, the loop has been divided into a dark and a light half. Notice that in part (A), the dark half is moving along (parallel to) the lines of force. Consequently, it is cutting none of these lines. The same is true of the light half, moving in the opposite direction. Since the conductors are cutting no lines of force, no emf is induced. As the loop rotates toward the position shown in part (B), it cuts more and more lines of force per second because it is cutting more directly across the field (lines of force) as it approaches the position shown in (B). At position (B), the induced voltage is greatest because the conductor is cutting directly across the field.

As the loop continues to be rotated toward the position shown in part (C), it cuts fewer and fewer lines of force per second. The induced voltage decreases from its peak value. Eventually, the loop is once again moving in a plane parallel to the magnetic field, and no voltage (zero voltage) is induced. The loop has now been rotated through half a circle (one alternation, or 180 degrees). The curve shown in the lower part of the figure shows the induced voltage at every instant of rotation of the loop. Notice that this curve contains 360 degrees, or two alternations. Two alternations represent one complete circle of rotation.

Looking at the dark half of the loop in part (B), the direction of current flow is depicted by the heavy arrow. When the loop is further rotated to the position shown in part (D), the action is reversed. The dark half is moving up instead of down, and the light half is moving down instead of up. It is really apparent that the direction of the induced emf and its resulting current have reversed, as shown by the directional arrow in (D). The voltage builds up to maximum in this new direction, as shown by the sine-wave tracing. The loop finally returns to its original position (E), at which point volt-

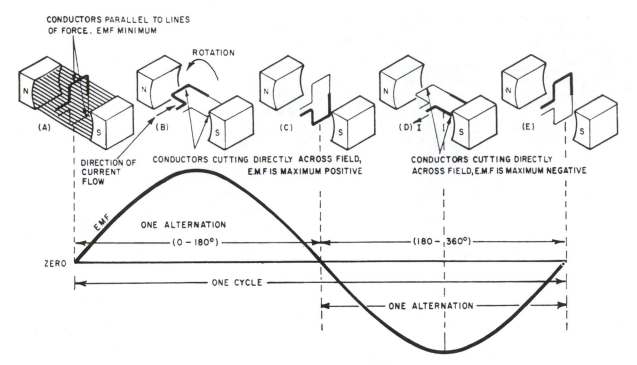

Fig. 7-2. Basic AC generator.

age is again zero. The wave of induced voltage has gone through one complete cycle.

As mentioned previously, the cycle is two complete alternations in a period of time. Recently, the Hertz (Hz) has been designated to be used in lieu of cycles per second. While it may seem confusing to the reader that in one place a cycle is used to designate two alternations per period of time and in another instance a Hertz is used to designate two alternations per second, the key to determine which is used is the time factor. One Hertz is one cycle per second. Therefore, throughout this manual, a cycle is used when no specific time element is involved, and a Hertz is used when the time element is measured in seconds.

If the loop is rotated at a steady rate, and if the strength of the magnetic field is uniform, the number of Hertz and the apparent output voltage will remain at fixed values. Continuous rotation will produce a series of sine-wave voltage cycles or, in other words, an AC voltage. In this way mechanical energy is converted into electrical energy.

The rotating loop in Fig. 7-2 is called an armature. The armature may have any number of loops of coils.

If the reader will study the wave illustrating the current flow in Fig. 7-2, it will readily be seen that the current starts the cycle with a value of zero, increases steadily to a peak value, drops off to zero again, then reverses direction and again increases to a peak, and drops off to zero again, to start another cycle. Thus, in each cycle, there are two brief moments when the current has reached a peak, and two briefer moments when the current flowing is actually zero. It happens so fast, however, that the effect in most electrical apparatus appears to us as a continuous flow of electricity. The moments of zero flow do not affect the light given off by an ordinary light bulb, or a fluorescent lamp, as long as the current is being supplied at a cycle speed fast enough that our eyes do not attempt to differentiate between the minute fluctuations which are produced as a result of the alternating current. In this country, most of the alternating current furnished by our public utilities is supplied at a rate of 60 cycles per second, or 60Hz. In many parts of the world, the basic cycle speed is 50Hz.

If one were to attempt supplying electricity at cycle speeds under 50Hz, then some people would find that for some unknown reason, they become tired and they would experience difficulty with their eyes. This happens because at slower cycle speeds, there may develop an interference between the natural frequency of the sensory nerves in the eyes and the fluctuations of the light output from the electric lamps. You may not notice the fluctuations, but the eyes may still be affected by them.

However, there would probably be no noticeable flicker detected until the cycle speed drops to about 25Hz.

One further note here which may be of interest to the electrical worker, concerning the voltages generated by the sine wave forms shown here, and prevalent in all of our AC systems, relates to the varying voltages throughout the cycle. The voltage which the system receives, and which is indicated on the instruments, is not that generated at the peaks of the sine wave curves, but is the resultant effective voltage produced by the ever-changing voltages throughout the cycle. The relationship between the peak voltages and the effective, usable, voltage need not concern the electrician who is working with and maintaining the electrical equipment in the plant, as it has no discernible effect on his activities.

C. Frequency

The frequency (f) of an alternating current or voltage is the number of complete cycles occurring in each second of time. Hence, the speed of rotation of the loop determines the frequency. For a single loop rotating in a two-pole field, as in Fig. 7-2, you can see that each time the loop makes one complete revolution, the current reverses direction twice. A single Hertz will result if the loop makes one revolution each second. If it makes two revolutions per second, the output frequency will be 2Hz. In other words, the frequency of a two-pole generator happens to be the same as the number of revolutions per second. As the speed is increased, the frequency is increased.

If an alternating-current generator has four pole pieces, as in Fig. 7-3, every complete mechanical revolution of the armature will produce 2 cycles. The more poles that are added, the higher the frequency per revolution becomes. To find the output frequency of any AC generator, the following formula can be used:

$$f = \frac{P \times rpm}{120}$$

where f is frequency in Hertz, rpm is revolutions per minute, and P is the number of poles.

A generator made to deliver 60Hz, having two field poles, would need an armature designed to rotate at 3,600 rpm. If it had four field poles, it would need an armature designed to rotate at 1,800 rpm. In either case, frequency would be the same. In actual practice, a generator designed for low-speed operation generally has a greater number of pole pieces, while high-speed machines will have relatively fewer pole pieces, if both are to deliver power at the same frequency.

Most of the electricity in this country is generated by steam turbines which operate most efficiently at higher speeds. Therefore, the generators have two poles, and operate at 3,600 rpm to produce 60Hz. Fifty Hz is produced by two-pole generators operating at 3,000 rpm. The reader may check those figures by use of the above formula.

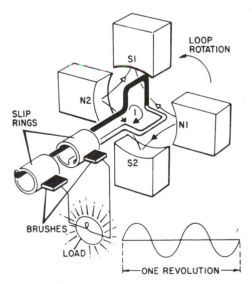

Fig. 7-3. Four-pole basic AC generator.

D. Period

An individual cycle of any sine wave represents a finite amount of TIME. Figure 7-3 shows 2 cycles of a sine wave which has a frequency of 2Hz. Since 2 cycles occur each second, 1 cycle must require one-half second of time. The time required to complete 1 cycle of a waveform is called the PERIOD of the wave. In this example, the period is one-half second.

The period of a wave is inversely proportional to its frequency. Thus, the higher the frequency (greater number of Hz), the shorter the period. In terms of an equation:

$$t = \frac{1}{f}$$

where t = period in seconds
f = frequency in Hz.

E. Single-Phase AC Circuits

The AC generator shown in Fig. 7-3 and described in the text is producing what is termed a "Single-Phase AC Circuit." By Single-Phase is meant one pair of conductors carrying the electricity to and from a load in an AC circuit.

There are generators producing more than one-phase circuits. In fact, the majority of them in use today produce three-phase circuits. This means that there are three pairs of conductors coming from the generator. However, this does not mean that there are six conductors to be wired to the loads, but only three, due to the internal wiring in the construction of the generator system. This will be shown later in this chapter, after we have covered the basics a little more thoroughly.

There are two-phase circuits, also, which we shall touch on slightly in due course.

First, we shall briefly describe the single-phase circuit, the one used most for lighting, small motors, small appliances, small heating units, and control circuits. To do so, we refer you first to Fig. 7-4, which depicts a typical single-phase AC circuit (as designated by the symbol used for the source of the electrical power).

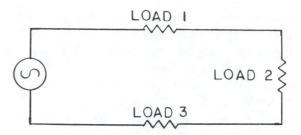

Fig. 7-4. Single-phase series circuit.

This circuit contains three resistance loads, and by studying this diagram it is readily seen that the same current flows in all parts of the circuit. The voltage drop across each load may be easily measured, as we shall illustrate later in this book.

As we stated earlier, the voltage drop across each resistance and amperage through the circuit may be measured. The electrical power, in watts, consumed by each load may thus be ascertained according to the following equations:

Power = I^2R = Volts × Amps
(For circuits containing only resistance).

We now refer the reader to Fig. 7-5, which depicts a single-phase AC circuit with multiple resistance loads connected in parallel. Again, we shall cover only the simple aspects of the circuit.

In a circuit of this type, the voltage across each of the loads is the same, being the voltage imposed across the entire system, as supplied by the leads or conductors. In this case, the total amperage draw, or current flow, is the sum of the individual amperages drawn across the individual loads. The maintenance man should keep this fact in mind when using voltmeters or wattmeters to check a circuit.

In the parallel circuit, the total power con-

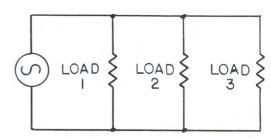

Fig. 7-5. Single-phase parallel circuit.

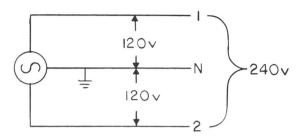

Fig. 7-6. Single-phase dual-voltage system.

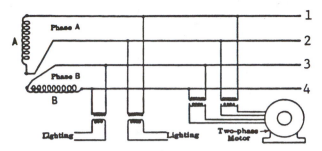

Fig. 7-7. Two-phase four-wire system.

sumed is the sum of the individual circuit powers, as measured by a wattmeter. However, the total amperage draw, if the circuit contains components other than resistors, is not the sum of the individual amperages, and the total amperage cannot be determined by any of the equations for power, amperage, or resistance as given in this book. This, too, we shall let slide, for reasons already stated.

Figures 7-2 and 7-3 show a simple type of single-phase AC generator, one which is found only in the smaller sizes, such as in portable units, and for isolated locations. Larger, stationary single-phase generators usually produce two single-phase circuits through three conductors, or wires. Such a system is known as a single-phase three-wire system. We show the wiring diagram for this system in Fig. 7-6. You will note that two different voltages are available from one generator without the use of a transformer, depending upon which wires or leads are used.

Note that the neutral conductor is either white or gray.

F. Two-Phase Systems

The two-phase AC systems described here are very seldom encountered. However, we shall cover them briefly, as the reader may be forced to deal with them in isolated locations. We shall not, however, cover them in sufficient detail to enable one to construct such a system, but merely to be able to identify it.

Figure 7-7 illustrates what is known as a two-phase four-wire system. It is important that it

not be confused with a three-phase four-wire system, which also has four leads.

Notice that the motors in this system have four terminals, and it is important that the correct wires be connected to the proper terminals. Small single-phase loads may be connected across either pair of leads constituting a phase. Do not connect any load across leads 2 and 3, or trouble will result.

Figure 7-8 is a two-phase three-wire AC system, and is also not to be confused with another circuit, this time the three-phase three-wire system, which also has three leads or conductors.

Notice that there is a common wire in this system, the middle one in the diagram. It is common to both Phase A and Phase B. That common wire must be capable of taking 1.41 times the current that either of the other two leads carry.

In Figure 7-2 we show a basic AC wave form. This is typical of the single-phase wave form as produced by a single-phase generator. Single phase simply means that only one pulse of electrical power is sent out through only one pair of leads from the generator.

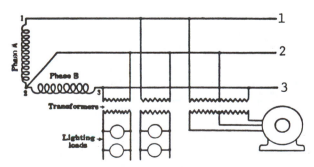

Fig. 7-8. Two-phase three-wire system.

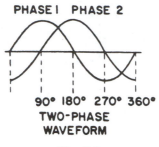

Fig. 7-9.

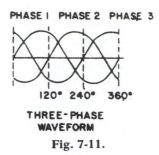

Fig. 7-11.

Now look at Fig. 7-9 and notice the difference. Superimposed over the single-phase wave form is another wave form, but one which starts its cycle later. This is a typical two-phase wave form, as produced by the generator circuit of either Fig. 7-7 or 7-8. In Fig. 7-7, the first pulse of power would be produced across terminals 1 and 2, for instance. One-fourth of a cycle later, another pulse of power would be produced across terminals 3 and 4, and this offbeat relationship would continue indefinitely, as long as the generator was operating. In both circuits, of course, the power pulse goes through the complete cycle of flowing first in one direction, then reversing direction for another pulse of power, all as described in Section B of this chapter.

In Fig. 7-8, the same condition exists, with terminals 1 and 2 making one phase, and terminals 2 and 3 making the second phase of the two-phase system.

G. Three-Phase Systems

Figure 7-10 shows the most common multiple-phase circuit in use today.

In Section E of this chapter, we mentioned

that three-phase systems had three pairs of leads coming from the generator which produces the three-phase system. Referring to Fig. 7-10, we see only three leads for this three-phase system. As can readily be seen, terminals 1 and 2 make up one phase, terminals 2 and 3 make up the second phase, and terminals 1 and 3 make up the third phase. Thus, each wire is made to do double duty in the system.

Figure 7-11 shows the resultant wave form for this three-phase system. As can be seen from the diagram, each phase is one-third of a cycle out of synchronization with the previous phase, so that the power pulses sent out over the wires in a three-phase system are not synchronized together, but are offset an equal amount in a continuous repetition of power pulses.

Figure 7-12 shows another three-phase system, one which uses four wires to produce three-phase circuits and single-phase circuits from the same generator. This system is produced by bringing the fourth wire out from the generator or transformer as shown. This is a neutral point, and is a common line for all single-phase circuits in the system. The main advantage of this system is that the total weight

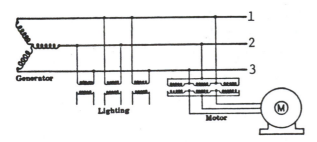

Fig. 7-10. Three-phase three-wire system.

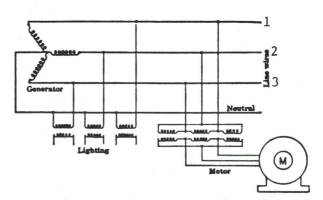

Fig. 7-12. Three-phase four-wire system.

of copper in the wiring system is much less than for the three-wire three-phase system, based on equal voltage losses in transmission.

There are several reasons why three-phase systems are preferred over the single-phase system.

1. The generators are cheaper for the same amount of power generated, as they are more efficient in the use of copper and steel.

2. It is easier to regulate the voltage from a three-phase generator.

3. Three-phase motors are smaller, more efficient, and have better operating characteristics than single-phase motors. More about this later, however.

4. A three-phase generator can produce both single-phase and three-phase power very easily, but a single-phase generator cannot be made to produce three-phase power.

H. Types of AC Generators

AC generators are made in many different sizes, depending on their intended use. For example, any one of the generators at Boulder Dam can produce millions of watts, while generators used for residential emergency produce only a few hundred watts.

Regardless of size, however, all generators operate on the same basic principle—a magnetic field cutting through conductors, or conductors passing through a magnetic field. Thus, all generators have at least two distinct sets of conductors. They are (1) a group of conductors in which the output voltage is generated, and (2) a second group of conductors through which direct current is passed to obtain an electromagnetic field of fixed polarity. The conductors in which the output voltage is generated are always referred to as the armature windings. The conductors in which the electromagnetic field originates are always referred to as the field windings.

In addition to the armature and field, there must be motion between the two. To provide this, AC generators are built in two major assemblies, the stator and the rotor. The rotor rotates inside the stator. It may be driven by any one of a number of commonly used power sources, such as gas or hydraulic turbines, electric motors, steam, or internal-combustion engines.

There are various types of alternating-current generators utilized today. However, they all perform the same basic function. The types discussed in the following paragraphs are typical of the more predominant ones encountered in electrical equipment.

In the revolving-armature AC generator, the stator provides a stationary electromagnetic field. The rotor, acting as the armature, revolves in the field, cutting the lines of force producing the desired output voltage. In this generator, the armature output is taken through sliprings and thus retains its alternating characteristic.

For a number of reasons, the revolving armature AC generator is seldom used. Its primary limitation is the fact that its output power is conducted through sliding contacts (sliprings and brushes). These contacts are subject to frictional wear and sparking. In addition, they are exposed, and thus liable to arc-over at high voltages. Consequently, revolving-armature generators are limited to low-power, low-voltage applications.

The revolving-field single-phase AC generator (Fig. 7-13) is a widely used type. In this type of generator, direct current from a separate source is passed through windings on the rotor by means of sliprings and brushes. This maintains a rotating electromagnetic field similar to a rotating bar magnet. The rotating magnetic field, following the rotor, extends outward and cuts through the armature windings embedded in the surrounding stator. As the rotor turns, alternating voltages are induced in the windings since magnetic fields of first one direction and then the other cut through them. Since the output power is taken from stationary windings, the output may be connected through fixed terminals T1 and T2 in Fig. 7-13. This is advantageous, in that there are no sliding con-

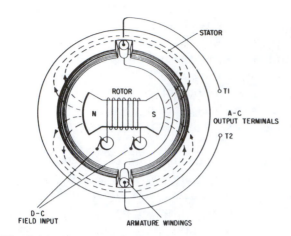

Fig. 7-13. Diagram of a rotating field AC generator.

tacts, and the whole output circuit is continuously insulated, thus minimizing the danger of arc-over.

Sliprings and brushes are adequate for the DC field supply because the power level in the field is much smaller than in the armature circuit.

The rating of an AC generator pertains to the load it is capable of supplying. The normal-load rating is the load it can carry continuously. Its overload rating is the above-normal load which it can carry for specified lengths of time only. The load rating of a particular generator is determined by the internal heat it can withstand. Since heating is caused mainly by current flow, the generator's rating is identified very closely with its current capacity.

The maximum current that can be supplied by an AC generator depends upon (1) the maximum heating loss (I^2R power loss) that can be sustained in the armature and (2) the maximum heating loss that can be sustained in the field. The armature current varies with the load. AC generators are rated in terms of armature load current and voltage output, or kilovolt-ampere (kva) output, at a specified frequency and power factor.* The specified power

---

*Power Factor: A term used to give the approximate efficiency with which the equipment utilizes the electrical power available to it. Perfect is usually taken as 1.0, and any decimal figure below that, such as .80, signifies that the plant is using only about 80% of the electrical power properly, the remainder disappearing as heat.

factor is usually 80 percent. For example, a single-phase AC generator designed to deliver 100 amperes at 1,000 volts is rated at 100 kva. This machine would supply a 100-kw load at unit power factor or an 80-kw load at .80 power factor.

A typical rotating-field AC generator consists of an AC generator and a smaller DC generator built into a single unit. The output of the AC generator section supplies alternating current to the load for which the generator was designed. The DC generator's only purpose is to supply the direct current required to maintain the AC generator field. This DC generator is referred to as the exciter. A typical three-phase generator is shown in Fig. 7-14(A); Fig. 7-14 (B) is a simplified schematic of the generator.

Any rotary generator requires a prime moving force to rotate the AC field and exciter armature. This rotary force is transmitted to the generator through the rotor drive shaft (Fig. 7-14 (A)) and is usually furnished by a combustion engine, turbine, or electric motor. The exciter shunt field (2, Fig. 7-14 (B)) creates an area of intense magnetic flux between its poles.

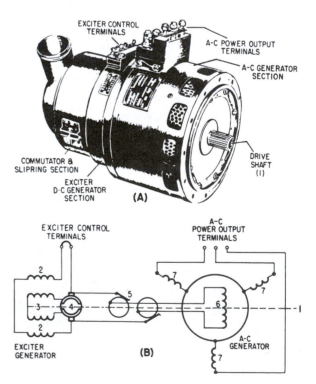

Fig. 7-14. AC generator and schematic.

When the exciter armature (3) is rotated in the exciter field flux, voltage is induced into the exciter armature windings. The exciter output commutator and brushes (4) connects the exciter output directly to the AC generator field input sliprings and brushes (5). Since these sliprings, rather than a commutator, are used to supply current through the AC generator field (6), current always flows in one direction through these windings. Thus, a fixed magnetic field is maintained at all times in the AC generator field windings. When the AC generator field is rotated, its magnetic flux is passed through and across the AC generator armature windings (7). Remember, a voltage is induced in a conductor, the same as if the field is stationary and the conductor is moved. The alternating voltage induced in the AC generator armature windings is connected through fixed terminals to the load.

AC generators may be divided into three classes according to the type of prime mover. These classes are as follows:

1. Low-speed engine driven

2. High-speed turbine driven

3. High-speed engine driven.

The stator, or armature, of the revolving-field AC generator is built up from steel punchings, or laminations. The laminations of an AC generator stator form a steel ring that is keyed or bolted to the inside circumference of a steel frame. The inner surface of the laminated ring has slots in which the stator winding is placed.

The low-speed engine-driven AC generator has a large diameter revolving field with many poles, and a stationary armature relatively short in axial length. The stator contains the armature windings, and the rotor consists of separated poles, on which are mounted the DC field windings. The exciter armature is a smaller unit and mounted on an extension of the AC generator shaft.

The high-speed turbine-driven AC generator is connected either directly or through gears to a steam turbine. The enclosed metal structure is a part of a forced ventilation system that carries away the heat by circulation of air or hydrogen through the stator and rotor. The exciter is a separate unit. The enclosed stator directs the paths of the circulating cooling currents and also reduces windage noise.

The high-speed engine-driven generator (Fig. 7-15) may be driven directly by an engine, a hydraulic constant-speed drive, or a gas turbine.

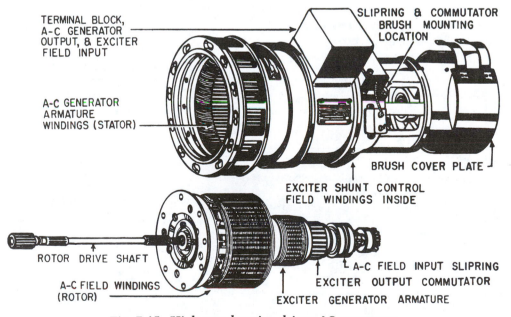

Fig. 7-15. High-speed engine-driven AC generator.

I. Troubleshooting

1. Noisy Operation

Probable Cause	Remedy
Unbalanced load	Balance load.
Coupling loose or misaligned	Realign coupling and tighten.
Improper air gap	Check for bent shaft, loose or worn bearings. Straighten and realign shaft. Replace bearings.
Loose laminations	Tighten bolts. Dip in varnish and bake.

2. Overheating

Probable Cause	Remedy
Overloaded	Check meter reading against nameplate ratings. Reduce load.
Unbalanced load	Balance load.
Open load-line fuse	Replace fuse.
Restricted ventilation	Clean, and remove obstructions
Rotor winding short-circuited, open-circuited, or grounded	Check, and replace defective coil or coils.
Stator winding short-circuited, open-circuited, or grounded	Check, and replace defective coil or coils.
Bearings	Check for worn, loose, dry, or overlubricated bearings. Replace worn or loose bearings, lubricate dry bearings, relieve overlubrication.

3. No Output Voltage

Probable Cause	Remedy
Stator coils open- or short-circuited	Check and replace defective coil or coils.
Rotor coils open- or short-curcuited	Check, and replace defective coil or coils.
Shorted sliprings	Disconnect field coils and check ring-insulation resistance with megger. Repair.
Internal moisture	Check with megger and dry windings.

Probable Cause	Remedy
No voltage at slipring brushes (no DC exciter voltage)	Check for defective switch or blown fuse in exciter feeder lines. Repair switch or replace fuses. Check feeder cables for opens or shorts. Repair connections or replace cables. Refer to "Failure to Build Up Voltage" (Chapter 13, Section J, Item 1).
Voltmeter defective	Check with a voltmeter known to be working properly. Replace.
Ammeter shunt open	Replace ammeter and shunt.

4. Output Voltage Unsteady

Probable Cause	Remedy
Poor commutation at sliprings	Clean sliprings and brushes. Reset brushes.
Loose terminal connections	Clean and tighten all connections and contacts.
Maladjusted voltage regulator and speed governor	Readjust speed governor and voltage regulator.

5. Output Voltage Too High

Probable Cause	Remedy
Overspeeding	Adjust speed-governing device.
Overexcited	Adjust voltage regulator.
Delta-connected stator open on one leg	Remake connection, repair or replace defective coil or coils.

6. Frequency Incorrect or Fluctuating

Probable Cause	Remedy
Speed incorrect or fluctuating	Adjust speed-governing device.
DC excitation fluctuating	Adjust belt tension of exciter generator.

7. Voltage Hunting

Probable Cause	Remedy
External field resistance in total "out" position	Readjust resistance.
Voltage regulator contacts dirty	Clean and reset contact points.

8. Stator Overheats in Spots

Probable Cause	Remedy
Short-circuitated phase winding	Check and replace defective coils.
Rotor off center (improper air gap)	Check for bent shaft, loose or worn bearings. Straighten and realign shaft. Replace bearings.
Unbalanced winding circuits	Balance winding circuits.
Loose winding connections	Tighten winding connections
Wrong phase polarity connections	Correct connections for proper phase polarity.

9. Field Overheating

Probable Cause	Remedy
Shorted field coil or coils	Check and replace defective coil or coils.

DC excitation current too high	Reduce exciter current by adjusting DC voltage regulator.
Clogged air passages (poor ventilation)	Clean equipment. Remove obstructions.

10. Alternator Produces Shock when Touched

Probable Cause	Remedy
Reversed stator field coil	Check polarity. Make correction to connections.
Static charges or grounded stator field coil	Check generator frame-ground connection or connections, clean and tighten. Repair or replace stator field coil.

C H A P T E R 8
INDUCTANCE AND CAPACITANCE

A. Inductance

Inductance is the characteristic of an electrical circuit that makes itself evident by opposing the starting, stopping, or changing of current flow. The above statement is of such importance to the study of inductance that it bears repeating in a simplified form. Inductance is the characteristic of an electrical conductor which opposes a CHANGE in current flow.

One does not have to look far to find a physical analogy of inductance. Anyone who has ever had to push a heavy load (wheelbarrow, car, etc.) is aware that it takes more work to start the load moving than it does to keep it moving. This is because the load possesses the property of inertia. Inertia is the characteristic of mass which opposes a CHANGE in velocity. Therefore, inertia can hinder us in some ways and help us in others. Inductance exhibits the same effect on current in an electric circuit as inertia does on velocity of a mechanical object. The effects of inductance are sometime desirable—sometimes undesirable.

Michael Faraday started to experiment with electricity around the year 1805 while working as an apprentice bookbinder. It was in 1831 that Faraday performed experiments on magnetically coupled coils. A voltage was induced in one of the coils due to a magnetic field created by current flow in the other coil. From this experiment came the induction coil, the theory of which eventually made possible many of our modern conveniences such as the automobile, doorbell, auto, radio, etc. Two months later, based on these experiments, Faraday constructed the first direct current generator. At the same time Faraday was doing his work in England, Joseph Henry was working independently along the same lines in New York. The discovery of the property of self-induction of a coil was actually made by Henry a little in advance of Faraday, and it is in honor of Joseph Henry that the unit of inductance is called the HENRY.

It was from the experiments performed by these, and many other men, that the laws and theories of inductance grew.

Even a perfectly straight length of conductor has some inductance. Current in a conductor always produces a magnetic field surrounding, or linking with, the conductor. When the current changes, the magnetic field changes, and an emf is induced in the conductor. This emf is called a SELF-INDUCED EMF because it is induced in the conductor carrying the current. The direction of the induced emf has a definite relation to the direction in which the field that induces the emf varies. When the current in a circuit is increasing, the flux linking with the circuit is increasing. This flux cuts across the conductor and induces an emf in the conductor in such a direction as to oppose the increase in current and flux. Likewise, when the current is decreasing, an emf is induced in the opposite direction and opposes the decrease in current. These effects are summarized by Lenz's law, which states that THE INDUCED EMF IN ANY CIRCUIT IS ALWAYS IN A DIRECTION TO OPPOSE THE EFFECT THAT PRODUCED IT.

The inductance is increased by shaping a conductor so that the electromagnetic field around each portion of the conductor cuts across some other portion of the same conductor. This is shown in its simplest form in Fig. 8-1. A length of conductor is looped so that two portions of the conductor lie adjacent and parallel to one another. These portions are labelled conductor 1 and 2. When the switch is closed, electron flow through the conductor establishes a typical concentric field around ALL portions of the conductor. For simplicity, however, the field is shown in a single plane that is a perpendicular to both conductors. Although the field originates simultaneously in both conductors, it is considered as originating in conductor 1 and its effect on conductor 2 will be noted. With increasing current, the field expands outward, butting across a portion of conductor 2. The resultant induced emf in conductor 2 is shown by the dashed arrow. Note that it is in OPPOSITION to the battery current and voltage, according to Lenz's law.

Many things affect the self-inductance of a circuit. An important factor is the degree of linkage between the circuit conductors and its electromagnetic flux. In a straight length of conductor, there is very little flux linkage between one part of the conductor and another. Therefore, its inductance is extremely small.

Conductors become much more inductive when they are wound into coils, as shown in Fig. 8-2. This is true because there is maximum flux linkage between the conductor turns, which lie side by side in the coil.

The coil is made more inductive by winding it in three layers, and providing a highly permeable core, as in Fig. 8-2. Note that some turns, such as (b), lie directly adjacent to six other turns (shaded). The magnetic properties of the iron core increase the total coil flux strength many times that of an air core coil of the same number of turns.

When a circuit containing a coil is energized with direct current, the coil's effect in the circuit is evident only when the circuit is energized, or when it is de-energized. For instance, when the switch in Fig. 8-3 is placed in position 1, the inductance of coil L will cause a delay in the time required for the lamps to attain normal brilliance. After they have attained normal brilliance, the inductance has no effect on the circuit as long as the switch remains closed. When the switch is opened, an electric spark will jump across the opening switch contacts. The emf which produces the spark is caused by the collapsing magnetic field cutting the turns of the inductor.

When the inductive circuit is supplied with alternating current, however, the inductor's effect is continuous and much greater than when it was supplied with direct current. For equal supplied voltages, the current through the circuit is less when AC is applied, as may be demonstrated by the circuits of Fig. 8-3. The al-

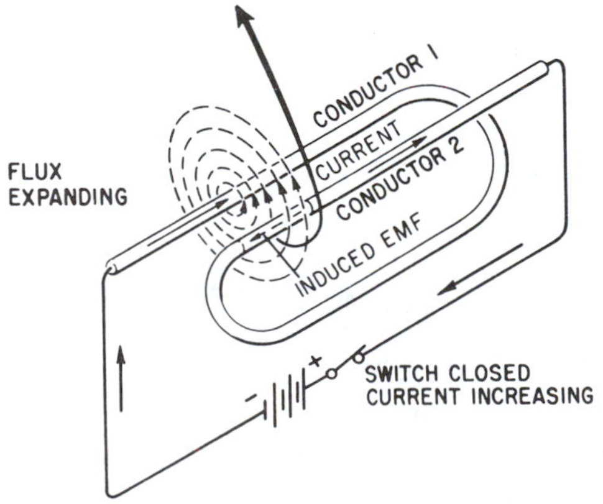

Fig. 8-1. Simple inductive circuit.

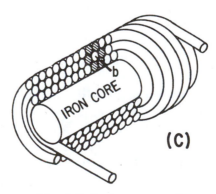

Fig. 8-2. Inductance coil.

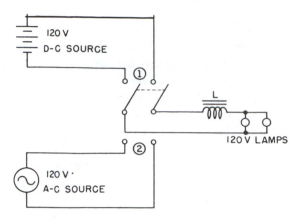

Fig. 8-3. AC versus DC inductance.

ternating current is accompanied by an alternating magnetic field around the coil, which cuts through the turns of the coil. This action induces a voltage in the coil that always opposes the changing current, which causes the lamps to glow dimly.

We have shown that an inductor in which there is a changing current becomes a source of emf, and that the direction of this emf is such that it tends to oppose the change in current producing it. As a result of this action, the current in the inductor does not rise to its full value the instant the switch is closed. Likewise, when the switch is opened, the source removed, and a short placed across the circuit, the current does not instantaneously fall to zero.

The circuits so far considered have been composed of ideal components. However, components are not perfect, therefore, no switch could be manufactured that would be capable of being moved from one position to another instantaneously. This leads to an explanation of the action of the induced voltage at the INSTANT a switch is opened. Because the magnitude of self-induced voltage can be extremely great even though the source voltage is very low, the development of high induced voltages when an inductive circuit is suddenly opened can be put to constructive use.

The energy contained in the collapsing magnetic field must be dissipated somewhere within the circuit. The voltage developed in the conductor is sufficient to create an arc across the switch contacts. The energy in the magnetic field is dissipated in the heat of this arc. The energy expended in the arc can seriously burn

an individual, damage the switch contacts, or break down the insulation of the coil. For these reasons, care should be taken in the abrupt interruption of DC current in any inductive circuit.

The development of a large voltage pulse from a low voltage source (INDUCTIVE KICK) is not always a disadvantage, but is commonly used in the spark coil circuits (ignition system) of most gasoline engines.

When used in AC circuits, the effect is different than when used in DC circuits. As we discovered in the previous chapter, the current and voltage flow in an AC circuit are constantly changing, from zero to maximum, to zero in the reverse direction, and back to zero again. If an inductance coil is inserted in this constantly changing electrical field of force, the result will be an induced flow of current and voltage of similar pattern, but in the reverse direction in each portion of the cycle.

What this all reduces to in the case of an inductance coil in an AC circuit is the following.

1. The characteristics of the resulting or induced voltage and current may be altered to suit a specific need, depending upon the design of the inductance coil.

2. Inductance is the principle upon which transformers operate; and without transformers, much of our use of AC electricity would not be possible.

3. It is possible to pass an electric current through an AC circuit without a continuous closed path, the power being made to jump across a gap in the circuit by means of an induced field of force.

4. It may be used as a current-limiting device in an AC circuit.

The last item above is one of the most common applications in an industrial plant. When an AC circuit carrying a large amount of power is short-circuited, the resultant current flow may be many times the normal flow. This can cause heavy strains upon the equipment, such as motors and transformers, due to the induced voltages caused by the heavy surge of current, as explained earlier in this section. The electrical equipment can very easily be damaged by the forces which result.

By placing an inductive reactor in the circuit, the sudden surge of current through the reactor will cause a counter force, which in turn resists or opposes the flow of current to ground. This protection may save the equipment, if it is properly designed. An inductive reactor installed in each line of an AC circuit may permit the use of moderately sized distribution equipment, such as transformers, bus bars, circuit breakers, fuses, and switches.

This protection is not obtained entirely without cost, however, as it uses some power; but if properly designed and chosen, the amount of power it consumes is a very small percentage of the total which it passes.

B. Capacitance

CAPACITANCE is defined as the property of an electrical device or circuit that tends to oppose a CHANGE in VOLTAGE. Capacitance is also a measure of the ability of two conducting surfaces, separated by some form of non-conductor, to store an electric charge. For the present time, air will be used as the insulating material between the conducting surfaces.

The device used in electrical circuits to store a charge by virtue of an electrostatic field is called a CAPACITOR.* The larger the capacitor, the larger the charge that can be stored.

The simplest type of capacitor consists of two metal plates separated by air. It has been discovered that a free electron inserted in an electrostatic field will move. The same is true, with qualifications, if the electron is in a bound state.

The material between the two charged surfaces of Fig. 8-4 (air in this case) is composed of atoms containing bound orbital electrons.

---

*A more commom term, and one which the reader will probably recognize more readily, is CONDENSER. We shall use the scientific term, CAPACITOR, in our discussion.

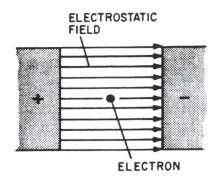

Fig. 8-4. Principle of capacitance, #1.

Since the electrons are bound, they cannot travel to the positively charged surface. Therefore, the resultant effect will be a distorting of the electron orbits. The bound electrons will be attracted toward the positive surface, and repelled from the negative surface. This effect is illustrated in Fig. 8-5. In Fig. 8-5 (A), there is no difference in charge placed across the plates; and the structure of the atom's orbits is undisturbed. If there is a difference in charge across the plates as shown in Fig. 8-5 (B), the orbits will be elongated in the direction of the positive charge.

As energy is required to distort the orbits, energy is transferred from the electrostatic field to the electrons of each atom between the charged plates. Since energy cannot be destroyed, the energy required to distort the orbits can be recovered when the electron orbits are permitted to return to their normal positions. This effect is analogous to the storage of energy in a stretched spring. A capacitor can thus "store" electrical energy.

Figure 8-6 (A) depicts a capacitor in its simplest form. It consists of two metal plates sepa-

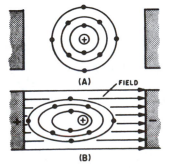

Fig. 8-5. Principle of capacitance, #2.

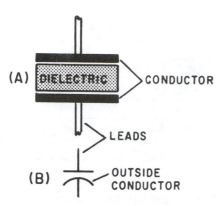

Fig. 8-6. Basic capacitor.

rated by a thin layer of insulating material (dielectric). When connected to a voltage source (battery), the voltage forces electrons onto one plate, making it negative, and pulls them off the other, making it positive. Electrons cannot flow through the dielectric. Since it takes a definite quantity of electrons to "fill up," or charge, a capacitor, it is said to have a CAPACITY. This characteristic is referred to as CAPACITANCE.

Various materials differ in their ability to support electric flux or to serve as dielectric material for capacitors. This phenomenon is somewhat similar to permeability in magnetic circuits. Dielectric materials, or insulators, are rated in their ability to support electric flux in terms of a figure called the DIELECTRIC CONSTANT. The higher the value of the dielectric constant (other factors being equal), the better is the dielectric material.

A vacuum is the standard dielectric for purposes of reference, and it is assigned the value of unity (or one). Dielectric constants for some common materials are given in Table 8-1.

Table 8-1.

Material	Dielectric Constant
Vacuum	1.0000
Air	1.0006
Paraffin paper	3.5
Glass	5–10
Mica	3–6
Rubber	2.5–35
Wood	2.5–8
Glycerine (15° C)	56
Petroleum	2
Pure water	81

Notice the dielectric constant for a vacuum. Since a vacuum is the standard of reference, it is assigned a constant of one; and the dielectric constants of all materials are compared to that of a vacuum. Since the dielectric constant of air has been determined experimentally to be approximately the same as that of a vacuum, the dielectric constant of AIR is also considered to be equal to one.

The capacitance of a capacitor depends on the three following factors:

1. The area of the plates

2. The distance between the plates

3. The dielectric constant of the material between the plates.

C. Capacitor Types

Capacitors may be divided into two major groups—fixed and variable.

Fixed capacitors are constructed in such a manner that they posses a fixed value of capacitance which cannot be adjusted. They may be classified according to the type of material used as the dielectric, such as paper, oil, mica, and electrolyte.

A PAPER CAPACITOR is one that uses paper as its dielectric. It consists of flat, thin strips of metal foil conductors, separated by the dielectric material. In this capacitor, the dielectric used is waxed paper. Paper capacitors are sealed with wax to prevent the harmful effects of moisture and to prevent corrosion and leakage.

Many different kinds of outer covering are used for paper capacitors, the simplest being a tubular cardboard. Some types of paper capacitors are encased in a mold of very hard plastic; these types are very rugged and may be used over a much wider temperature range than the cardboard-case type. Figure 8-7 (A) shows the construction of a tubular paper capacitor; part (B) shows a completed cardboard-encased capacitor.

A MICA CAPACITOR is made of metal foil plates

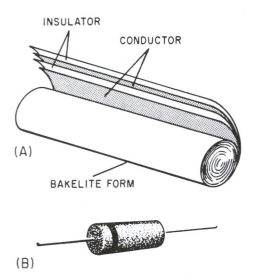

Fig. 8-7. Paper capacitor.

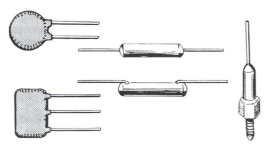

Fig. 8-9. Ceramic capacitor.

that are separated by sheets of mica, which form the dielectric. The whole assembly is covered in molded plastic. Figure 8-8 shows a cutaway view of a mica capacitor. Mica is an excellent dielectric and will withstand higher voltages than paper without allowing arcing between the plates.

A CERAMIC CAPACITOR is so named because of the use of ceramic dielectrics. One type of ceramic capacitor uses a hollow ceramic cylinder as both the form on which to construct the capacitor and as the dielectric material. The plates consist of thin films of metal deposited on the ceramic cylinder. Typical capacitors are shown in Fig. 8-9.

ELECTROLYTIC CAPACITORS are used where a large amount of capacitance is required. As the name implies, electrolytic capacitors contain an electrolyte. The electrolyte can be in the form of either a liquid (wet electrolytic capacitor) or a paste (dry electrolytic capacitor). Wet electrolytic capacitors are no longer in popular use due to the care needed to prevent spilling of the electrolyte.

Dry electrolytic capacitors consist essentially of two metal plates between which is placed the electrolyte. In most cases, the capacitor is housed in a cylindrical aluminum container which acts as the negative terminal of the capacitor (Fig. 8-10). The positive terminal (or terminals if the capacitor is of the multi-section type) is in the form of a lug on the bottom end of the container. The size and voltage rating of the capacitor is generally printed on the side of the aluminum case.

An example of a multi-section type of electrolytic capacitor is depicted in Fig. 8-10. The cylindrical aluminum container will normally enclose four electrolytic capacitors into one can. Each section of the capacitor is electrically independent of the other sections, and one sec-

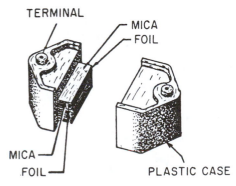

Fig. 8-8. Mica capacitor.

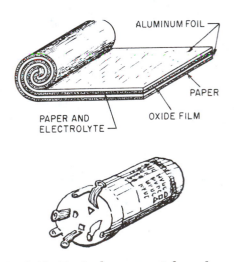

Fig. 8-10. Typical commercial condenser (capacitor).

tion may be defective while the other sections are still good. The can is the common negative connection with separate terminals for the positive connections identified by an embossed mark as shown in Fig. 8-10. The common identifying marks on electrolytic capacitors are the half moon, triangle, square, and no identifying mark. By looking at the bottom of the container and the identifying sheet pasted to the side of the container, the maintenance man can identify each section.

OIL CAPACITORS are often used in radio transmitters where high output power is desired. Oil-filled capacitors are nothing more than paper capacitors that are immersed in oil. The oil-impregnated paper has a high dielectric constant which lends itself well to the production of capacitors that have a high value. Many capacitors will use oil with another dielectric material to prevent arcing between the plates. If an arc should occur between the plates of an oil-filled capacitor, the oil will tend to reseal the hole caused by the arc. These types of capacitors are often called SELF-HEALING capacitors.

D. Capacitor Applications

Capacitors are used in many ways in electrical and electronic equipment and circuits. A few of the more common applications are blocking direct current, filtering, and spark suppression.

The blocking capacitor in Fig. 8-11 is used in a circuit where both direct and alternating currents flow at the same time and it is necessary or desirable to pass the AC and block the DC. This can be done by using a blocking capacitor

as in Fig. 8-11. The capacitor C in the diagram allows alternating current to flow and blocks the flow of direct current.

The filter capacitor in Fig. 8-12 is used to maintain a steady DC voltage by filtering out or removing undesired AC or ripple voltages by capacitor action opposing any change in voltage. Filter capacitors are commonly used to filter power supply voltages.

Spark suppression is obtained by placing a capacitor across the contacts of relays and other movable points subject to electrical sparking when opening and closing. The capacitor minimizes the effects of the sparking and extends the life of the relay contacts or points. The buffer capacitor across the breaker points of an automobile ignition system is a fine example.

However, probably the most common use of the capacitor, and the one which the reader will be most likely to see in the average industrial plant, is the type of capacitor used to improve the power factor of the plant's electrical system.

POWER FACTOR is a term used to denote the efficiency of utilization of the electrical power available to the plant or system, as defined in Chapter 7, Section H.

The use of capacitors wired into the circuits of inductive equipment, such as induction motors, is a cheap and efficient method of raising the power factor. Another method is to drive one or more pieces of equipment with what is known as a synchronous motor, although this is not a cheap way of accomplishing this end.

These two methods will not be explored in this discussion, as they are generally selected and sized by the engineering staff, an outside consultant, or suppliers of the equipment. The plant maintenance staff may have to install

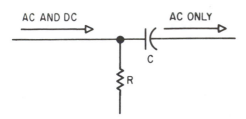

Fig. 8-11. Blocking capacitor.

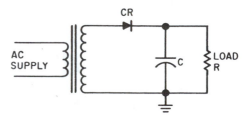

Fig. 8-12. Filtering capacitor.

them, and will most certainly be expected to maintain them. In this matter, they present no particular problem, as all that is usually required is to follow the supplier's instructions carefully, along with the safety rules given in Chapter 1.

However, should the reader feel it within his ability to improve the power factor of the plant, then we refer to Section G in the Appendix in which is given the various sizes of capacitors required to raise the plant's power factor. It must be used with care, as the values given are approximate only, but will give sufficient information for preliminary evaluation of the project. For final confirmation, we suggest that the supplier of the capacitor be consulted.

C H A P T E R 9
ALTERNATING CURRENT TRANSFORMERS

A. Introduction

We stated in Chapter 7 that one of the chief advantages of alternating current is the possibility of generating and transmitting it at one voltage, then changing the voltage to any desired value at the point of use. This flexibility is highly desirable, and has resulted in its adoption almost universally throughout the world for commercial, residential, and industrial use. In this chapter, we shall explain how the high voltages transmitted across country and delivered into the industrial plant are than reduced to those common voltages required to drive the plant's machinery and provide all of the many other power and light uses found in the average industrial plant.

First, a few words on the common voltages mentioned in the last paragraph are in order at this point. The following list covers most of the more common ones, but this is not meant to be all-inclusive. The reader may find others used in specific applications, simply because the usual voltages are not satisfactory for the results desired. The principles given in this book are still applicable, however.

Single-Phase Voltages	Three-Phase Voltages	
110/120	208	416
220/240	216	440
208/216	220	480
	240	4160

The reason we have listed the voltages in the single-phase category with a slash between them (110/120, etc.) is to illustrate the fact that these voltages are often used interchangeably. There appears to be some difference of opinion in various parts of the country over the best basic voltage to use. Some utilities prefer a basic single-phase voltage of 110 volts, while others prefer 120 volts. The basic decision made by the utility company serving your particular plant sets the design of most of the large power using equipment in the plant, with some exceptions, of course.

As a practical matter, equipment made for the 110 volt services can be used on 120 volt service, and usually equipment made for 220 volt service may be used on 240 volt lines. The plant maintenance engineer should use this advice with caution, however, as there are pitfalls for the unwary which we shall attempt to explain, briefly.

Most equipment made to operate on, say, 220 volts, will operate with a voltage of plus or minus 10% of that amount. Thus, it may be safely used on voltages varying from 198 volts to 242 volts. Now for the pitfall to guard against.

We learned in an earlier chapter that in order for a piece of electrical equipment to produce a certain amount of output, a definite amount of electrical power has to be put into the equipment. This is a fundamental fact of all machinery or power using apparatus, regardless of whether it uses electricity, steam, or diesel fuel. In the case of electrical equipment, the power is determined by the following formula:

Power in Watts = Volts × Amperes = E × I.

Obviously, to maintain a definite amount of power input to an electrical device, if the voltage is reduced, then the amperage draw must be increased proportionately. For instance, if an electric motor designed to be used on 240 volts is wired into a circuit providing only 220 volts, the amperage draw on the circuit may possibly be greater than the motor was designed to take. The trouble will come from the increased heat produced from the extra amperage draw, as shown by the formula, given earlier in this book:

Heat Produced = I² × Resistance.

The resistance of the motors has not changed, therefore, if the amperage draw increases, the heat produced increases very rapidly with the increase in the amperage draw. This increased heat may burn out the insulation and cause the motor to burn up due to a short in the windings. The maintenance engineer should keep this in mind when the decision is made to install an old motor which has been in the warehouse for a few years, and there is no history on record for it.

Fortunately, most motors for the past 30 or 40 years have been made with ample insulation to resist a higher heat release in the windings than the electrical code requires. This has kept many a plant engineer out of trouble. Of course, you can also get into trouble if you rely on that practice too blindly. Check the nameplate ratings of every motor or piece of electrical equipment being installed, regardless of whether it is new or used.

There is one other thing to keep in mind before hooking up any piece of electrical apparatus, and that is the manufacturer's guarantee on it. If you install a motor designed for operation on 220 volts on a circuit rated at 240 volts, the 240 volts is within the plus or minus guarantee on voltage made by the manufacturer of the motor, as 220 volts plus 10% is equal to 242 volts, just within the guarantee. But how certain are you that the voltage will not climb to much higher than 242 volts, especially during periods of low power draw on the system? No power supply is that stable, and variations of at least plus or minus 5% may be expected in any commercial power supply. In short, a power supply rated at 240 volts may climb to 252 volts (240 volts plus 5% = 252 volts), which in the case in our example would be 10 volts over the guarantee. There are areas in this country where the voltages may vary considerably more than the guaranteed amount, due to unbalanced conditions during periods of low loads. It is well to keep in mind that all electrical supply systems in the country are not perfectly designed and balanced, and wide voltage swings are common.

For some time, there has existed considerable confusion over the appropriate split between high voltage and low voltage. Several agencies, industries, etc., have each attempted to codify the difference between the two voltages, usually within the framework of their own operations, with little or no regard to other electrical fields.

As a practical matter, we see no reason why each industrial plant cannot adopt its own split between the high voltage systems and the low voltage systems in the plant.

B. Principles of Transformers

A transformer is just what the name implies. It is a device for taking an electrical input and transforming it to a more convenient output. In the case of electrical distribution and supply systems, the desirable output is usually based upon the voltage required for the plant equipment. In this chapter, we shall describe the principles involved in the barest essentials only, to permit the plant maintenance man to know what he is working with, and to permit him to know fairly well what to expect when he makes a change in the wiring, or is required to perform any maintenance upon it.

The basic equation of power upon which the transformer operates is as follows:

Power Into the Transformer = Power Out Plus Losses.

The losses are minimal in the transformer, and in most cases are actually ignored in preliminary considerations. They generally constitute about 2% or 3% of the input power, so that the efficiency of the average transformer is between 97% to 98%. Transformers are thus one of the most efficient pieces of electrical or mechanical apparatus you will find.

A transformer is a device that has no moving parts and that transfers energy from one circuit to another by electromagnetic induction. The energy is always transferred without a change in frequency, but usually with changes in voltage and current. A stepup transformer receives electrical energy at one voltage and delivers it at a higher voltage. Conversely, a stepdown transformer receives energy at one voltage and delivers it at a lower voltage. Transformers require little care and maintenance because of their simple, rugged, and durable construction. The efficiency of transformers is high. The conventional constant voltage transformer is designed to operate with the primary connected across a constant voltage source and to provide a secondary voltage that is fairly constant from no load to full load, within the limits stated in Section A.

The various types of small single-phase transformers are used in electrical equipment. In many installations, transformers are used on switchboards to step down the voltage for indicating lights. Low-voltage transformers are included in some motor control panels to supply control circuits or to operate overload relays.

Instrument transformers include voltage transformers and current transformers. Instrument transformers are commonly used with AC instruments when high voltages or large currents are to be measured.

Transformers are rated in Kilovolt-Amperes, and this rating is determined thus:

$$\text{Single-Phase Kva} = \frac{\text{Volts} \times \text{Amperes}}{1000}$$

$$\text{Three-Phase Kva} = \frac{\text{Volts} \times \text{Amperes} \times 1.73}{1000}.$$

Small transformers, usually less than 1000 VA, are rated in volt-amperes.

C. Construction

The typical transformer has two windings insulated electrically from each other. These windings are wound on a common magnetic core made of laminated sheet steel. The principal parts are: (1) the core, which provides a circuit for the magnetic flux; (2) the primary windings, which receive the energy from the AC source; (3) the secondary windings, which receive the energy by mutual induction from the primary and delivers it to the load; and (4) the enclosure.

When a transformer is used to step up the voltage, the low-voltage winding is the primary. Conversely, when a transformer is used to step down the voltage, the high-voltage winding is the primary. The primary is always connected to the source of the power; the secondary is always connected to the load. It is common practice to refer to the windings as the primary and secondary rather than the high-voltage and low-voltage windings.

The principal types of transformer construction are the core type and the shell type as illustrated, respectively, in Fig. 9-1 (A) and (B). The cores are built of thin stampings of silicon steel.

In the core type transformer, the copper windings surround the laminated iron core. In the shell type transformer, the iron core surrounds the copper windings. Distribution transformers are generally of the core type; whereas some of the largest power transformers are of the shell type.

The windings are sub-divided, and half of each winding is placed on each leg of the core. The windings may be cylindrical in form and placed one inside the other with the necessary insulation, as shown in Fig. 9-1(A).

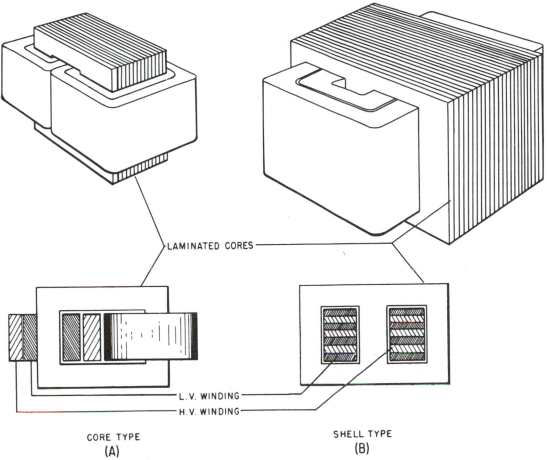

LAMINATED CORES

L.V. WINDING

H.V. WINDING

CORE TYPE
(A)

SHELL TYPE
(B)

Fig. 9-1. Types of transformer construction. (A) Core type. (B) Shell type.

In another method, the windings are built up in thin flat sections called pancake coils. These pancake coils are sandwiched together, with the required insulation between them, as shown in Fig. 9-1 (B).

The complete core and coil assembly (Fig. 9-2 (A)) is placed in a steel tank. In some transformers, the complete assembly is immersed in a special mineral oil to provide a means of insulation and cooling, while in other transformers they are mounted in drip-proof enclosures as shown in Fig. 9-2 (B).

Transformers are built in both single-phase and polyphase units. A three-phase transformer consists of separate insulated windings for the different phases, wound on a three-legged core capable of establishing three magnetic fluxes displaced 120 degrees in time phase.

D. Voltage and Current Relationships

The operation of the transformer is based on the principle that electrical energy can be transferred efficiently by mutual induction from one winding to another. When the primary winding is energized from an AC source, an alternating magnetic flux is established in the transformer core. This flux links the turns of both primary and secondary, thereby induc-

Fig. 9-2. Single-phase transformer. (A) Coil and core assembly. (B) Enclosure.

ing voltages in them. Because the same flux cuts both windings, the same voltage is induced in each turn of both windings. Hence, the total induced voltage in each winding is proportional to the number of turns in that winding, that is,

$$\frac{E_1}{E_2} = \frac{N_1}{N_2}$$

where E_1 and E_2 are the induced voltages in the primary and secondary windings, respectively, and N_1 and N_2 are the number of turns in the primary and secondary windings, respectively. Hence, in ordinary transformers, the applied primary voltage and the secondary induced voltage are approximately proportional to the respective number of turns in the two windings.

Figure 9-3 shows the voltage and winding relationships for an average stepdown transformer, ignoring any losses within the transformer. The common equations pertaining to the transformer are included, and a study of the figure and the equations will acquaint the student with most of the mathematics involved in day-to-day maintenance and operation.

Many transformers are rated in terms of the kva load that they can safely carry continuously without exceeding a temperature rise of 80° C when maintaining rated secondary voltage and when operating with an ambient (surround atmosphere) temperature of 40° C. The actual temperature rise of any part of the transformer is the difference between the total temperature of that part and the temperature of the surrounding air.

It is possible to operate transformers on a higher frequency than that for which they are designed, but it is not permissable to operate them at more than 10 percent below their rated frequency, because of the resulting overheating. At reduced frequency, the exciting current becomes excessively large and the accompanying heating may damage the insulation and the windings.

Probably the best method of explaining the design and use of a typical transformer is to use as an example the style found on the utility poles in any ordinary residential neighborhood distribution system. A typical distribution voltage is 7200 volts, and this is dropped to two single phase, 120 volt circuits as shown in Fig. 9-4. Also, by connecting terminals X_2 and X_3 to a common ground, 240 volts will be obtained across terminals X_1 and X_4.

In the above diagram, the terminals are marked in the manner usually found on commercial transformers. The terminals H_1, H_2,

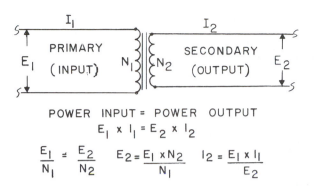

Fig. 9-3. Basic transformer principle.

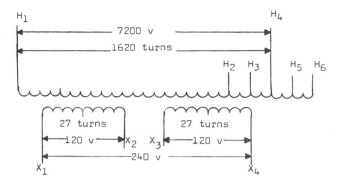

Fig. 9-4. Typical distribution transformer.

etc., are on the high voltage side, on the input side, as this is a stepdown transformer. Terminals X_1, X_2, etc., are on the outlet, or low voltage, side.

You will notice that there are several extra terminals on the high voltage side, and these are known as "taps." The purpose of these is to permit the ratio of windings to be altered to compensate for voltages that are either lower or higher than the nominal required voltages. If the voltage in the particular area where the transformer is to be installed remains fairly steady at 7200 volts, then the terminal H_4 will produce the proper turns ratio to give required outlet voltages of 120 and 240 volts, as described above.

However, if the supply voltage at this location varies from the 7200 volts, then the outlet voltage connections may be altered to compensate for this altered voltage. Ordinarily, H_3 and H_2 will each vary the outlet voltage by $2\frac{1}{2}\%$ in this manner; if the inlet voltage is low, then changing the H_4 connection to H_3 will compensate for an inlet voltage that is $2\frac{1}{2}\%$ low. Changing the H_4 connection to H_2 will compensate for an inlet voltage that is 5% low. The aim in either case is to ensure that the outlet voltage will be the 120/240 volts required at the customer's premises.

Conversely, if the distribution voltage is high at this location, the connection H_4 may be changed to either H_5 or H_6 in the same manner to compensate. This does not happen too frequently, however, as the tendency in today's expanding power demand is to overload the systems and thus lower the distribution voltages.

Such a transformer is known generally as a 2-$2\frac{1}{2}\%$FCBN and 2-$2\frac{1}{2}\%$FCAN, which means that is has 2, $2\frac{1}{2}\%$ taps below normal and 2, $2\frac{1}{2}\%$ taps above normal voltage, the FCBN standing for "Full Capacity Taps Below Normal," and the FCAN standing for "Full Capacity Taps Above Normal."

You will notice that the changes required to compensate for varying inlet voltages follow a distinctive direction. If the inlet voltage is low, the connection is made to a lower tap(H_3 or H_2). If the inlet voltage is high, the connection is made to a higher tap(H_5 or H_6). This should make it very easy to remember which direction the connections must be moved to compensate for changes in inlet voltages, and also outlet voltages, as the outlet varies in the same direction as the inlet voltages.

What, you may ask, has happened to the amperages in the system after all this has taken place? To answer this question, we refer you back to the Introduction, Section A, of this chapter. There we gave you the formula for power as being equal to the volts times the amperes. The relationship carries through from the primary side of the transformer to the secondary side.

In the example just given for our distribution transformer having an input voltage of 7200 volts and an output of 120 volts, the stepdown ratio is 7200 divided by 120, which is a ratio of 60 to 1. Thus, if the voltage is dropped by a ratio of 60, the amperage on the secondary side must increase to 60 times that on the primary side. Remember, the following equation must be satisfied:

Primary volts × Amps = Secondary volts × Amps.

It is important for you to think this through to ensure that you have it well in mind.

E. Large Power Transformers

So far we have described only the smaller single-phase transformers. Now we shall give you a few pointers concerning the larger transformers, the type found in substation and distribution station installations. The principles are the same, so it will not be necessary for us to go into that part of them again.

The three things that set the power transformers apart from the smaller single-phase transformers are the size, the requirement for more extensive cooling, and the three-phase arrangement.

Because of their size, and due to the enormous amounts of power that they handle, the substation and distribution transformers are usually isolated from all other equipment, with either high walls or high fences around them. This is not only to protect unauthorized people from getting hurt or killed, but also to prevent sabotage and vandalism, as these units are usually in a vital portion of the electrical distribution system, either for the plant or for the area being served. There are very strict rules set down by the **National Electric Code**® defining the use and installation of these transformers, and these rules should be very carefully followed. To deviate from them could lead to disastrous results, as most of the rules have been written as a result of others having trouble in the past.

Because of their size, three-phase power transformers are available in either one casing or three separate casings. It is cheaper to install all three transformers of a three-phase unit inside one case, as there is a decided saving in

material, cooling system, and manufacturing labor. It makes for a simpler installation, also.

However, there is a limit as to the size that can be transported from the factory to the installation location. Transformers are very heavy pieces of equipment, and highway and railroad limitations have to be considered when ordering a large power transformer.

Therefore, in the higher capacity units, it is quite common to install three separate, single-phase transformers, and wire them on the jobsite to meet the requirements of the specifications. The main advantage in using three separate units is reliability, because if one phase causes trouble, it is cheaper and easier to replace the single-phase transformer than to replace an entire three-phase unit. Also, in many cases, it is possible to continue operation with only two phases on the line, so service may not have to be curtailed as much. We refer you to the next section, where we describe the open Delta system, which is an example of this situation.

We mentioned earlier in this chapter that transformers operate at very high efficiencies. The losses, as small as they may be from a percentage viewpoint, can result in large amounts of heat released in the larger units, for the losses appear as heat, and this heat could damage the insulation and cause a massive short, if not controlled. For this reason, the large power transformers will have a method of cooling, either by natural convection or by forced circulation of a cooling medium. It is not unusual to see small fans mounted at the base of these transformers, sending cooling air over the fins built into the housing of the transformer sections.

Many of the transformers in this size range will be submerged in a cooling oil, with circulating systems to circulate the coolant around the core and the coil sections. This oil should be handled with care, as some of the transformers furnished in past years have contained an oil that is highly toxic, and therefore dangerous to personal health. Be sure that you are aware of the type of coolant in your transformers.

One of the common PCB oils, which are very

toxic, is "askarel." Great care should be used in working with this material, so as not to spill any of it. Follow the instructions given with the transformer to prevent accidents.

The oil used for cooling also serves partially as insulation between the windings and the core, therefore, it must not contain any moisture. Water in the oil, even in minute amounts, will cause the insulating value to break down. For this reason, the oil that is charged into the cooling system must be dehydrated, and the cooling chamber evacuated with a vacuum pump for several hours before the oil is charged into the system. Also, it may be necessary to periodically circulate the oil through a dryer to remove accumulated water. This is when the maintenance worker must be very careful in handling the oil.

Reconditioning of the PCB oils is often done by outside contractors who have the equipment and the training to perform this task with a minimum of trouble and danger to the plant personnel.

The rating on any transformer is based upon one set ambient, or surrounding, temperature, and the rating plate will list the temperature as being so many degrees Centigrade above ambient. It is very important that the location of the unit be selected with care, and the operating temperatures watched very carefully.

It is sometimes possible to operate a transformer over its rated capacity, by careful attention to the cooling system. By increasing the circulation of the coolant, or lowering its temperature, or by setting the transformer in a cooler area, it may be possible to push the unit above its normal rating. Winter time in cold areas often permits operation at above rated capacity, also.

There is an inherent fire hazard with oil cooled transformers, and if they are installed indoors, a fire or smoke detector should be installed above them to sound an alarm, and take any other action necessary. It is sometimes an excellent idea to have a fire smothering system installed which automatically smothers the transformer with a chemical upon detecting an elevated temperature or the presence of smoke.

Such chemicals are often very messy, and should be used only with careful advice of the supplier.

Probably the only maintenance which will be required of the plant maintenance electrician is to clean the outside surfaces of the transformer once a year under ordinary circumstances, or more often if ambient conditions require it. The cooling surfaces in particular should receive special attention, along with any cooling fans and their electrical systems.

As you can readily see from the above, transformers, due to their large power through-put, are very important affairs, and must be handled very considerately if the best service is to be obtained from them. But then, this is true of most of the equipment in your plant, isn't it?

F. Three-Phase Combinations

In Chapter 7 we introduced you to three-phase systems, as well as the common single-phase systems, and the uncommon two-phase systems. Before proceeding any further in this discussion of transformers and electrical distribution systems, it is necessary that you become familiar with the various methods of connecting three-phase systems. It is these connections and their voltage relationships that allow the use of so many different voltages in today's electrical equipment and appliances.

In all of the succeeding illustrations and diagrams, it will be noticed that the transformers are grounded on the output side. This is for protection, as well as serving as another power leg in some cases.

Transformers rated 1 KVA and larger are always grounded through a separate cable or bus to a grounding electrode such as a cold water pipe, driven rod, building steel, etc. Transformers smaller than 1 KVA may be grounded to the cabinet or the conduit serving it. The rules stated in Chapter 11, Section G, apply in either case, regardless of the size of the unit.

Going back to Fig. 7-10, we see that there are three wires bringing power from the generator.

In the usual power distribution system, these three wires may be connected into the distribution transformers in two basic configurations. These two basic methods are known as the Y, or Star, system, and the Delta system. The Delta system is named after the Greek letter, Delta, which is in the shape of a triangle. These two configurations are illustrated in Figs. 9-5 and 9-6 on the secondary, or output side, of the transformer.

The wires leading from the transformers may, in turn, be connected in several variations of both basic methods, and these will now be explained briefly, giving the common voltages which are obtainable by these wiring systems.

We shall start with a description of the basic three-wire Y system, as shown in Fig. 9-5. In all of the following diagrams, the wiring shown inside the boxed portion is considered to be inside the transformer, and is not accessible. The transformers are usually ordered to the customer's specifications, and there are various combinations of the basic types available, depending upon which distribution system the transformer will be used. Most transformer terminals are marked thus; X_0, X_1, X_2, etc. We shall dispense with the "X" and use only the numerical markings.

The three-wire transformer shown in Fig. 9-5 is the type most often used to produce the following voltages:

480-volt, single phase, two-wire; terminals 1 & 2, 2 & 3, 1 & 3

480-volt, three-phase, three-wire; terminals 1, 2, & 3

277-volt, single-phase; any terminal to ground.

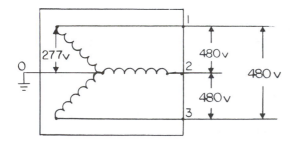

Fig. 9-5. Y-connected secondary side of transformer.

Figure 9-5 shows that it is sometimes possible to ground the common or neutral terminal, 0, which is often done for protection. When this is the case, the grounding is usually done right at the transformer location. The voltages commonly obtained are in the following combinations:

Combination 1—480-volt circuits; same as above

Combination 2—277-volt, single-phase; terminals 0 to 1, 2, or 3.

This system is also used in higher voltages for distribution systems on long rural lines. In this case, the grounded junction, 0, is not only grounded at the central transformer, but is also carried along the entire route with a white covered fourth wire, which grounds every step-down transformer along the system. These stepdown, or substation, transformers are also grounded individually at each location. This permits isolation of sections of the system in case of trouble.

In Fig. 9-5, the 277-volt circuits are often used for the plant lighting systems.

Turning now to the other basic system, the Delta system, we refer you to Fig. 9-6, which illustrates a corner grounded Delta circuit. The 240-volt system is the one most commonly used in this method, and we have shown it here. Not all Delta transformers are grounded, however. The voltages ordinarily supplied with the Delta system are as follows:

240-volt, single-phase, two-wire; terminals 1 & 2, 2 & 3, 1 & 3

240-volt, three-phase, three-wire; terminals 1, 2, & 3

0-volts; terminal 3 to ground.

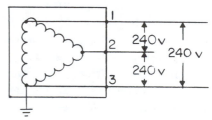

Fig. 9-6. Delta three-wire secondary system.

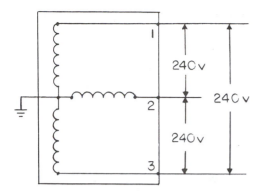

Fig. 9-7. T connected secondary transformer.

Table 9-1.

Voltage Across the Terminals	Terminal-to-Ground Voltage[a]
208 volts	179 volts
240 volts	206 volts
480 volts	413 volts
4160 volts	3578 volts
12,470 volts	10,724 volts

[a] 86% of voltages in first column.

This system is often used for 480 volt and 600 volt circuits, with the same terminal–voltage relationships as given for the 240 volt systems.

Figure 9-7 shows a variation of the Y system, known as the T system, the reason being quite obvious from the diagram. These transformers are sometimes known as "Scott" connected units, and are available only in smaller sizes. They are easily confused with other three-phase, three-wire transformers, and the maintenance engineer should be very careful not to mix them up when running wires from them.

This system usually supplies loads rated as follows:

240-volt, single-phase, two-wire; terminals 1 & 2, 2 & 3, 1 & 3

240-volt, three-phase, three-wire; terminals 1, 2, & 3.

If the neutral terminal is accessible on the transformer, and is solidly grounded, then a voltage of 206 volts will be obtained between terminal and ground. This fact should be kept in mind, as the same relationship holds true for any of the three-phase, three-wire systems discussed above, when the transformer is firmly grounded. We shall give this relationship in the form of Table 9-1, for easy reference.

Another variation of the T system transformer is shown in Fig. 9-8 which is similar to Fig. 9-7, except that in this case the transformer is not only grounded, but the grounded conductor, which is considered "neutral," is carried along with the other three conductors. The resultant four-wire, three-phase system is avail-

able only in the smaller sizes with voltages as follows:

120-volt, single-phase, two-wire; terminals 0 to 1, 2, or 3

208-volt, single-phase, two-wire; terminals 1 & 2, 2 & 3

208-volts, three-phase, three-wire; terminals 1–3.

Figure 9-9 shows another variation of the Y system, one which is used for high-rise apartment complexes and similar installations. The voltages supplied are:

120-volt, single-phase, two-wire; terminals 1 & 2, 2 & 3

208-volt, three-phase, three-wire; terminals 1, 2, & 3.

Figure 9-10 is a variation of the Delta system, known as the open Delta system, which may or not be grounded. It is used on systems which are to be enlarged in the future, as only two transformers are required to complete the circuit, as shown. In the future, the third transformer may be added, which increases the capacity, for the capacity of the open Delta

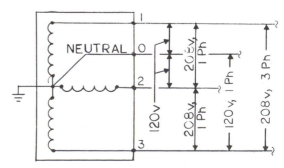

Fig. 9-8. Four-wire T connected secondary.

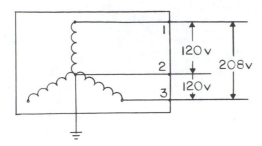

Fig. 9-9. Single-phase dual-voltage Y secondary system.

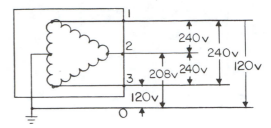

Fig. 9-11. Delta four-wire secondary system.

arrangement is only 57.7% of the capacity of the three-legged Delta arrangement. Typical voltage-phase combinations are available as follows:

240-volt, single-phase, two-wire; terminals 1 & 2, 2 & 3

240-volt, three-phase, three-wire; terminals 1, 2, & 3.

These transformers are also available in 480 volt and 600 volt units, with the same terminal relationships as for the 240 volt units.

Figure 9-11 is a more popular version of the Delta system, in which one of the phase transformers is grounded in the center, and a ground, or neutral, wire is carried along with the other three wires to carry the loads. The 208 volt circuit between terminals 0 and 2 is almost never used in practice, as there are other systems which are better for use on 208 volt circuits.

The one thing to be kept in mind with the center grounded Delta four-wire system is the possibility of any load or equipment which is connected to terminal 2 becoming accidentally grounded. When this happens, the voltage

drops to 208 volts, as shown in Fig. 9-11, and the amperage draw by the load increases to maintain the power requirement of the load. If you remember, the formulas are:

Power = Volts × Amps, and Power = Amperage² × Resistance.

Thus, as the amperage draw for the load increases, the heating effect in the conductors and on the load increases rapidly. This increased heating may cause the insulation to burn up, shorting the system. So the maintenance engineer should keep in mind that loads on the 240 volt circuit in this system should be insulated for the 208 volt system, and so also should the wires to the load.

The line coming from terminal 2 is commonly known as the "stinger" leg, for the reasons given in the previous paragraph. It is always of a very distinctive orange color, as a warning to anyone working on a circuit of this configuration to be very careful how it is used and handled.

As you can see by Fig. 9-11, the loads which may be supplied with this system are:

120-volt, single-phase, two-wire; terminals 0 & 1, 0 & 3

208-volt, single-phase, two-wire; terminals 0 & 2

240-volt, single-phase, two-wire; terminals 1 & 2, 2 & 3, 1 & 3

240-volt, three-phase, three-wire; terminals 1, 2, & 3.

Table 9-2 contains recommendations for instrumentation of various common sizes of commercial power transformers.

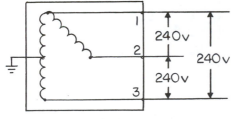

Fig. 9-10. Open delta secondary system.

Table 9-2. Minimum Instrumentation for Transformers

Devices	Low-voltage secondary		Medium-voltage secondary — Unit Capacity											
	500 kVA and below	500 kVA to 1,500 kVA[1]	2,500 kVA and below				3,750 kVA to 7,500 kVA				10,000 kVA and above			
	Secondary main	Secondary main	Incoming line[1]	Secondary main	Feeder	Parallel sources	Incoming line[2]	Secondary main	Feeder	Parallel sources	Incoming line	Secondary main	Feeder	Parallel sources
Relays														
Synchronism check (25)[3]	None required	None required				X				X				X
Overcurrent, phase & ground (51)(51N)			X	X	X	X	X	X	X	X	X	X	X	X
Directional overcurrent (67)(67N)						X				X				X
Reclosing (79)				X				X					X	
Differential current (87)							X				X			
Meters														
Ammeter	None required	X	X	X			X	X	X		X	X	X	
Voltmeter		X	X	X			X	X			X	X		
Frequency meter						X				X				X
Wattmeter				X				X				X	X	
Varmeter				X				X				X		
Watthour meter		X	X				X				X			
Miscellaneous														
Synchroscope	—	—				X				X				X

[1]For transformers 2,000 kVA and above, provide instrumentation as shown for medium-voltage secondaries.

[2]Relays apply only when circuit breakers are provided.

[3]Numbers in parentheses are ANSI device numbers.

G. Other Types of Transformers

There are many uses for transformers in today's industrial system. The electronics industry uses thousands of them in many sizes, styles, and materials. The electric power field is the one we are primarily interested in, so we are limiting our discussion to the ones which the maintenance engineer in the average industrial plant will be most likely to encounter.

The vast majority of the transformers used industrially are what is known as "stepdown" transformers, as they reduce the incoming voltage to a lower value for use in the plant or the equipment. This is the style we have been involved with so far in this discussion.

There is also the "stepup" transformer, in which output voltage is increased from the input voltage. This style uses the same system of terminal markings as the "stepdown" unit. The terminals marked with an H_1, H_2, etc., are the input terminals, and the terminals marked X_1,

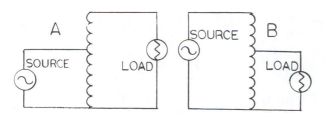

Fig. 9-12. Autotransformers.

X_2, etc., are the output terminals. They may also have taps on the input side for voltage adjustment, as described earlier in this chapter for the stepdown units.

Another style is the isolating transformer, in which the input and the output voltages are identical. This is used to isolate the circuit for protection, as there is no direct electrical contact between the two portions of the system. However, when this situation is encountered, it is important that very close attention be paid to the forces which are applied to the transformer when a short occurs.

Figure 9-12 illustrates another type of transformer, the autotransformer. In this transformer, one winding does double duty, and it may be a stepup or a stepdown transformer. In Fig. 9-12 (A), we show a typical stepup transformer, and in (B) we show a stepdown unit. The main advantages of this type are lower price, steady voltages under varying loads, and low power losses. One of the chief disadvantages is the lack of electrical isolation between the input side and the output side.

A word of caution here regarding autotransformers. They may not be permitted under your local codes, so be sure and check before you decide to install one.

The autotransformer is also known as a "Buck and Boost" transformer. Whether it "Bucks" or "Boosts" depends upon how it is wired into the circuit. One of the more common uses of this transformer is to alter the voltage in small amounts to a particular piece of equipment, when the plant supply at that point is not correct or varies too much. We suggest you contact the supplier for proper use of the autotransformer.

Two other types of transformers—current and potential transformers—are described in the next chapter.

CHAPTER 10
MEASURING

A. Introduction

In the field of electricity, as in all the other physical sciences, accurate quantitative measurements are essential. This involves two important items—numbers and units. Simple arithmetic is used in most cases, and the units are well defined and easily understood. The standard units of current, voltage, and resistance, as well as other units, are defined by the National Bureau of Standards. At the factory, various instruments are calibrated by comparing them with established standards.

The maintenance man commonly works with ammeters, voltmeters, ohmmeters, wattmeters, and watt-hour meters.

Electrical equipment is designed to operate at certain efficiency levels. To aid the maintenance man in maintaining the equipment, technical instruction books and sheets containing optimum performance data, such as voltages and resistances, are available from the equipment manufacturer.

To the maintenance man, a good understanding of the functional design and operation of electrical instruments is important. In electrical service work, one or more of the following methods are commonly used to determine if the circuits of the plant equipment are operating properly.

1. Use an ammeter to measure the amount of current flowing in a circuit.

2. Use a voltmeter to determine the voltage existing between two points in a circuit.

3. Use an ohmmeter or megger (megohmmeter) to measure circuit continuity and total or partial circuit resistance.

The maintenance man may also find it necessary to employ a wattmeter to determine the total POWER being consumed by plant equipment. If he wishes to measure the ENERGY consumed, a watt-hour or kilowatt-hour meter is used.

A thorough understanding of the construction, operation, and limitations of electrical measuring instruments, coupled with the theory of circuit operation, is most essential in servicing and maintaining electrical equipment.

The discussion and explanations in this chapter will be based primarily on those basic instruments which have evolved over the 100 or more years since electricity became a major source of useful power in our lives. Today's electrical worker will find an increasing number of more sophisticated instruments on the market, not only to do the same work as those described herein, but also many which will make the daily tasks easier.

These new instruments are designed around electronic circuits, and usually require a high degree of skill and care in their use. They must be maintained and calibrated properly, according to the instructions which accompany them. An excellent example is the Autoranging Digital Multimeter shown in Fig. 10-24.

B. Direct Current Instruments

In this chapter, a discussion of both the DC and AC meters and measuring instruments is presented. The DC type meters will be discussed first since they are the simpler type. Some of the DC meters may also be used for measuring AC potentials.

D'ARSONVAL METER

The stationary permanent-magnet moving-coil meter is the basic movement used in most measuring instruments for servicing electrical equipment. This type of movement is commonly called the D'Arsonval movement and consists of a stationary permanent magnet and a movable coil. When current flows through the coil, the resulting magnetic field reacts with the magnetic field of the permanent magnet and causes the coil to rotate. The greater the amount of current flow through the coil, the stronger the magnetic field produced; and the stronger this field, the greater the rotation of the coil.

The principle of the D'Arsonval movement may be shown by the use of the simplified diagram (Fig. 10-1) of the D'Arsonval movement commonly used in DC instruments.

While D'Arsonval type galvanometers are useful in the laboratory for measurements of extremely small currents, they are not portable, compact, or rugged enough for use in the

maintenance of industrial equipment. The Weston meter movement is used instead.

The Weston meter uses the principles of the D'Arsonval galvanometer, but it is portable, compact, rugged, and easy to read. In the Weston meter, the coil is mounted on a shaft fitted between two permanently mounted jewel bearings. A lightweight pointer is attached and turns with the coil; the pointer indicates the amount of current flow. Figure 10-2 illustrates this movement.

The amount of current required to turn the meter pointer to full-scale deflection depends upon the magnet's strength and the number of turns of wire in the moving coil. This amount of current is the ammeter's MAXIMUM ALLOWABLE AMOUNT. A further increase of meter current would damage the meter.

AMMETER

The small size of the wire with which an ammeter's movable coil is wound places severe limits on the current that may be passed through the coil. Consequently, the basic D'Ar-

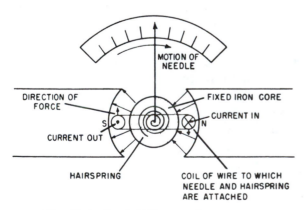

Fig. 10-1. Basic D'Arsonval movement.

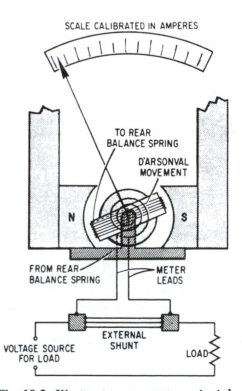

Fig. 10-2. Weston type ammeter principle.

sonval movement discussed this far may be used to indicate or measure only very small currents.

To measure a larger current, a shunt must be used with the meter. A shunt is a heavy low-resistance conductor connected across the meter terminals to carry most of the load current. This shunt has the correct amount of resistance to cause only a small part of the total circuit current to flow through the meter coil. The meter current is proportional to the load current. If the shunt is of such a value that the meter is calibrated in milliamperes, the instrument is called a MILLIAMMETER. If the shunt is of such a value that the meter is calibrated in amperes, it is called an AMMETER.

A single type of standard meter movement is generally used in all ammeters, no matter what the range of a particular meter. For example, meters with working ranges of 0 to 10 amperes, 0 to 5 amperes, or 0 to 1 ampere all use the same galvanometer movement. The designer of the ammeter simply calculates the correct shunt resistance required to extend the range of the 1-milliampere meter movement to measure any desired amount of current. This shunt is then connected across the meter case (internal shunt) or somewhat away from the meter (external shunt), with leads going to the meter. An external shunt arrangement is shown in Fig. 10-2.

To ensure accurate readings, the meter leads for a particular ammeter should not be used interchangeably with those for a meter of a different range. Slight changes in lead length and size may vary the resistance of the meter circuit and thus its current, and may cause an incorrect meter reading. External shunts are generally used where currents greater than 50 amperes must be measured.

It is important to select a suitable shunt when using an external shunt ammeter so that the scale indication is easily read. For example, if the scale has 150 divisions and the load current to be measured is known to be between 50 and 100 amperes, a 150-ampere shunt is suitable.

If the scale deflection is 75 divisions, and the load current is 75 amperes, the needle will deflect half-scale when using the same 150-ampere shunt.

A shunt having exactly the same current rating as the estimated normal load current should never be selected because any abnormally high load would drive the pointer off scale and might damage the movement. A good choice would bring the needle somewhere near the mid-scale indication when the load is normal.

For limited current ranges (below 50 amperes), internal shunt ammeters are most often employed. In this manner, the range of the meter may be easily changed by selecting the correct internal shunt having the necessary current rating.

Various values of shunt resistances may be used, by means of a suitable switching arrangement, to increase the number of current ranges that may be covered by the meter. Two switching arrangements are shown in Fig. 10-3. Figure 10-3 (A) is the simpler of the two arrangements. However, it has two disadvantages.

1. When the switch is moved from one shunt resistor to another, the resultant surge of current could easily damage the coil.

2. The contact resistance—that is, the resistance between the blades of the switch when they are in contact—is of a variable nature. Thus, the ammeter indication may not be accurate.

As in the use of any instrument, always follow the instructions which accompany it very carefully.

A more generally accepted method of range switching is shown in Fig. 10-3 (B). Although

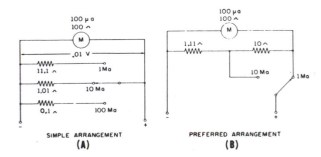

Fig. 10-3. Internal shunt connections.

only two ranges are shown, as many ranges as needed can be used. In this type of circuit, the range selector switch contact resistance is external to the shunt and meter in each range position, and therefore has no effect on the accuracy of the current measurement.

CURRENT-MEASURING INSTRUMENTS MUST ALWAYS BE CONNECTED IN SERIES WITH A CIRCUIT AND NEVER IN PARALLEL WITH IT. If an ammeter were connected across a constant-potential source of appreciable voltage, the meter would burn out.

If the approximate value of current in a circuit is not known, it is best to start with the highest range of the ammeter and switch to progressively lower ranges until a suitable reading is obtained.

Most ammeter needles indicate the magnitude of the current by being deflected from left to right. If the meter is connected with reversed polarity, the needle will be deflected backwards, and this action may damage the move-

ment. Hence, the proper polarity should be observed in connecting the meter in the circuit. That is, the meter should always be connected so that the electron flow will be into the negative terminal and out of the positive terminal.

Figure 10-4 shows various circuit arrangements; the ammeter or ammeters are properly connected for measuring current in various portions of the circuits.

VOLTMETER

The D'Arsonval system used as the basic meter for the ammeter may also be used to measure voltage if a high resistance is placed in series with the moving coil of the meter. For low-range instruments, this resistance is mounted inside the case with the D'Arsonval movement, and typically consists of resistance wire having a low temperature coefficient and wound either on spools or card frames. For higher voltage ranges, the series resistance may be connected

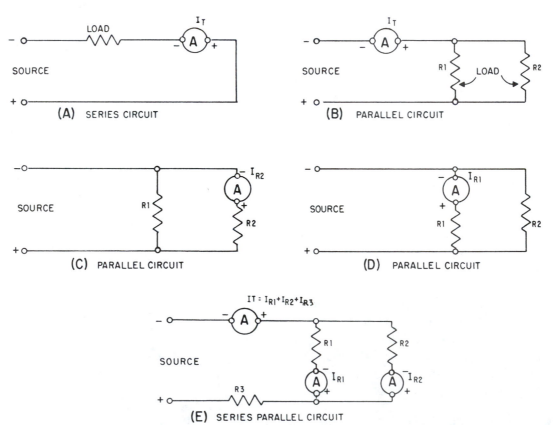

Fig. 10-4. **Proper basic ammeter connections. (A) Series circuit. (B) Parallel circuit. (C) Parallel circuit. (D) Parallel circuit. (E) Series parallel circuit.**

externally. When this is done, the unit containing the resistance is commonly called a MULTIPLIER.

A simplified diagram of a voltmeter is shown in Fig. 10-5.

MULTIRANGE VOLTMETERS

The value of the necessary series resistance is determined by the current required for full-scale deflection of the meter and by the range of voltage to be measured. The meter scale is calibrated directly in volts for each fixed series resistance.

Multirange voltmeters utilize one meter movement with the required resistance connected in series with the meter by a convenient switching arrangement. A multirange voltmeter with three ranges is shown in Fig. 10-6.

VOLTAGE-MEASURING INSTRUMENTS ARE CONNECTED ACROSS (IN PARALLEL WITH) A CIRCUIT. If the approximate value of the voltage to

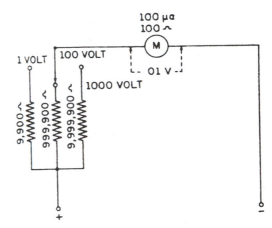

Fig. 10-6. Multirange voltmeter.

be measured is not known, it is best to start with the highest range of the voltmeter and progressively lower the range until a suitable reading is obtained.

In many cases, the voltmeter is not a central-zero indicating instrument. Thus, it is necessary to observe the proper polarity when connecting the instrument to the circuit, as is the case in the DC ammeter. The positive terminal of the voltmeter is always connected to the positive terminal of the source, and the negative terminal to the negative terminal of the source when the source voltage is being measured. In any case, the voltmeter is connected so that electrons will flow into the negative terminal and out of the positive terminal of the meter.

The function of a voltmeter is to indicate the potential difference between two points in a circuit. When the voltmeter is connected across a circuit, it shunts the circuit. If the voltmeter has low resistance, it will draw an appreciable amount of current. The effective resistance of the circuit will be lowered and the voltage reading will consequently be lowered.

When voltage measurements are made in high-resistance circuits, it is necessary to use a high-resistance voltmeter to prevent the shunting action of the meter. The effect is less noticeable in low-resistance circuits because the shunting effect is less.

The accuracy of a meter is generally expressed in percent. For example, a meter that has an accuracy of 1 percent will indicate a value that is within 1 percent of the correct

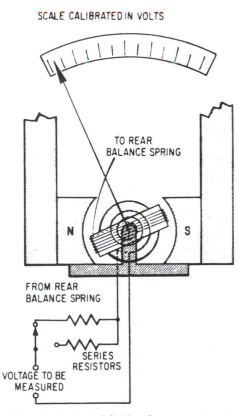

Fig. 10-5. Simplified voltmeter circuit.

value. The statement means that if the correct value is 100 units, the meter indication may be anywhere within the range of 99 to 101 units.

MEASURING RESISTANCE

The two instruments most commonly used to check the continuity, or to measure the resistance of a circuit or circuit element, are the OHMMETER and the MEGGER (megohmmeter). The ohmmeter is widely used to measure resistance and check the continuity of electrical circuits and devices. Its range usually extends to only a few megohms. The megger is widely used for measuring insulation resistance, such as between a wire and the outer surface of its insulation, and insulation resistance of cables and insulators. The range of a megger may extend to more than 1,000 megohms.

OHMMETER

The ohmmeter consists of a DC milliammeter, which was discussed earlier in this chapter, with a few added features. The added features are:

1. a DC source of potential (usually two $1\frac{1}{2}$ volt batteries)

2. one or more resistors (one of which is variable).

A simple ohmmeter circuit is shown in Fig. 10-7.

The ohmmeter's pointer deflection is controlled by the amount of battery current passing through the moving coil. Before measuring the resistance of an unknown resistor or electrical circuit, the test leads of the ohmmeter are first shorted together, as shown in Fig. 10-7. With the leads shorted, the meter is calibrated for proper operation on the selected range. (While the leads are shorted, meter current is maximum and the pointer deflects a maximum amount, somewhere near the zero position on the ohms scale.) When the variable resistor (rheostat) is adjusted properly, with the leads shorted, the pointer of the meter will come to rest exactly on the zero graduation. This indicates ZERO RESISTANCE between the test leads,

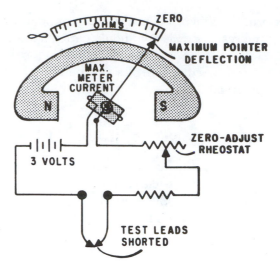

Fig. 10-7. Simple ohmmeter circuit.

which in fact are shorted together. The zero readings of series type ohmmeters are sometimes on the right-hand side of the scale, whereas the zero reading for ammeters and voltmeters is generally to the left-hand side of the scale. When the test leads of an ohmmeter are separated, the pointer of the meter will return to the left side of the scale.

After the ohmmeter is adjusted for zero reading, it is ready to be connected in a circuit to measure resistance. A typical circuit and ohmmeter arrangement is shown in Fig. 10-8.

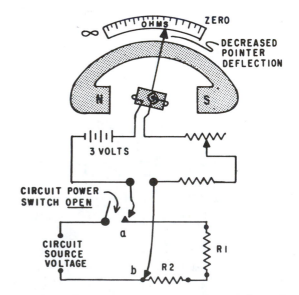

Fig. 10-8. Measuring circuit resistance with an ohmmeter.

The power switch of the circuit to be measured should always be in the OFF position. This prevents the circuit's source voltage from being applied across the meter, which could cause damage to the meter movement.

The test leads of the ohmmeter are connected across (in series with) the circuit to be measured. (See Fig. 10-8.) This causes the current produced by the meter's 3 volts to flow through the circuit being tested. (Assume that the meter test leads are connected at points a and b of Fig. 10-8. The amount of current that flows through the meter coil will depend on the resistance of resistors, R_1 and R_2, plus the resistance of the meter. Since the meter has been preadjusted (zeroed), the amount of coil movement now depends solely on the resistance of R_1 and R_2. The pointer will now come to rest at a scale figure indicating the combined resistance of R_1 and R_2. Movement of the moving coil is proportional to the amount of current flow. The scale reading of the meter, in ohms, is inversely proportional to current flow in the moving coil.

The amount of circuit resistance to be measured may vary over a wide range. In some cases it may be only a few ohms, and in others it may be as great as 1,000,000 ohms. To enable the meter to indicate any value being measured, with the least error, scale multiplication features are incorporated in most ohmmeters. For example, a typical meter will have four test lead jacks, marked as follows: COMMON, R × 1, R × 10, and R × 100. The jacks are connected to three different size resistors located within the ohmmeter. This is shown in Fig. 10-9.

Some ohmmeters are equipped with a selector switch for selecting the multiplication scale desired, so that only two test jacks are necessary. Other meters have a separate jack for each range, as shown in Fig. 10-9. The range to be used in measuring any particular unknown resistance (R_x in Fig. 10-9) depends on the approximate ohmic value of the unknown resistance. For instance, assume the ohmmeter scale in Fig. 10-9 is calibrated in divisions from 0 to 1,000. If R_x is greater than 1,000 ohms, and the R × 1 range is being used, the ohmmeter cannot measure it. The test lead would have to be plugged into the next range, R × 10. With

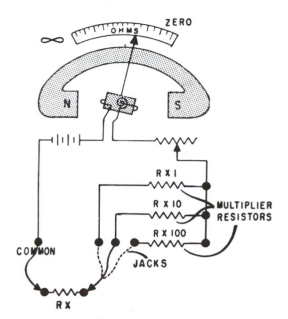

Fig. 10-9. Ohmmeter with multiplication jacks.

this done, assume the pointer deflects to indicate 375 ohms. This would indicate that R_x has 375 × 10 = 3,750 ohms resistance. The change of range caused the deflection because resistor R × 10 has only 1/10 the resistance of resistor R × 1. Thus, selecting the smaller series resistance permitted a battery current of sufficient amount to cause a useful pointer deflection. If the R × 100 range were used to measure the same 3,750-ohm resistor, the pointer would deflect still further, to the 37.5 ohm position. This increased deflection would occur because resistor R × 100 has only 1/10 the resistance of resistor R × 10.

The foregoing circuit arrangement allows the same amount of current to flow through the meter's moving coil whether the meter measures 10,000 ohms on the R × 1 scale, or 100,000 ohms on the R × 100 scale, or 1,000,000 ohms on the R × 100 scale.

It always takes the same amount of current to deflect the pointer to a certain position on the scale (mid-scale position, for example), regardless of the multiplication factor being used. Since the multiplier resistors are of different values, it is necessary to ALWAYS "zero" adjust the meter for each multiplication factor selected.

The operator of the ohmmeter should select

the multiplication factor that will result in the pointer coming to rest as near as possible to the midpoint of the scale. This enables the operator to read the resistance more accurately, because the scale readings are more easily interpreted at or near midpoint.

MEGGER

An ordinary ohmmeter cannot be used for measuring resistance of multimillions of ohms, such as conductor insulation. To adequately test for insulation breakdown, it is necessary to use a much higher potential than is furnished by an ohmmeter's battery. This potential is placed between the conductor and the outside surface of the insulation.

An instrument called a MEGGER (megohm-meter) is used for these tests. The megger (Fig. 10-10) is a portable instrument consisting of two primary elements—(1) a hand-driven DC

generator, G, which supplies the necessary voltage for making the measurement, and (2) the instrument portion, which indicates the value of the resistance being measured. When the generator is not operated, the pointer floats freely and may come to rest at any position on the scale.

If the test leads are open-circuited, the pointer moves to infinity, which indicates a resistance too large to measure. When a resistance such as R_x is connected between the test leads, the pointer indicates the value of resistance R_x. The instrument is not injured under the circumstances because the current is limited by R_3.

To avoid excessive test voltages, most meggers are equipped with friction clutches. When the generator is cranked faster than its rated speed, the clutch slips and the generator speed and output voltage are not permitted to exceed their rated values. For extended ranges, a 1,000-volt generator is available. When extremely high resistances—for example, 10,000 megohms or more—are to be measured, a high voltage is needed to cause sufficient current flow to actuate the meter movement.

MULTIMETER

The MULTIMETER is a multipurpose instrument that can measure resistance, voltage, or current. It contains one milliammeter. The face of the instrument has separate graduated scales to indicate the three values that can be measured. Fig. 10-11 shows a multimeter that is widely used today.

The front panel of the multimeter is constructed and labelled in such a way that all functions are self-explanatory. One pin jack marked − DC ± AC OHMS is common to all functions and ranges. One test lead is always plugged into this common jack, and the remaining lead into the jack marked for the particular function or range desired. The internal arrangement of the meter is controlled by means of the FUNCTION switch. This switch selects the proper circuit elements for the type of measurement desired. The desired voltage, current, or ohmmeter range is determined by the

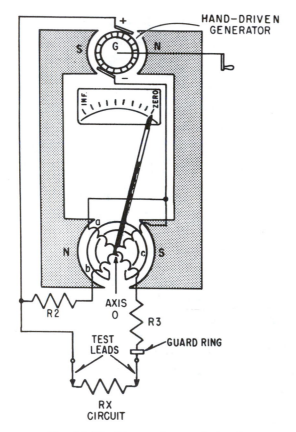

Fig. 10-10. Megger internal circuit.

Fig. 10-11. Commercial multimeter. (courtesy of Simpson Electric Co.)

combined action of positioning the function and range switches and selecting certain pin jacks. The range switch selects the resistance or current range, and the pin jacks select the voltage range. The function switch selects the type of operation to be performed. Separate pin jacks are provided for ohms and for the 10-ampere range. A CASE GROUND jack is connected directly to the cast aluminum case.

The meter panel contains the following operating controls.

1. FUNCTION SWITCH. This is a six position rotary switch which is clearly marked with the type of measurement to be taken.

2. SELECTOR SWITCH. This is a twelve position switch used in selecting one of five resistance ranges or one of the seven current ranges.

3. OHMS ZERO ADJ. This control is a rheostat used to zero the pointer on the ohmmeter. With the test leads shorted, the rheostat is turned until the pointer comes to rest at zero. The rheostat also serves to compensate for variations in battery voltage.

This is a complicated piece of measuring equipment, and the instructions which accompany it must be followed very carefully, as is true for all other instruments described in this chapter.

MOVING IRON-VANE METER

The moving iron-vane meter is another basic type of meter. Unlike the D'Arsonval type meter, which employs permanent magnets, the moving iron-vane meter depends on induced magnetism for its operation. It employs the principle of repulsion between two concentric iron vanes, one fixed and one movable, placed inside a coil, as shown in Fig. 10-12. A pointer is attached to the movable vane.

Portable voltmeters are made with self-contained series resistance for ranges up to 750 volts. Higher ranges are obtained by the use of additional external multipliers.

The moving iron-vane instrument may be used to measure direct current, but has an error due to residual magnetism in the vanes. The error may be minimized by reversing the meter connections and averaging the readings. When used on AC circuits, the instrument has an accuracy of 0.5 percent. Because of its simplicity, its relatively low cost, and the fact that no current is conducted to the moving element, this type of movement is used extensively to measure current and voltage in AC power circuits.

However, the moving iron-vane meter requires much more power to produce full-scale deflection than is required by a D'Arsonval me-

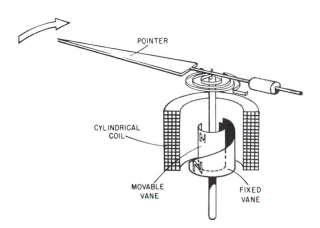

Fig. 10-12. Diagram of a moving iron-vane meter.

ter of the same range. Therefore, the moving iron-vane meter is seldom used in high-resistance low-power circuits.

INCLINED-COIL IRON-VANE METER

The principle of the moving iron-vane mechanism is applied to the inclined-coil type of meter shown in Fig. 10-13. The inclined-coil iron-vane meter has a coil mounted at an angle to the shaft. Attached obliquely to the shaft, and located inside the coil, are two soft-iron vanes. When no current flows through the coil, a control spring holds the pointer at zero and the iron vanes lie in planes parallel to the plane of the coil. When current flows through the coil, the vanes tend to line up with magnetic lines passing through the center of the coil at right angles to the plane of the coil. Thus, the vanes rotate against the spring action to move the pointer over the scale.

The iron vanes tend to line up with the magnetic lines regardless of the direction of current flow through the coil. Therefore, the inclined-coil iron-vane meter can be used to measure either alternating current or direct current.

Like the moving iron-vane meter, the inclined-coil type requires a relatively large amount of current for full-scale deflection and hence is seldom used in high-resistance low-power circuits. It is available as either a voltmeter or as an ammeter.

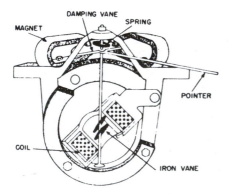

Fig. 10-13. Inclined-coil iron-vane meter.

C. Alternating Current Instruments

So far, the simpler electrical indicating instruments were discussed. Only a few could measure AC. Most were used to measure only direct voltage or current. It will be necessary for the maintenance man to become familiar with additional, more advanced AC indicating instruments. Those that will be discussed in this section are (1) rectifier type AC meters; (2) wattmeters and watt-hour meters; (3) instrument transformers; and (4) frequency meters.

RECTIFIERS

In Chapter 7, we described, both in words and in symbolic diagrams, how alternating current flows in one direction through the circuit in one half of a cycle. In Chapter 5, we explained that direct current flows always in one direction, which is why it is termed "direct current."

There are applications where it is desirable to convert alternating current to direct current, without the use of batteries or direct current generators. This is mainly because alternating current is the most readily available in our society, for reasons already covered in previous chapters. We solve this problem by the use of rectifiers, which are devices for taking an AC input and converting it to a usable form of DC.

The resultant DC may not be as pure or true a supply of DC as is available from either a battery or generator, but it often is close enough to the pure form as is necessary for the required purposes. Some rectifiers alter the form of AC input very close to a pure form of DC, depending upon the design of the rectifier.

The simplest form of rectifier is the semiconductor, or diode, rectifier. Figure 10-14 shows diagrammatically how this type of rectifier alters the alternating current input by blocking the flow of current in one direction, and produces a type of DC output as a result of the remaining current being allowed to proceed in one direction only. In effect, the rectifier is a switch, which opens to prevent the flow of alternating current when it comes into the switch

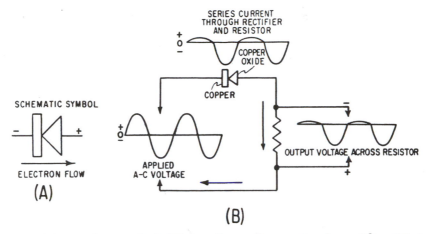

Fig. 10-14. (A) Rectifier symbol; (B) waveforms from a simple rectifier AC circuit.

from one direction, but closes to allow current to flow from the opposite direction during the second half of the cycle. The resulting wave form is shown in Fig. 10-14 (B), consisting of current flowing only in one direction for half the cycle of the input AC, and no current at all through the second half of the cycle. This is known as half-wave rectification.

The electrical circuits using this distorted DC wave form do not usually detect the altered wave form, but see only a wavy one-direction wave form at a reduced voltage and current. Keep in mind that if the AC input is the standard 60 cycles per second, the DC output flows for 1/120th of a second, then stops flowing for 1/120th of a second, then resumes for the next 1/120th of a second, and so on.

The rectifiers we are interested in here are generally of two types at the present. They are the copper oxide rectifier and the selenium rectifier. The copper oxide must operate at temperatures below 140°F, and they operate at efficiencies between 60% and 70%, the losses appearing in the form of heat, which must be removed by some means of cooling, either by ample air space surrounding them or by circulating air or water, depending upon their current rating and location.

The selenium rectifier operates at higher efficiencies, in the range of 65% to 86%, and can tolerate higher temperatures.

Rectifiers for other uses in the plant, aside from those built into the different instruments we will cover here, are of various designs and sizes, but they all have the same general common characteristics. They are designed for a specific DC output wave form, voltage, and amperage, and they all are heat sensitive to varying degrees.

Many of the instruments discussed in the remainder of this chapter utilize semiconductor rectifiers. Since these units are common to a number of different instruments, they will be discussed first.

RECTIFIER TYPE AC INSTRUMENTS

It is possible to connect a D'Arsonval direct current type instrument and a rectifier so as to measure AC quantities. The rectifier is usually of the semiconductor type and is arranged in a bridge circuit, as shown in Fig. 10-15. By the use of four rectifiers in the bridge, current flow through the meter is always in one direction. When the voltage being measured has a wave form as shown in Fig. 10-15, the path of current flow will be from the lower input terminal through rectifier No. 3 through the instrument, and then through rectifier No. 2, thus completing its path back to the source's upper terminal. The next half cycle of the input voltage (indicated by dotted sine wave) will cause the current to pass through rectifier No. 1, through the instrument, and through rectifier No. 4, completing its path back to the source.

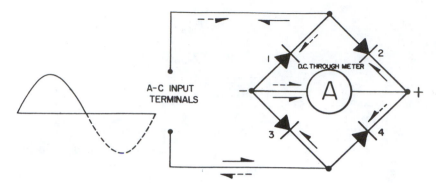

Fig. 10-15. Typical full wave rectifier type instrument.

This type of instrument generally is characterized by errors due to wave form and frequency. Allowances must be made according to data furnished by the manufacturer. It is possible, however, to add corrective networks to the instrument which will make it practically free from error up to 100kHz. An instrument of this type requires a current from the line of only about one milliampere for full scale deflection. It is widely used for AC voltmeters, especially of the lower ranges.

There is some "aging" of the rectifier with a corresponding change in the calibration of the instrument. Because of this aging, such instruments must be recalibrated from time to time. A rectifier type instrument reads the average value of the AC quantity. However, because the EFFECTIVE values are more useful, AC meters are generally calibrated to read effective, or equivalent, values.

WATTMETER

Electric power is measured by means of a wattmeter. This instrument is of the electro-dynamometer type. It consists of a pair of fixed coils, known as current coils, and a movable coil, known as the potential coil. (See Fig. 10-16.)

The current coil (stationary coil) of the wattmeter is connected in series with the circuit (load), and the potential coil (movable coil) is connected across the line.

Thus, we see that the wattmeter consists of two circuits, either of which will be damaged if too much current is passed through them. This fact is to be especially emphasized in the case

of wattmeters, because the reading of the instrument does not serve to tell the user that the coils are being overheated. If an ammeter or voltmeter is overloaded, the pointer will be indicating beyond the upper limit of its scale. In the wattmeter, both the current and potential circuits may be carrying such an overload that their insulation is burning, and yet the pointer may be only part way up the scale. The safe rating is generally given on the face of the instrument.

A wattmeter is always distinctly rated, not in watts but in volts and amperes.

Figure 10-17 shows the proper way to connect a wattmeter in various circuits.

WATT-HOUR METER

The watt-hour meter is an instrument for measuring energy. As energy is the product of power and time, the watt-hour meter must take into consideration both of these factors.

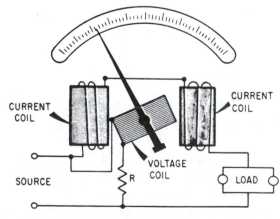

Fig. 10-16. Simplified wattmeter circuit.

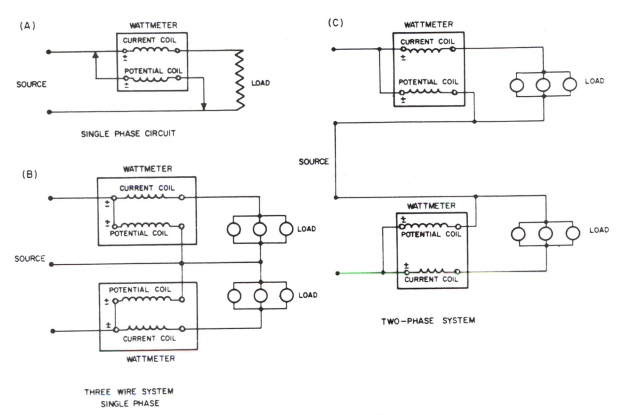

Fig. 10-17. Correct wattmeter applications.

In principle, the watt-hour meter is a small motor whose instantaneous speed is proportional to the POWER passing through it. The total revolutions in a given time are proportional to the total ENERGY or watt-hours consumed during that time.

Referring to Fig. 10-18, the lines are connected to two terminals on the left-hand side of the meter. The upper terminal is connected to two coils FF in series. The armature A rotates in the field produced between the coils FF. The outer line wire goes directly to the load.

The following directions should be followed when reading the dials of a watt-hour meter. The meter, in this case, is a four-dial type.

The pointer on the right-hand dial (Fig. 10-19) registers 1 kw-hr or 1,000 watt-hours for each division on the dial. A complete revolution of the hand on this dial will move the hand of the second dial one division and register 10 kw-hr or 10,000 watt-hours. A complete revolution of the hand of the second dial will move the third hand one division and register 100 kw-hr or 100,000 watt-hours, and so on.

Accordingly, you must read the hands from left to right, and add three zeros to the reading of the lowest dial to obtain the reading of the meter in watt-hours. The dial hands should always be read as indicating the figure which they have LAST PASSED, and not the one they are approaching.

We will not go into the repair and maintenance of electrical power meters, as they are usually the property and responsibility of the

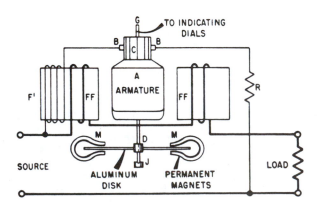

Fig. 10-18. Watt-hour meter internal diagram.

Fig. 10-19. Commercial watt-hour meter dials.

public utility furnishing the electric power. However, this does not remove the user from the responsibility of taking reasonable care in its operating environment. This means that the plant's maintenance personnel are expected to keep the area surrounding the meter in reasonably clean and uncluttered condition, enabling the public utility's service and recording personnel to easily gain access to the meter. This responsibility is usually spelled out in the contract agreement with the utility.

INSTRUMENT TRANSFORMERS

It is not usually practical to connect instruments and meters directly to high-voltage circuits. Unless the high-voltage circuit is grounded at the instrument, a dangerously high potential-to-ground voltage may exist at the instrument or switch board. Further, instruments become inaccurate when connected directly to a high voltage, because of the electrostatic forces that act on the indicating element. Specially designed instruments may be constructed so that they can be connected di-

rectly to high-voltage circuits, but these instruments are usually expensive.

By means of instrument transformers, instruments may be entirely insulated from the high-voltage circuit and yet indicate accurately the current, voltage, and power in the circuit. Low-voltage instruments having standard current and voltage ranges may be used for all high-voltage circuits, irrespective of the voltage and current ratings of the circuits, if instrument transformers are utilized.

POTENTIAL TRANSFORMERS

Potential transformers do not differ materially from the constant-potential power transformers, as discussed elsewhere. An exception is that their power rating is small, and they are designed for minimum error. For taking measurements below 5,000 volts, potential transformers are usually of the dry type; between 5,000 and 13,800 volts they may be either the dry type or oil immersed; and above 13,800 volts they are oil immersed.

Since only instruments, meters, and sometimes indicator lights are ordinarily connected to the secondaries of potential transformers, they have ratings from 40 to 500 watts. For primary voltages of 34,500 volts and higher, the secondaries are rated at 115 volts. For primary voltages less than 34,500 volts, the secondaries are rated at 120 volts. Figure 10-20 shows a simple connection for measuring voltage in a 14,000 volt circuit by means of a potential transformer.

The secondary should always be grounded at one point to eliminate static electricity from the instrument and to ensure the safety of the operator.

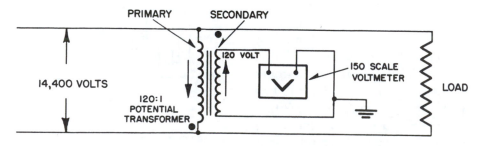

Fig. 10-20. Diagram of a 14,400 volt potential transformer circuit.

CURRENT TRANSFORMERS

To avoid connecting instruments directly into high-voltage lines, current transformers are used. In addition to insulating from high voltage, they step down the line current in a known ratio. This permits the use of a lower-range ammeter than would be required if the instrument were connected directly into the primary line. Figure 10-21 shows a simple connection for measuring current in a 14,400 volt circuit by means of a current transformer.

The secondary windings of practically all current transformers are rated at 5 amperes regardless of the primary current rating. For example, a 2,000 ampere current transformer has a ratio of 400 to 1, and a 50 ampere transformer has a ratio of 10 to 1.

The current transformer differs from the ordinary constant potential transformer in that its primary current is determined entirely by the load on the system and not by its own secondary load. If its secondary becomes open-circuited, a high voltage will exist across the secondary, with a large increase in the flux, producing excessive core loss and heating, as well as a dangerously high voltage across the secondary terminals. **THEREFORE, THE SECONDARY OF A CURRENT TRANSFORMER SHOULD NOT BE OPEN-CIRCUITED UNDER ANY CIRCUMSTANCES**.

Figure 10-22 shows the method of connecting a complete instrument system, through instru-

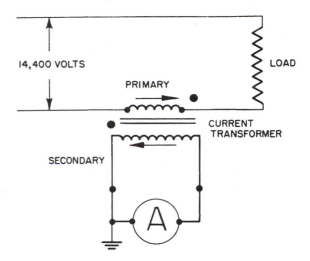

Fig. 10-21. Typical application of a current transformer.

ment transformers, to a high-voltage line. The load on the instrument transformers includes an ammeter A, a voltmeter V, a wattmeter, and a watt-hour meter.

POLARITY MARKING

Instruments, meters, and relays must be connected so that the correct phase relations exist between their potential and current. In instrument transformers, it is important that the relation of the instantaneous polarities of the secondary terminals to the primary terminals be known. It has become standard to designate or mark the primary terminals and the secondary

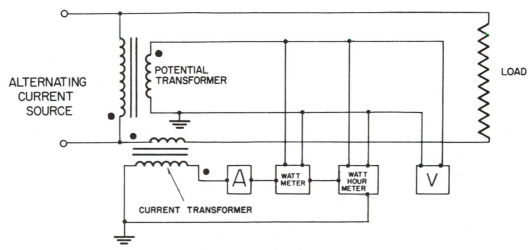

Fig. 10-22. Combination single-phase measuring system.

terminals that have the same instantaneous polarity.

In a wiring diagram, primary and secondary terminals having the same instantaneous polarity are marked with a dot. This is shown in Fig. 10-22.

HOOK-ON TYPE VOLTAMMETER

The hook-on AC ammeter consists essentially of a current transformer with a split core and a rectifier type instrument connected to the secondary. The primary of the current transformer is the conductor through which the current to be measured flows. The split core permits the instrument to be "hooked on" the conductor without disconnecting it. Therefore, the current flowing through the conductor may be measured safely with a minimum of inconvenience, as shown in Fig. 10-23.

The instrument is usually constructed so that voltages also may be measured. However, in order to read voltage, the meter switch must be set to VOLTS, and leads must be connected from the voltage terminals on the meter to the terminals across which the voltage is to be measured.

A recent addition to the field of electrical measurement is the Autoranging Digital Multimeter, shown in Fig. 10-24. This has the ad-

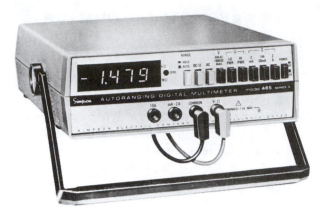

Fig. 10-24. Autoranging digital multimeter. (Courtesy of Simpson Electric Co.)

vantage of not requiring the user to experiment in locating the correct range of voltage, amperes, or resistance, which eliminates the possibility of destroying the instrument. The autoranging feature consists of a built-in ability to adjust internally for the correct range in making the lighted digital read-out.

As yet, the instrument is designed primarily for bench use, but it is generally portable enough to use for on-the-spot locations throughout the plant. Its cost is more than the conventional design of meters, but the elimination of the danger of damaging the meter makes the cost very attractive.

D. Applications

In this section we shall give a number of typical applications for the meters described in this chapter. These applications are taken from various sources, some of which are listed in the ACKNOWLEDGMENTS at the front of this book. Both portable and permanently mounted installations will be included.

The installations shown are typical only, as the final method will depend to a certain extent on the make and design of the instrument. Here, as in other cases throughout this book, the reader is cautioned to follow the specific instructions furnished with the equipment.

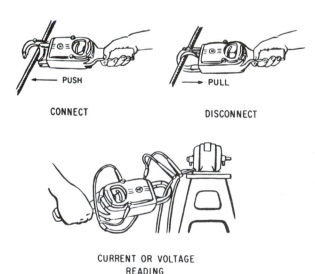

PUSH   PULL

CONNECT   DISCONNECT

CURRENT OR VOLTAGE
READING

Fig. 10-23. Hook-on type voltammeter.

HOW TO CHECK CAPACITORS

Some appliance motors have a capacitor in series with the starting winding. They are called capacitor-start motors. When operating properly, the capacitor provides the necessary phase shift to give the motor a high starting torque. When defective, the capacitor severely degrades motor performance, and sometimes causes damage to the motor. Determining that a capacitor is defective is an easy process with the aid of an ohmmeter. Replacing a capacitor is simple.

A capacitor consists of two conducting plates separated by an insulator, or dielectric. Large conducting surfaces separated by a thin dielectric give a capacitor a high capacity. The electrolytic capacitor (Fig. 10-25) on capacitor-start motors has a high rating due to its extremely thin chemical-layer dielectric. In an AC circuit, the capacitor provides a phase shift in the starting winding which exerts a rotational force on the motor armature.

Excessive voltage, heat drying, and age can cause a capacitor to become defective. This means that the capacitor is either shorted internally, that it has a broken lead-in wire, or that it has lost its ability to polarize. All three conditions can be checked with the Simpson Model 372, or similar ohmmeter set on the Rx100 scale, as shown in Fig. 10-25. Following are the indications to be expected.

A shorted capacitor registers instantly as zero ohms or very low resistance.

A good capacitor will cause the pointer to move toward zero at the instant the meter leads are attached. The pointer will then slowly return to the high-resistance end of the scale.

A capacitor with a broken wire will register ∞ (infinity) on the meter.

A capacitor that can no longer polarize properly will cause the pointer to deflect slightly, then stabilize very quickly at a high reading, close to ∞.

The "good" capacitor indication deserves an explanation. At the start of the test, an empty

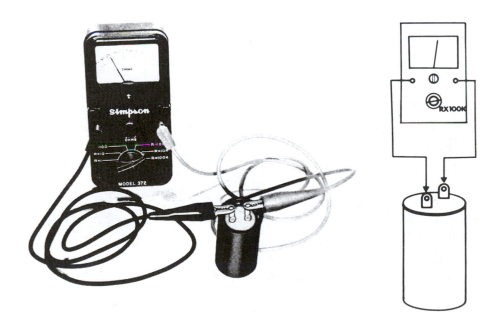

Fig. 10-25. Testing capacitor on capacitor-start motor. (Courtesy of Simpson Electric Co.)

capacitor charges heavily from the battery in the ohmmeter and the pointer responds by moving quickly across the scale, indicating current flow. Then, as the capacitor becomes charged, the current diminishes, causing the pointer to drop back. In a rough way, this capacitor check also indicates capacity value by the length of time the needle takes to return to zero.

On the capacitor-run motor shown in Fig. 10-26, the capacitor remains in the starting winding circuit at all times. The dielectric in this type of capacitor is paper, oil, or a plastic material. This type of capacitor is used to provide a minor phase shift, offsetting the reverse shift caused by the inductance of the motor winding. The result is a reduction in the current drawn by a motor for a given horsepower.

Since the phase shift required is small (compared with the shift necessary on a capacitor-start motor), the capacitor rating is lower than that of the capacitor used in a motor-starting application.

Capacitors from both types of motors are checked in the same way.

HOW TO CHECK SPLIT-PHASE MOTORS

Electrical faults in the split-phase motors used to operate washers, irons, dishwashers, and disposers become immediately apparent because either the motor smokes, a fuse blows, or both. Such a motor is overheated due to an overload. It should be tested after the overload is removed, to make sure that no permanent electrical damage has occurred.

As there is no insulated winding on the rotor, the field coils are the only ones to be tested. Relatively large wire is used on both windings; however, the starting winding wire may be slightly smaller than the running winding wire. Both are quite low in resistance and require a low-reading ohmmeter.

The pocket-size Simpson Model 362, or equivalent, which reads only 2 ohms at mid-scale, is recommended. Several tests can be made (see Figs. 10-27 and 10-28).

1. A resistance measurement can be used to identify lead wires belonging to the same winding.

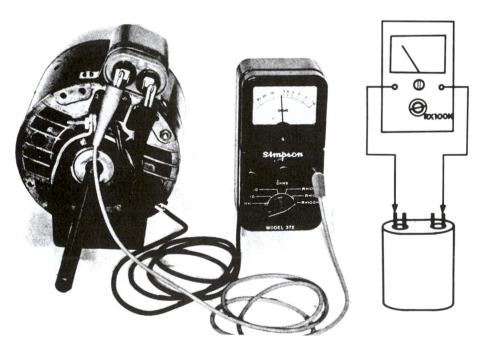

Fig. 10-26. Testing capacitor on capacitor-run motor. (Courtesy of Simpson Electric Co.)

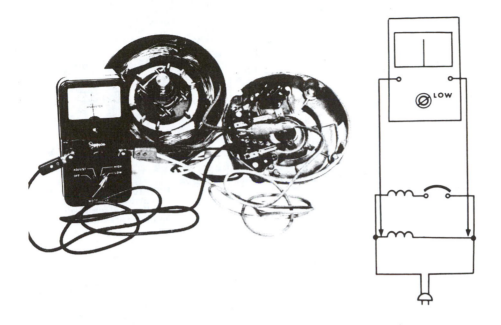

Fig. 10-27. Testing split-phase motor running winding. (Courtesy of Simpson Electric Co.)

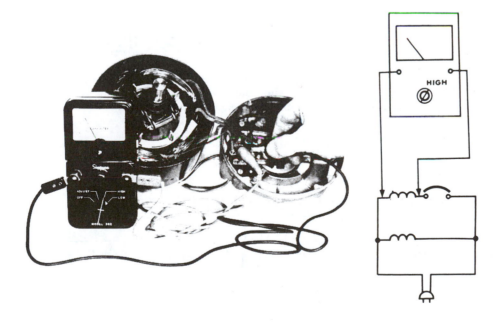

Fig. 10-28. Testing split-phase motor starting winding. (Courtesy of Simpson Electric Co.)

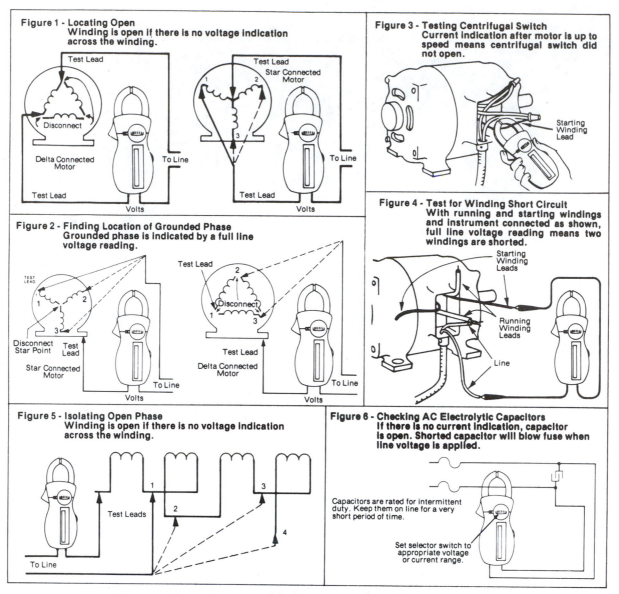

Figure 1 - Locating Open
Winding is open if there is no voltage indication across the winding.

Test Lead
Test Lead
Star Connected Motor
Disconnect
Delta Connected Motor
To Line
Test Lead
Test Lead
Volts
Volts

Figure 2 - Finding Location of Grounded Phase
Grounded phase is indicated by a full line voltage reading.

TEST LEAD
Test Lead
Disconnect Star Point
Test Lead
Star Connected Motor
To Line
Volts

Test Lead
Disconnect
Test Lead
Delta Connected Motor
To Line
Volts

Figure 3 - Testing Centrifugal Switch
Current indication after motor is up to speed means centrifugal switch did not open.

Starting Winding Lead

Figure 4 - Test for Winding Short Circuit
With running and starting windings and instrument connected as shown, full line voltage reading means two windings are shorted.

Starting Winding Leads
Running Winding Leads
Line

Figure 5 - Isolating Open Phase
Winding is open if there is no voltage indication across the winding.

Test Leads
To Line

Figure 6 - Checking AC Electrolytic Capacitors
If there is no current indication, capacitor is open. Shorted capacitor will blow fuse when line voltage is applied.

Capacitors are rated for intermittent duty. Keep them on line for a very short period of time.

Set selector switch to appropriate voltage or current range.

Fig. 10-29. Troubleshooting motors with an amprobe clamp-on instrument. (Courtesy of Amprobe Instrument Co.)

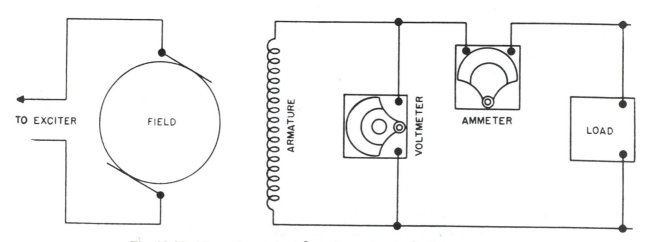

TO EXCITER
FIELD
ARMATURE
VOLTMETER
AMMETER
LOAD

Fig. 10-30. Measuring output from two-wire single-phase generator.

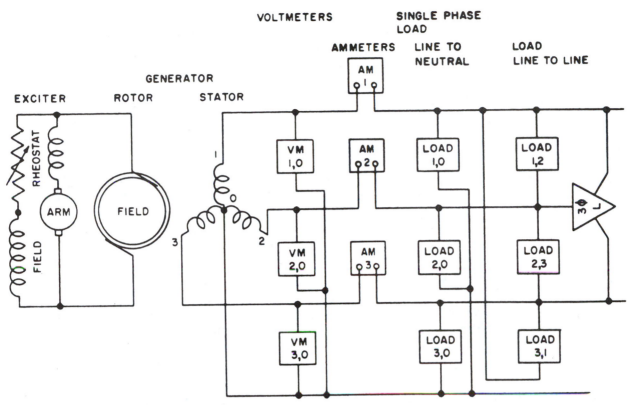

Fig. 10-31. Measuring output from four-wire three-phase generator.

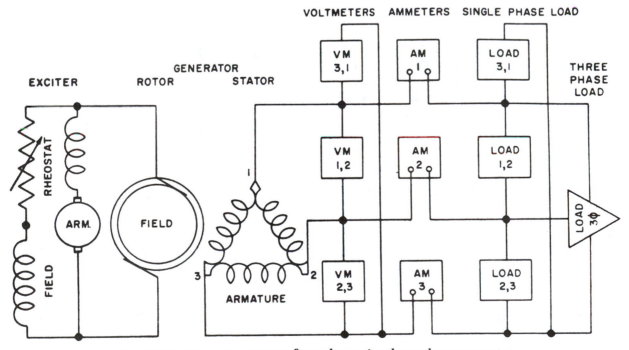

Fig. 10-32. Measuring output from three-wire three-phase generator.

2. If only three wires are brought out from the field coils, a resistance measurement will locate the wire common to both windings.

3. Often it is possible to distinguish the starting winding from the running winding by the difference in measured resistance.

If no defect exists, the low resistance of a running winding will be measured between one pair of leads, while the somewhat higher resistance of a starting winding will be measured between the other pair. As this identification is made, an open-circuited coil will show up since only one wire pair will measure a typical wind-

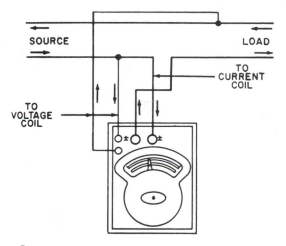

① CONNECTIONS FOR MEASURING WATTS IN SINGLE PHASE A-C CIRCUIT

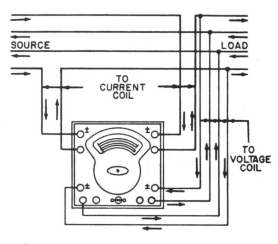

② CONNECTIONS FOR MEASURING WATTS IN TWO PHASE FOUR WIRE A-C CIRCUIT

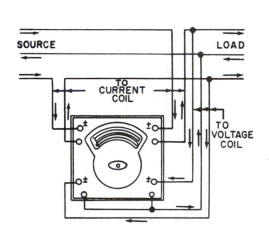

③ CONNECTIONS FOR MEASURING WATTS IN TWO PHASE AND THREE PHASE THREE WIRE A-C CIRCUITS

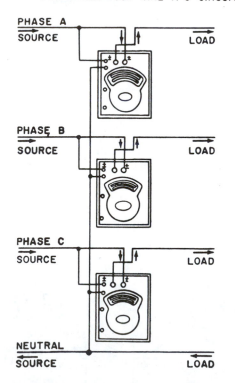

④ CONNECTIONS FOR MEASURING WATTS IN THREE PHASE FOUR WIRE A-C CIRCUITS

Fig. 10-33. Typical AC wattmeter applications.

ing resistance; the other pair will measure very high resistance.

If the correct resistance of each winding is known, or can be found in the manufacturer's literature, a comparison can be made with the values measured. Too low a resistance will result if several turns or an entire section of the winding is shorted; however, a single-turn short is difficult to identify because the variation in resistance is too small.

Actual tests of a typical $\frac{1}{3}$-hp motor field are illustrated. The running winding (Fig. 10-27)

measures 2 ohms; the starting winding (Fig. 10-28) measures $3\frac{1}{2}$ ohms.

A motor can be checked for ground with the Simpson 362 using the high-ohms scale. If either winding is grounded, the resistance between the wire lead and the motor frame will measure less than the highest reading on the meter.

Figures 10-29, 10-30, 10-31, 10-32, and 10-33 cover numerous methods of taking electrical measurements which may be of help to the electrical maintenance man.

C H A P T E R 11
ELECTRICAL CIRCUIT PROTECTION

A. Introduction

We have constantly stressed the safety aspects of electrical energy in this book, and shall continue to do so when the subject appears appropriate to the discussion at hand. Of course, safety is not the only thing to be considered, but it is the most important to you, the electrical maintenance mechanic. There is also the matter of conserving plant equipment which usually represents a huge investment by the plant owners who have employed you to keep it in first class running condition.

There is a natural law concerning power that goes something like this: Whenever a source of power is created, discovered, or developed by man's efforts, an equal amount of effort must be expended to devise means to safely control that power. Otherwise, the end result could be just that—the end!

As we stated very early in this book, electricity is one of nature's sources of energy in the universe. In this chapter, we shall explore and explain the various methods of controlling it. You will no doubt find that most of your maintenance work on electrical equipment in your plant is directed towards keeping this control equipment in proper working order.

Most of the discussion in this chapter is based upon equipment produced and marketed for many years by the Square D Company, who have kindly consented to the use of their material. Controls produced by other firms may have some minor variations, but the principles are the same. The reader is, of course, advised to follow the instructions issued by the maker of his particular brand of controls, using the material given here purely as a guide.

When the electrical unit is built, the greatest care is taken to ensure that each separate electrical circuit is fully insulated from all others so that the current in a circuit will follow its intended individual path. Once the unit is placed into service, however, there are many things that can happen to alter the original circuitry. Some of these changes can cause serious troubles if they are not detected and corrected in time.

Perhaps the most serious trouble we can find in a circuit is a dead short. Recall that this term is used to describe a situation in which some point in the circuit, where full system voltage is present, comes in direct contact with the ground or return side of the circuit. This establishes a path for current flow that contains no resistance other than that present in the wires carrying the current, and these wires have very little resistance.

According to Ohm's Law, if the resistance in a circuit is extremely small, the current will be extremely great. When a dead short occurs, then there will be an extremely heavy current flowing through the wires. Suppose, for instance, that the two leads from a battery to a motor came in contact with each other. Not only would the motor stop running, because of the current going through the short, but the battery would become discharged quickly (perhaps ruined), and there would also be danger of fire.

The battery cables in our example would be very large wires, capable of carrying very heavy currents. Most wires used in electrical circuits are considerably smaller, and their current-carrying capacity is quite limited. The size of the wires used in any given circuit is determined by the amount of current the wires are expected to carry under normal operating conditions. Any current flow greatly in excess of normal, such as there would be in case of a dead short, would cause a rapid generation of heat.

If the excessive current flow caused by the short is left unchecked, the heat in the wire will continue to increase until something gives way. Perhaps a portion of the wire will melt and open the circuit so that nothing is damaged other than the wires involved. The probability exists, however, that much greater damage would result. The heat in the wires could char and burn their insulation and that of other wires bundled with them, which could cause more shorts. If any combustible material is near any of the hot wires, a disastrous fire might be started.

To protect electrical systems from damage and failure caused by excessive current, several kinds of protective devices are installed in the system. Fuses, circuit breakers, and thermal protectors are used for this purpose.

Circuit protective devices, as the name implies, all have a common purpose: to protect the units and the wires in the circuit. Some are designed primarily to protect the wiring. These open the circuit in such a way as to stop the current flow when the current becomes greater than the wires can safely carry. Other devices are designed to protect a unit in the circuit by stopping current flow to it when the unit becomes excessively warm.

B. Fuses

The simplest protective device is a fuse. All fuses are rated according to the amount of current that is safely carried by the fuse element at a rated voltage. Usually, the current rating is in amperes, but some instrument fuses are rated in fractions of an ampere. When a fuse blows, it should be replaced with another of the same rated voltage and current capacity, including the same current-versus-time characteristic.

The most important fuse characteristic is its current-versus-time or "blowing" ability. Three time ranges for existence of overloads can be broadly defined as fast, medium, and delayed. FAST may range from 5 microseconds through $\frac{1}{2}$ second; MEDIUM, $\frac{1}{2}$ to 5 seconds; DELAYED, 5 to 25 seconds.

Normally, when the circuit is overloaded, or a fault develops the fuse element melts and opens the circuit that it is protecting. However, all fuse openings are not the result of current overload or circuit faults. Abnormal production of heat, aging of the fuse element, poor contact due to loose connections, oxides of corrosion forming within the fuse holder, and the heated condition of the surrounding atmosphere will alter the heating conditions and the time required for the element to melt.

Some equipment, such as an electric motor, requires more current during starting than for normal running. Thus, a fast-time or medium-time fuse rating that will give running protection might blow during the initial period when high starting current is required. Delayed-action fuses are used to handle these situations.

One type of delayed-action fuse has a heater element connected in parallel with the fuse element in order to get the delayed action. During normal operation, the heat developed in the fuse link is not great enough to melt the link. The melting, or opening, of the fuse link depends on the transfer of heat to the link from the heater. Therefore, more time is needed to melt the link than would be required if the link were directly heated.

Because the heater and fuse element are in parallel, the opening of the fuse element will cause the total circuit current to flow through the heater. The high current will cause the heater to burn out and completely open the circuit.

Another type of delayed-action fuse has the

fuse element and heater connected in series. Current above that of the rated value for a short time will have no effect on the fuse or heater. However, prolonged overloads cause the heater section to become hot enough to melt the junction between the elements; this action opens the circuit.

Delayed-action fuses are sometimes called Time-Delay fuses; and three trade names—"Slo Blo," "Fusestat," and "Fusetron"—are in common use. See Fig. 11-1 for two common styles in use in industrial plants.

The plug fuse is constructed so that it can be screwed into a socket mounted on the control panel or distribution center. The fuse link is enclosed in an insulated housing of porcelain or glass. The construction is so arranged that the fuse link is visible through a window of mica or glass. Therefore, an open element may be located by visual examination. When found to be defective, the fuse is discarded and a new fuse installed in its place. The plug fuse is used primarily to protect low-voltage, low-current circuits. The operating ratings range from 0.5 to 30 amperes up to 150 volts.

In operation, the cartridge fuse is exactly the same as the plug fuse. In construction, the fuse link is enclosed in a tube of insulating material with metal ferrules at each end (for contact with the fuse holder). The dimensions of cartridge fuses vary with the current and voltage ratings.

It is not always possible to detect a blown fuse by a visual examination. Hence, fuses are often equipped with a device that will provide a visual indication so that a blown-fuse condition can be readily detected (Fig. 11-1). These devices consist of the spring-loaded and the neon-lamp types of blown-fuse indicators.

In the spring-loaded type, when the link opens, it releases a spring that is held under tension. This action exposes an indicator, which makes the visual location of the blown fuse possible.

The neon-lamp type is designed to be mounted on the fuse. When the link opens, a neon lamp glows to show a blown fuse.

Most fuse panels and switchboards are of the enclosed panel type. The term "dead-front" means that all fuses and bus connections are enclosed in a metal cabinet when the cover is closed. The use of this type of construction reduces the possibility of equipment damage and danger to personnel. Modern switchboards are of the "dead-front" type.

However, the complete enclosure of the equipment makes it less accessible for test purposes. Therefore, most fuses used on "dead-front" switchboards have indicators that show when a fuse is blown.

When an indicator is used, it is necessary to test the fuse continuity with a megger, ohmmeter, or voltmeter. Two methods of testing will be discussed in Section C.

C. Troubleshooting Fused Circuits

From the main power supply, the total electrical load is divided into several feeder circuits and each feeder circuit is further divided into several branch circuits. Each final branch circuit is fused to safely carry only its own load, while each feeder is safely fused to carry the total current of its several branches. This reduces the possibility of one circuit failure interrupting the power for the entire system.

The distribution wiring diagram showing the connections that might be used in a lighting

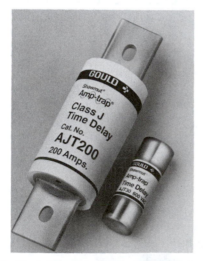

Fig. 11-1. Cartridge fuses. (Courtesy of Gould Inc., Circuit Protection Div.)

system is illustrated in Fig. 11-2. An installation might have several feeder distribution boxes, each supplying six or more branch circuits through distribution boxes.

Fuses F1, F2, and F3 (Fig. 11-2) protect the main feeder supply from heavy surges, such as may be caused by short circuits or overloads on the feeder cable. Fuses A-A1 and B-B1 protect branch NO. 1. If trouble develops and work is to be done on branch NO. 1, switch S1 may be opened to isolate this branch. Branches 2 and 3 are protected and isolated in the same manner by their respective fuses and switches.

Usually, receptacles for portable equipment and fans are on branch circuits separate from lighting branch circuits. Test procedures are the same for any branch circuit. Therefore, a description will be given of the steps necessary to (1) locate the defective circuit and (2) follow through on that circuit and find the trouble.

Assume that, for some reason, several of the lights are not working in a certain section. Because several lights are out, it will be reasonable to assume that the power supply has been interrupted on one of the branch circuits.

To verify this assumption, first locate the distribution box feeding the circuit that is inoperative. Then make sure that the inoperative circuit is not energized. Unless the circuits are identified in the distribution box, the voltage at the various circuit terminations will have to be measured. For the following procedures, use the circuits shown in Fig. 11-2 as an example circuit.

To pin down the trouble, connect the voltage tester to the load side of each pair of fuses in the branch distribution box. No voltage between these terminals indicates a blown fuse or a failure in the supply to the distribution box. To find the defective fuse, make certain S1 is closed, then connect the voltage tester across A-A1, and next across B-B1 (Fig. 11-2). The full-

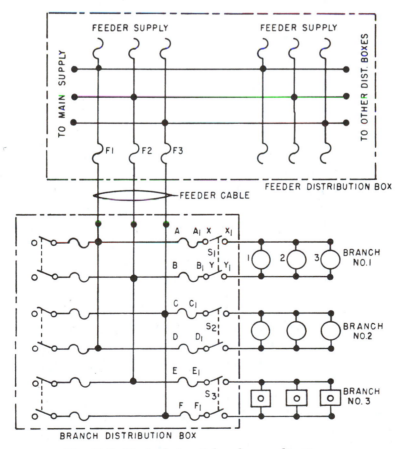

Fig. 11-2. Typical circuit breaker application.

phase voltage will appear across an open fuse, provided circuit continuity exists across the branch circuit. However, if there is an open circuit at some other point in the branch circuit, this test is not conclusive. If the load side of a pair of fuses does not have the full-phase voltage across its terminals, place the tester leads on the supply side of the fuses. The full-phase voltage should be present. If the full-phase voltage is not present on the supply side of the fuses, the trouble is in the supply circuit from the feeder distribution box.

Assume that you are testing at terminals A-B (Fig. 11-2) and that normal voltage is present. Move the test lead from A to A1. Normal voltage between A1 and B indicates that fuse A-A1 is in good condition. To test fuse B-B1, place the tester leads on A and B, and then move the lead from B to B1. No voltage between these terminals indicates that fuse B-B1 is open. Full-phase voltage between A and B1 indicates that the fuse is good.

This method of locating blown fuses is preferred to the method in which the voltage tester leads are connected across the suspected fuse terminals, because the latter may give a false indication if there is an open circuit at any point between either fuse and the load in the branch circuit.

In sizing a replacement fuse, it is necessary to obtain the running current of the motor being protected. This is known as the Full-Load Current, or FLC, and the fuses are then chosen according to the code letter on the motor, with the following being maximum allowable percentages permitted by the **NEC**®:

Non-Time-Delay	300% FLC
Time-Delay	175% FLC
Circuit Breakers	200% FLC

Sizing a replacement fuse may not be as simple as the above brief descriptions would indicate, as there are many types, styles, and operating characteristics available. If any particular circuit is causing trouble, we suggest that a supplier be contacted, and let him make his recommendation, based on your problem. Once the proper fuse for each location has been de-termined, be sure and record the complete specification in the maintenance or equipment log, and order replacements for spares.

Table 11-1 lists the major types available through one well-known supplier. In Section C in the Appendix, there are selection tables for AC motor protection fuses. These tables are to be used as a starting point and a guide only.

At this point, we shall select the fuses for protecting the motor and its circuit proposed in Chapter 3.

The 50 hp 460 V motor draws 63 amps at full load, and we can estimate that it will take between 2 and 5 seconds to attain full speed under its normal load, requiring a time delay fuse. Page 320 in the Appendix contains selection tables for UL Class RK5 and Class J fuses, and we shall start with the Class RK5 tables on page 320. Under the "Typical 5 Second" column, we see that for the 50 hp motor, 100 amp fuses are recommended. A call to the supplier would determine whether the RK5 or the J class would be the best choice.

As it is a three-phase circuit, three fuses will be required, plus spares.

Fuses are also used in conjunction with disconnect, or safety, switches which are described fully in Chapter 17. The disconnect switches are available in various sizes, and each size is designed to handle a definite prescribed fuse capacity range.

Troubles with fuses usually involve loose holding clips, dirt collecting on the clips or fuse contact surfaces, or arcing across the contacts. If the clips become loose, it is useless to attempt to tighten them, as to do so will only make the situation worse. There are adapters available on the market, which will solve the problem very well, known as "Clip Clamps." There are also fuse reducers for installing fuses in disconnect switches which are designed to take larger fuses.

Never attempt to remove or replace cartridge fuses by hand or with ordinary pliers. Always use insulated fuse pullers when performing this task, and never allow anyone to stand in front of the fuse box or switch when the power switch is closed on the circuit.

Table 11-1. Fuse Availability Schedule

Voltage	Fuse Type	Ampere Rating	Interrupting Rating— KA[a]	Notes
A. UL Classifications				
125	Plug	0–30	10	
250	Class H	0–600	10	Includes Renewables
	Class K	0–600	50, 100 or 200	Interchangeable with Class H
	Class RK1	0–600	200	One-end rejection
	Class RK5	0–600	200	One-end rejection
	Midget	0–30	10	$\frac{13}{32}'' \times 1\frac{1}{2}''$
300	Class G	0–60	100	$\frac{13}{32}''$ diameter
	Class T	0–1200	200	Very small dims.
600	Class H	0–600	10	Includes Renewables
	Class J	0–600	200	600V dims. only
	Class K	0–600	50, 100 or 200	Interchangeable with Class H
	Class RK1	0–600	200	One-end rejection
	Class RK5	0–600	200	One-end rejection
	Class T	0–800	200	Very small dims.
	Class CC	0–30	200	Midget one-end rejection
	Midget	0–30	10, 50 or 100	$\frac{13}{32}'' \times 1\frac{1}{2}''$
	Class L	601–6000	200	Bolt-in
B. Other Types				
130–2500	Rectifier	0–2000	200	Many sizes UL component recognized
6–240	Glass	0–30	Varies to 10	Automotive and electronic, $\frac{1}{4}''$ dia. Many sizes UL Listed
600	Cable	4/0–750 MCM Cu or Al Cables	200	Crimp type, bolt type or solid stud
600	Capacitor	25–225	200	Variety of mountings
250, 600	Welder	70–600	200	Class H or Class J dims.

Courtesy of Gould Electronics, Inc.
[a]KA = Kiloamps.

D. Circuit Breakers

A circuit breaker is designed to break the circuit and stop the current flow when the current exceeds a predetermined value. It is commonly used in place of a fuse and may sometimes eliminate the need for a switch. A circuit breaker differs from a fuse in that it "trips" to break the circuit and it may be reset, while a fuse melts and must be replaced.

Several types of circuit breakers are commonly used. One is a magnetic type. When excessive current flows in the circuit, it makes an electromagnet strong enough to move a small armature which trips the breaker. Another type is the thermal overload switch or breaker. This consists of a bimetallic strip which, when it becomes overheated from excessive current, bends away from a catch on the switch lever and permits the switch to trip open.

Some circuit breakers must be reset by hand, while others reset themselves automatically. When a circuit breaker is reset, if the overload condition still exists, the circuit breaker will trip again to prevent damage to the circuit.

One common type of circuit breaker now being used is depicted in Fig. 11-3. This breaker is designed for front or rear connections as required, and may be mounted so as to be removable from the front without removing the circuit breaker cover. The voltage ratings of this breaker are 500 volts AC, 60 Hz, or 250 volts DC with a maximum current capacity of 250 amperes. Trip units (Fig. 11-4) for this breaker are available with current ratings of 125, 150, 175, 225, and 250 amperes.

1. Stationary contact.
2. Arc suppressors.
3. Terminal stud nuts and washers.
4. Trip unit line terminal screw-outer poles.
5. Trip unit line terminal screw-center pole.
6. Trip unit nameplate.
7. Terminal barriers.
8. Shunt trip.
9. Auxiliary switch.
10. Hole for shunt trip undervoltage release plunger.
11. Instantaneous trip adjusting wheels.

The trip unit houses the electrical tripping mechanisms, the thermal element for tripping the circuit breaker on overload conditions, and the instantaneous trip for tripping on short circuit conditions. The automatic trip devices of

Fig. 11-3. Typical circuit breaker. (Courtesy Square D Co.)

this circuit breaker are "trip free" of the operating handle; this means the circuit breaker cannot be held closed by the operating handle if an overload exists. When the circuit breaker has tripped due to overload or short circuit, the handle rests in a center position. To reclose after automatic tripping, the handle must be moved to the extreme OFF position which resets the latch in the trip unit; then the handle must be physically moved to the ON position. Metal locking devices are available that can be attached to the handles of circuit breakers to prevent accidental operation.

A thermal protector, or switch, is a device used to protect a circuit. It is designed to open the circuit automatically whenever the temperature of the element becomes excessively high. It has two positions—open and closed. The most common use for a thermal switch is to keep a motor from overheating. If some malfunction in the motor causes it to overheat, the thermal switch will break the circuit intermittently. If the trouble is a locked rotor, the intermittent opening and closing of the circuit may release the rotor and allow the motor to resume normal operation.

The electrical circuit for the internal arrangement of a circuit breaker is shown in Fig. 11-5. The magnetic element acts instantaneously to trip the breaker when a short occurs. The thermal element is a slower acting device, which trips the breaker when the current draw exceeds the normal design system requirement by a specified amount.

The thermal switch contains a bimetallic disk, or strip, which bends to break the circuit when it is heated. This happens because one of the metals expands more than the other when they are subjected to the same temperature. When the strip or disk cools, the metals contract and the strip returns to its original position and closes the circuit. These thermal switches are described more fully in Chapter 17, under the section on magnetic starters.

As mentioned previously, standard molded case circuit breakers usually contain: (1) a set of contacts; (2) a magnetic trip element; (3) a thermal trip element; (4) line and load terminals; (5) bussing used to connect these individ-

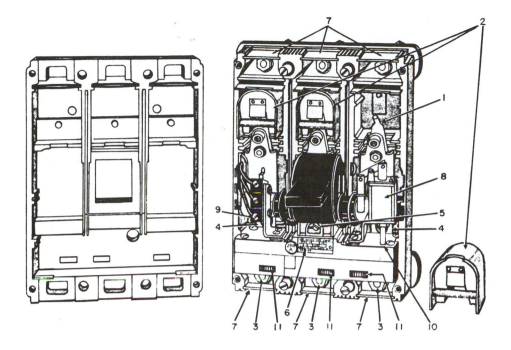

1. Stationary contact.
2. Arc suppressors.
3. Terminal stud nuts and washers.
4. Trip unit line terminal screw- outer poles.
5. Trip unit line terminal screw- center pole.
6. Trip unit nameplate.

7. Terminal barriers.
8. Shunt trip.
9. Auxiliary switch.
10. Hole for shunt trip undervoltage release plunger.
11. Instantaneous trip adjusting wheels.

Fig. 11-4. Circuit breaker details.

ual parts; and (6) an enclosing housing of insulating material. The circuit breaker handle manually opens and closes the contacts and resets the automatic trip units after an interruption. Some circuit breakers also contain a manually operated "push-to-trip" testing mechanism. Each of these elements will be discussed later along with their various important functions.

A circuit breaker can be rated for either alternating current or direct current system applications or for both. Single-pole circuit break-

ers, rated at 120/250 volts AC or 125/250 volts DC, can be used singly and in pairs on three-wire circuits having a neutral connected to the midpoint of the load. Single-pole circuit breakers rated at 120/240 volts AC or 125/250 volts DC also can be used in pairs on a two-wire circuit connected to the outside (ungrounded) wires of a three-wire system. Two-pole or three-pole circuit breakers rated 120/240 volts AC or 125/250 volts DC can be used *only* on a three-wire, direct current, or single-phase, alternating current system having a grounded neutral. Circuit breaker voltage ratings must be equal to or greater than the voltage of the electrical system on which they are used.

Circuit breakers have two types of current ratings. The first, and the one that is used most often, is the continuous current rating. The second is the short circuit current interrupting capacity.

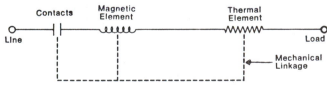

Fig. 11-5. Circuit breaker wiring diagram. (Courtesy of Square D Co.)

The rated continuous current of a device is the maximum current in amperes, which it will carry continuously without exceeding the specified limits of observable temperature rise. Continuous current ratings of circuit breakers are established based on standard UL ampere ratings. The ampere rating of a circuit breaker is located on the handle of the device, and the numerical value alone is shown.

General application requires that the circuit breaker current rating must be equal to or less than the load circuit conductor current carrying capacity (ampacity). However, as a safety measure, circuit breakers should not be loaded beyond 80% of their rated capacity.

The interrupting capacity (AIC) rating of a circuit breaker is the maximum short circuit current which the breaker will interrupt safely. The AIC rating is at rated voltage and frequency.

A circuit breaker must be selected with interrupting capacity equal to or greater than the available short circuit current at the point where the circuit breaker is applied in the system. The breaker interrupting capacity is based on tests to which the breaker is subjected. There are two such tests; one set up by UL and the other by NEMA.

The interrupting rating of a circuit breaker is as important in application as the voltage and current ratings, and should be considered each time a breaker is applied.

In residential applications, the available short circuit current is seldom higher than 10,000 amperes or even near this value.

Where still higher interrupting capacity than the standard ratings discussed above is required in the 15–100 ampere circuit breaker, higher interrupting capacity is available. One type has an interrupting rating of 65,000 amperes, symmetrical at 240 volts AC. The continuous current ratings are duplicated in these breakers (15–100 amperes) but the interrupting capacity has been increased to satisfy the need for greater interrupting capacity. This type of breaker is applied in installations where the higher short circuit currents are possible, such as large industrial plants.

When replacing an existing transformer, or any other piece of equipment, with a new one of recent manufacture, remember to check the AIC capacity of the existing circuit breakers in the circuit, as reduced resistance of the new equipment may require a higher AIC capacity.

The standard rated frequency for circuit breakers is 60 cycles per second. Frequencies below 60 cycles usually will not affect the trip setting of breakers, but molded case circuit breakers applied on systems with frequencies of 25 cycles or less should also be rated as suitable for use on DC systems. Frequencies above 60 cycles affect both thermal and magnetic characteristics. The supplier should be consulted before applying breakers on higher frequencies.

The rated ambient temperature of a circuit breaker is the temperature on which its continuous current rating is based. Ambient temperature is the temperature of the air immediately around the circuit breaker which affects its tripping point and must be considered for effective applications.

An overcurrent trip element is a device which, for a given pole of a circuit breaker, detects overcurrent and transmits the energy necessary to trip the breaker automatically. The overcurrent trip element in some circuit breakers is commonly called a thermal trip element, meaning it is sensitive to heat. These elements have an inverse time characteristic. Inverse time is a qualifying term indicating that a delayed action is purposely introduced, and this delay decreases as the magnitude of the current increases.

The thermal overload trip element is permanently fixed at the factory in all circuit breakers and cannot be adjusted in the field.

Working along with the thermal overcurrent trip element circuit breakers is the instantaneous magnetic trip element. An instantaneous magnetic trip is that part of a trip unit which contains an electromagnet which trips a breaker instantaneously at or above a predetermined value of the current. It is not affected by heat as is the thermal element just discussed.

In Chapter 3, we posed a typical problem involving the installation of a 50 hp motor, and proceeded to select some of the elements required to make the complete installation. At this point, we shall select a circuit breaker to handle the motor of 63 amps, by utilizing the data provided by Square D Company.

Table 11-2 contains a table of standard circuit breaker mechanisms for the various amperage and voltage combinations. We note that there are any number of styles that will meet our specifications for amperage and voltage available, and after studying the literature and possibly checking with the local distributor as confirmation of our selection we choose the FH-FHL in 600 volt size. This circuit breaker has an AIC rating of 65,000 amperes at 240 volts, which should do very nicely for our situation. As we have space available in the closest power distribution panel to take the new circuit breaker module, a separate cabinet will not be needed.

E. Auxiliary Devices

The advantage of the circuit breaker goes beyond permitting the disconnection and protection of the electrical circuit. By adding auxiliary devices, a breaker can do such things as sense voltage on its own system or other systems, alarm personnel of faulty conditions, disconnect or control associated equipment of completely different types of electrical systems as well as mechanical devices, and allow itself to be controlled from remote locations along with many other operations. We cannot list all the functions a circuit breaker can perform, but we can look at the auxiliary devices that help add many functions to the breaker.

A shunt trip is a mechanism which trips a circuit breaker by means of a trip coil energized from a separate circuit or source of power. The trip-coil circuit is closed by a relay, switch, or other means. The shunt trip is available in two- and three-pole breakers and is normally installed in the left pole. Shunt trip coils do not have a continuous current rating. A cut-off switch is included to break the coil circuit when the breaker opens.

An undervoltage trip device is one which trips a circuit breaker automatically when the main circuit voltage decreases to approximately 40% of its value. The breaker cannot be reset until the voltage returns to 80% of normal value.

A normally closed contact, such as those used in stop buttons, can be installed in the control circuit to open the breaker in a manner similar to a shunt trip.

An auxiliary switch is one which is mechanically operated by the main switching device for signaling, interlocking, or other purposes. An "A" type contact is one which is open when the breaker contacts are open. The "B" type contact is closed when the breaker contacts are open. Auxiliary switches are available in two- and three-pole breakers and are normally installed in the right pole.

The alarm switch is a single-pole device that is activated when the breaker is in the tripped position. It is used to actuate bell alarms and warning signals. The alarm switch is factory installed and is rated at least 1 ampere at 120 volts AC.

F. Enclosures

Up to this point in this discussion, the circuit breaker has been discussed without any consideration given to the enclosure in which it may be mounted.

The majority of circuit breakers are used in some type of enclosure (e.g., panelboards, switchboards, motor control centers, individual enclosures, etc.).

NEMA has established enclosure designations because electric apparatus is used in so many different types of locations, weather and water conditions, dust and other contaminating conditions, etc. The designation (i.e., NEMA 12) indicates an enclosure type to fulfill requirements for a particular application. Table 11-3 contains a brief description of the most commonly used enclosures as rated by NEMA.

Table 11-2. Circuit Breaker Group Identification

Catalog Number Prefix	No. Poles	Maximum AC Voltage Rating	Ampere Rating
QO-GFI, QOB-GFI	1 2	120 120/240	15-30 15-60
QO-VHGFI, QOB-VHGFI	1	120	15-30
QH, QHB	1 2 3	120/240 120/240 240	15-30 15-30 15-30
QO, QOB	1 2 3	120/240 120/240 240	10-70 10-125 10-100
QO-PL, QOB-PL	1 2 3	120/240 120/240 240	10-30 10-60 15-60
QO-H, QOB-H	2	240	15-100
QOT	1 2	120/240 120/240	15-20 15-20
QO-VH, QOB-VH	1 2 3	120/240 120/240 240	15-30 15-125 15-100
QOU	1 2 3	120/240 120/240 240	10-100 10-100 10-100
QOH, QOHB	2	120/240	35-125
Q2, Q2L	2 3	120/240 240	100-225 100-225
Q2-H, Q2L-H	2 3	240 240	100-225 100-225
Q4, Q4L	2 3	240 240	250-400 250-400
EH, EHB	1 2 3	277 480Y/277 480Y/277	15-60 15-100 15-100
EHB-PL	1 2 3	277 480Y/277 480Y/277	15-30 15-60 15-60
QOB-VH	2 3	240 240	150 110-150

Catalog Number Prefix	No. Poles	Maximum AC Voltage Rating	Ampere Rating
FY	1	277	15-30
FA, FAL	1 2 3	120 240 240	15-100 15-100 15-100
FA, FAL	1 2 3	277 480 480	15-100 15-100 15-100
FA, FAL	2 3	600 600	15-100 15-100
FH, FHL	1 2 3	277 600 600	15-100 15-100 15-100
FC, FCL	2 3	480 480	15-100 15-100
IF, IFL	2 3	600 600	20-100 20-100
KA, KAL	2 3	600 600	70-250 70-250
KH, KHL	2 3	600 600	70-250 70-250
KC, KCL	2 3	480 480	110-250 110-250
IK, IKL	2 3	600 600	110-250 110-250
LA, LAL	2 3	600 600	125-400 125-400
LH, LHL	2 3	600 600	125-400 125-400
IL-ILL	3	480	250-400
MA, MH	2 3	600 600	300-800 300-800
MAL, MHL	2 3	600 600	300-1000 300-1000
ME, MEL	3	600	50-800
NA, NAL	2 3	600 600	600-1200 600-1200
NC, NCL	2,3	600	600-1200
NE, NEL	3	600	300-1200
PAF	2 3	600 600	600-2000 600-2000
PHF	2 3	600 600	600-2000 600-2000
PEF, PEC	3	600	300-2500
PCF	2 3	600 600	1600-2500 1600-2500
SE	3	600	100-4000
PXF	3	600	600-2500
MX, MXL	3	600	125-800 125-800
NX, NXL	3	600	600-1200

G. Grounding Protection

One of the most common methods of protection, both for personnel and for plant equipment, is by the proper use of ground connections. By grounding, we mean the practice of running electrical conductors from the equipment or circuits into the ground. We shall now explain the reasoning and the systems in use for this important practice.

If you will look at Fig. 11-6, you will see a diagram of a typical conduit carrying three wires, or conductors, in a typical three phase supply system. Normally, of course, the three wires will be properly insulated from each other and from the conduit. Under normal conditions, then, the current for the process or systems being supplied will flow along the three wires.

But suppose that something should happen to damage the insulation on one or more of the wires in this conduit? Let us look at the possibilities.

There can be a possible "short" between A and B, between B and C, between A and C, and between any of the three wires and the conduit, which is grounded.

If this were a typical three phase system, then the possibilities for "shorts," as explained above, would be as follows:

Shorted Pair	Common Term for the Short
B & C, A & B, A & C	Phase to Phase Short
A & G, C & G, B & G	Phase to Ground Short

This "short" can be any degree, from a very slight leakage of electricity to what is called a "dead short," where there is an extreme flow of electricity.

The short may not occur in the conduit, but may be in any of the many electrical appliances, tools, or machines in the plant. In place of the conduit in Fig. 11-6, the frames, the cabinets, or enclosures of the electrical equipment may serve the same purpose as the conduit in this explanation or demonstration.

Notice that in Fig. 11-6 the conduit is grounded. Obviously, if there is a short between any of the wires and the conduit, the current will be bypassed from its normal path through the conduit into the ground. This is the primary function of the grounded conduit, machine, control, or cabinet whichever the case may be.

Now look at Fig. 11-7, where we have placed this same situation into the typical industrial plant supply system. Let us assume there is a slight short in the electrical wiring in the left-hand leg. The plant maintenance man unknowingly touches the frame and receives a shock. Why? To answer that, we refer you back to Chapter 2, where we explained that if an electrical current is flowing along a line and comes to a branching of the circuit, that current flow will split into two portions. The circuit which has the least resistance to the flow of electricity will naturally take the highest flow, while that circuit which has the highest resistance of the two paths, will take the least flow of electricity. That is what will happen there, when you or someone else in your plant touches that equipment in which a short to ground exists.

If the natural ground has a higher resistance than the body of the person making the contact between the shorted equipment and the ground, then the majority of the current flowing in the system will rush through that person's body and into the ground. This can easily happen if the feet are wet, or if the floor or ground is wet underneath the feet of the man taking the flow of current. We mentioned above that the man will receive a shock. That is all that will happen if he is lucky. If he isn't lucky, or if he is careless, then he could very well be electrocuted immediately. It happens too often in today's industrial establishments. This is why you will find that dry feet, hands, and clothing are *a must* when working around electrical equipment.

So now we have the unfortunate electrical maintenance man receiving a shock. What next? If he is lucky, and has managed to let go of the equipment which is shorted, then he will probably jump back and thank his guardian angel. The next step is to shut off that circuit and locate and repair the short.

Table 11-3. NEMA Enclosure Definitions

Definitions Pertaining to Nonhazardous Locations

Type 1 Enclosures are intended for indoor use primarily to provide a degree of protection against contact with the enclosed equipment.
NEMA Standard 1-10-1979.

Type 2 Enclosures are intended for indoor use primarily to provide a degree of protection against limited amounts of falling water and dirt.
NEMA Standard 1-10-1979.

Type 3 Enclosures are intended for outdoor use primarily to provide a degree of protection against windblown dust, rain, sleet, and external ice formation.
NEMA Standard 1-10-1979.

Type 3R Enclosures are intended for outdoor use primarily to provide a degree of protection against falling rain, sleet, and external ice formation.
NEMA Standard 1-10-1979.

Type 3S Enclosures are intended for outdoor use primarily to provide a degree of protection against windblown dust, rain, sleet, and to provide for operation of external mechanisms when ice laden.
NEMA Standard 1-10-1979.

Type 4 Enclosures are intended for indoor or outdoor use primarily to provide a degree of protection against windblown dust and rain, splashing water, and hose-directed water.
NEMA Standard 1-10-1979.

Type 4X Enclosures are intended for indoor or outdoor use primarily to provide a degree of protection against corrosion, windblown dust and rain, splashing water, and hose-directed water.
NEMA Standard 1-10-1979.

Type 5 Enclosures are intended for indoor use primarily to provide a degree of protection against dust and falling dirt.
NEMA Standard 1-10-1979.

Type 6 Enclosures are intended for indoor or outdoor use primarily to provide a degree of protection against the entry of water during occasional temporary submersion at a limited depth.
NEMA Standard 1-10-1979.

Type 6P Enclosures are intended for indoor or outdoor use primarily to provide a degree of protection against the entry of water during prolonged submersion at a limited depth.
NEMA Standard 1-10-1979.

Type 11 Enclosures are intended for indoor use primarily to provide, by oil immersion, a degree of protection to enclosed equipment against the corrosive effects of liquids and gases.
NEMA Standard 1-10-1979.

Type 12 Enclosures are intended for indoor use primarily to provide a degree of protection against dust, falling dirt, and dripping noncorrosive liquids.
NEMA Standard 1-10-1979.

Type 12K Enclosures with knockouts are intended for indoor use primarily to provide a degree of protection against dust, falling dirt, and dripping noncorrosive liquids other than at knockouts.
NEMA Standard 1-10-1979.

Type 13 Enclosures are intended for indoor use primarily to provide a degree of protection against dust, spraying of water, oil, and noncorrosive coolant.
NEMA Standard 1-10-1979.

Definitions Pertaining to Hazardous (Classified) Locations

Type 7 Enclosures are for use indoors in locations classified as Class I, Groups A, B, C, or D, as defined in the *National Electrical Code*.
NEMA Standard 1-10-1979.

Type 8 Enclosures are for indoor or outdoor use in locations classified as Class I, Groups A, B, C, or D, as defined in the *National Electrical Code*.
NEMA Standard 1-10-1979.

Type 9 Enclosures are for use in indoor locations classified as Class II, Groups E, F, or G, as defined in the *National Electrical Code*.
NEMA Standard 5-19-1986.

Type 10 Enclosures are constructed to meet the applicable requirements of the Mine Safety and Health Administration.
NEMA Standard 1-10-1979.

General Definitions Pertaining to Enclosures

Apparatus is the enclosure, the enclosed equipment, and the attached protruding accessories.
NEMA Standard 1-10-1979.

Design Tests demonstrate performance of a product designed to applicable standards; they are not intended to be production tests.
NEMA Standard 1-10-1979.

Flush Mounting means so constructed as to have a minimal front projection when set into a recessed opening and secured to a flat surface.
NEMA Standard 1-10-1979.

Hazardous (Classified) Locations are those areas which may contain hazardous (classified) materials in sufficient quantity to create an explosion. See article 500 of the *National Electrical Code.*

NEMA Standard 3-8-1985.

Hazardous Materials are those gases, vapors, combustible dusts, fibers, or flyings which are explosive under certain conditions.

NEMA Standard 1-10-1979.

Indoor Locations are those areas which are protected from exposure to the weather.

NEMA Standard 1-10-1979.

Knockout is a portion of the wall of an enclosure so fashioned that it may be removed readily by a hammer, screwdriver, and pliers at the time of installation in order to provide a hole for the attachment of an auxiliary device or raceway, cable, or fitting.

NEMA Standard 1-10-1979.

Nonventilated means so constructed as to provide no intentional circulation of external air through the enclosure.

NEMA Standard 1-10-1979.

Oil-Resistant Gaskets are those made of material which is resistant to oil or oil fumes.

NEMA Standard 1-10-1979.

Outdoor Locations are those areas which are exposed to the weather.

NEMA Standard 1-10-1979.

Surface Mounting means so constructed as to be secured to, and projected from, a flat surface.

NEMA Standard 1-10-1979.

Ventilated means so constructed as to provide for the circulation of external air through the enclosure to remove excess heat, fumes, or vapors.

NEMA Standard 1-10-1979.

The purposes for installing an adequate grounding system are as follows.

a) To limit the surge or flow of current in a shorted circuit.

b) To provide a path for excess surges to be conducted into the ground before they can cause damage.

c) To provide protection from lightning.

d) To provide protection for the plant personnel.

To accomplish the above, other devices are needed in addition to an adequate grounding system. In fact, the complete grounding system will usually consist of the following.

a) Amply sized connections into the grounding circuit from all current carrying apparatus.

b) Devices to detect and stop the flow of excess current once the flow to ground has

commenced, and before the flow reaches a dangerous level.

c) Devices which will isolate the circuit without shutting down too many other circuits in the plant or system.

Before we enter into the discussion of the various other control devices mentioned above, we shall go more deeply into the methods of effectively grounding the plant electrical systems. Most of the following material is covered in great detail in the **National Electrical Code®**, but we shall give the essentials here. You should understand that this discussion will be general in nature, and that specific cases should be handled by direct reference to the NEC book. A copy of that book should be in every maintenance office of any size where electrical maintenance work is to be performed.

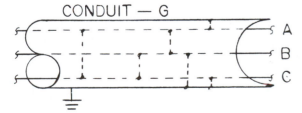

Fig. 11-6. Short circuit possibilities in a conduit.

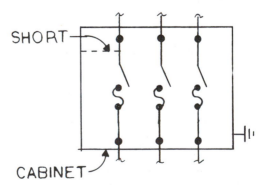

Fig. 11-7. Short circuit in a switch box.

First, let us examine the electrical service coming into the building or onto the plant premises. This service may, and usually does, come from distribution transformers, connected up in one of the systems covered in an earlier chapter. The primary sides of these transformers are under the control of the electrical utility serving the plant, and any wiring done to the secondary, or plant, sides of the transformers will have a definite effect upon the utility's service. Therefore, any attempt to ground the secondary side should be discussed first with the utility service department.

Chapters 9 and 16 contain several methods used to ground the secondary sides of the distribution transformers, and no further explanation is necessary. The principles covered later in this discussion will suffice to cover any grounding that may be permitted by the utility company.

What equipment needs to be grounded? We covered this briefly a few paragraphs ago, but will go more deeply into the subject here, by listing those more commonly encountered in the industrial plant.

a) All electrical distribution panels and control cabinets, including motor starters, fuse boxes, circuit breakers, and disconnect switches.

b) All machinery containing motors or electrical switches or devices.

c) All conduit, wire trays, or other metal systems containing electrical wires.

d) All metal structures which are subject to collecting stray currents from such other electrical machinery as may be in the vicinity. This includes wire fences under high tension power lines, and moving machinery subject to static charges, as covered in Chapter 2.

Some of the above grounding requirements are relatively simple (such as the ones mentioned in c). In this case, merely connecting a copper wire jumper from one end of a conduit to the other in a junction box, so as to carry the continuity on to the final grounding connec-

tion, is all that is required. This is assuming, of course, that the conduit is metal, or that it has a metal wire running within it for the purpose of grounding.

The size of the grounding wire is stipulated in the **National Electrical Code**® book. Generally speaking, the size is governed by the amount of current which the wire could conceivably ever be forced to handle under an extreme situation. In the case of electrical equipment, such as control panels, motors, etc., this translates to the current rating of the power input to the equipment. In any event, the equipment grounding wire is not to be smaller than size 14.

The equipment grounding wire is either bare or, if covered, it is green or green with yellow stripes. The system grounded line, running from control boxes, through conduits, between pull boxes, lighting fixtures, and such low voltage equipment, will be white or gray. However, there are times when the electrical installation crew may run out of the proper color wire, and will sometimes substitute another color. Such substitutions, not authorized by code, will confuse the maintenance worker who is not expecting it. If this happens, the electrician should look for any signs of tagging or colored tape on the wiring indicating the service for which it was intended.

Some of the other grounding methods may become somewhat more sophisticated, and may consist of attaching heavy copper cables to underground steel grounding rods or attaching the copper grounding cable to them. We shall now go into this latter method more thoroughly. The **National Electrical Code**® covers this in great detail, of course.

Ground rods (usually referred to as "grounding electrodes") are driven into the ground in sufficient depths and in sufficient quantities to produce a resistance between the system of rods and the ground of not over 25 ohms. As the ground, or earth, varies in resistance from place to place and from season to season, it is essential that great care be taken in providing the proper number, depth, and placing of the rods. Basically, the procedure consists of driving one rod into the ground, measuring the resistance between it and the ground, then driving more rods into the ground until the resistance is well below 25 ohms. These rods are driven as shown in Fig. 11-8, and the resistance is measured by means of a ground-testing megger, used as directed by the maker of the instrument, and as shown in Fig. 11-8.

It may not be possible to drive sufficient rods into the ground to produce the desired results. In this case, it is possible to drive one rod about 8 feet into the ground, and surround the rod with a ditch. Into this ditch, from 40 to 90 pounds of magnesium sulphate, copper sulphate, or rock salt is poured. The dimensions of this arrangement are shown in Fig. 11-9. In very dry soils, it may be necessary to pour water into the chemical trench occasionally to produce the required resistance. In fact, all grounding systems should be tested frequently to be sure that nothing has altered in this very important safety measure. Remember, it is your life that may be at stake.

Figure 11-10 shows several other methods of installing ground rods, and making the connections to them. The connection in all cases should be very tight, size 8 wire or larger, and not subject to corrosion, if at all possible. These connections should be checked often, along with the regular maintenance schedule which you, as an efficient maintenance electrician, will no doubt establish. At times, it may be worth making this connection by welding or

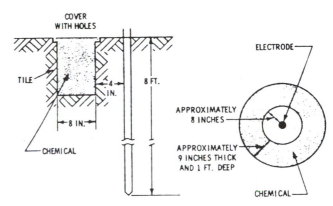

Fig. 11-9. Lowering the ground resistance.

brazing. Normally, a good tight clamp type connection made for this purpose will be sufficient, if properly made and maintained.

Fortunately, most of the electrical equipment now being supplied comes with grounding lugs for attaching to the grounding system. The instructions for this equipment usually contains directions for properly grounding the equipment. Follow them to the letter, check all connections and ground wiring, and regularly meter the resistance of the rods to ground, and you stand a good chance of staying out of the hospital.

So that you may not fall into the trap of complacency with your grounding system, we hasten to point out that this entire grounding system is a protection only in the case of shorts between the electrical wiring and the ground. Shorts between the wires themselves will not be protected with even the most efficient grounding system. However, shorts between the wires are usually protected by other means of overcurrent disconnect devices, some of which have already been covered in this chapter. We refer to fuses, thermal and magnetic overload disconnecting devices, etc.

H. Ground Fault Current Interrupters

These devices, often shortened to GFCI's, are ingenious safety devices which are designed to open a circuit when there is a difference in cur-

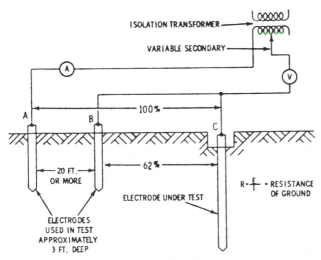

Fig. 11-8. Ground rod testing.

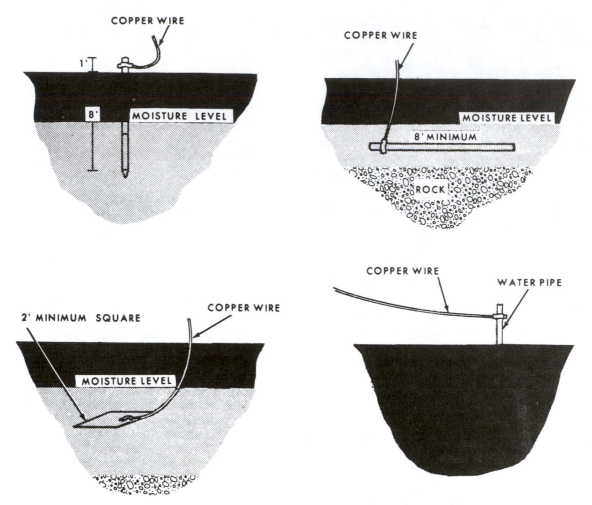

Fig. 11-10. Typical grounding fixtures.

rent flow between the hot and neutral conductors of a circuit. The current between these conductors should always be equal, unless there is a fault in the piece of equipment being used.

Please refer to Fig. 11-11, which shows a typical circuit involving an ordinary electric drill. In this case, there is assumed to be a break in the wiring inside the drill body, and there is a leakage of current to the drill body of 100ma (milliamps). If the grounding circuit is in good order, then the instant that this 100ma starts to flow, it is detected by the interrupter circuitry, which actuates a solenoid, which in turn breaks the circuit and shuts off the flow of current. This normally happens so fast that often the person holding the drill will not feel more than a slight tingle or shock.

Now suppose that there is a break in the grounding circuit, as shown in the ground wire coming from the drill. Instead of the short causing the flow to take place to the ground, it will go into the hand and body of the person holding the drill. In this case, 100ma will probably produce a fatal jolt. In the process, it will not shut off the flow of current by blowing the fuse or kicking out the breaker.

Figure 11-12 shows a diagram of the GFCI. Notice that it contains a test button, which is simply a bypass switch producing an artificial short between the wiring and the ground. Actuating this switch at frequent intervals will ensure that the GFCI is ready and able to protect the system. You must be sure, however, that when this switch is actuated that the resultant power outage in that part of the plant circuitry will not cause problems. Figure 11-13 contains

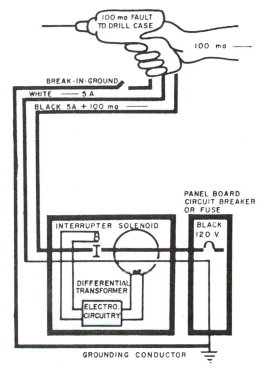

Fig. 11-11. Protecting a portable drill. (Courtesy of Square D Co.)

several typical GFCI applications to be found in industrial plant electrical systems.

The GFCI sensor continuously monitors the current balance in the ungrounded "hot" load conductor and the neutral load conductor. If the current in the neutral load wire becomes less than the current in the "hot" load wire, then a ground fault exists, since a portion of the current is returning to the source by some means other than the neutral load wire. When an imbalance in current occurs, the sensor, which is a differential current transformer, sends a signal to the solid state circuitry which activates the ground trip solenoid mechanism and breaks the "hot" load connection. A cur-

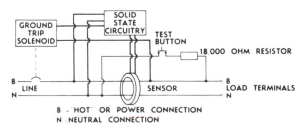

Fig. 11-12. Typical GFCI circuit. (Courtesy of Square D Co.)

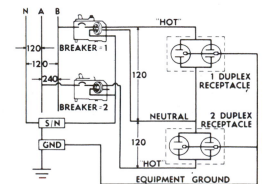

Fig. 11-13. Typical GFCI circuits. Note: S/N = solid neutral. (Courtesy of Square D Co.)

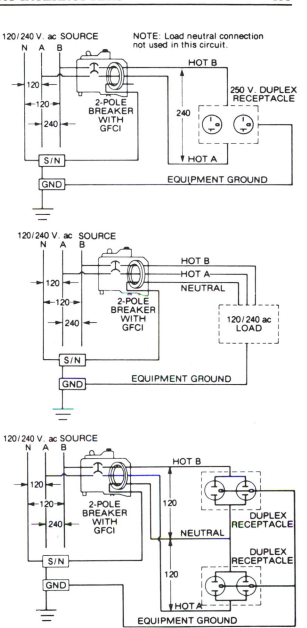

rent imbalance as low as 6ma will cause the circuit breaker to interrupt the curcuit.

Just to emphasize the importance of the GFCI and the grounding system as a whole, we shall give you the expected results when currents of varying amounts flow directly into the human body.

Hand-to-hand body resistance of an adult lies between 1,000 and 4,000 ohms, depending on moisture, muscular structure, and voltage. The average value is 2,100 ohms at 240V.AC and 2800 ohms at 120V.AC.

Using Ohm's law, the current resulting from the above average hand-to-hand resistance values is 114 milliamperes (.114 amps) at 240V.AC and 43 milliamperes (.043 amps) at 120V.AC. The effects of 60 Hz alternating current on a normal healthy adult are as follows (note that current is in millamperes):

More than 5ma—generally painful shock

More than 15ma—sufficient to cause "freezing" to the circuit for 50% of the population

More than 30ma—breathing difficult (possible suffocation)

50 to 100ma—possible ventricular fibrillaton*

100 to 200ma—certain ventricular fibrillation*

Over 200ma—severe burns and muscle contractions. The heart is more likely to stop than fibrillate.

Now, will a conventional overcurrent device open a circuit before irreparable harm is done? NO! Here's why.

The current that would flow from a defective electric drill, for example, through the metal housing and through the human body to ground would be 100 milliamperes, calculated using 1200 ohms as average body resistance.

---

*Ventricular fibrillation is defined as "very rapid uncoordinated contractions of the ventricles of the heart resulting in loss of synchronization between heartbeat and pulse beat." Once ventricular fibrillation occurs in man, it usually continues, and death will ensue within a few minutes.

100 milliamperes is only .68% of the current required to open a 15 ampere circuit breaker or fuse, and yet it approaches the current level which may produce ventricular fibrillation. Obviously, the standard circuit breaker or fuse will not open the circuit from such low levels of current flow.

I. Testing and Servicing

In the previous section on GFCI's, we mentioned a test button which may be found on some systems. Merely pressing this button will simulate a ground fault, and shut off the circuit if the GFCI is operating properly. Should this not happen, then the fault is probably in the solenoid coil. Usually replacing this will restore the system back to normal protective condition again. In addition, the GFCI circuit equipment must be kept clean, as in any similar protective device.

The following instructions are furnished by the General Electric Company, covering their line of circuit breakers. The same instructions are applicable to any other similarly constructed circuit breakers.

The following material covers the ordinary type of circuit breaker found in industrial plants, as used in ratings up to 600 volts. Above that rating, there is another type of circuit breaker, known as the oil-filled circuit breaker (OCB). Should you find one on your premises, then we suggest you contact the manufacturer of it and obtain complete servicing instructions before attempting to maintain it. There is only one thing we point out which we feel is within the scope of this book—NEVER OPEN OR CLOSE AN OIL-FILLED CIRCUIT BREAKER WHEN THERE IS A LOAD ON ITS CIRCUIT. This means that all switches, starters, etc., in the load centers downstream of the OCB should be open when the OCB is either opened or closed.

There are two general types of testing procedures used for inspecting and testing circuit breakers of the molded case design. The one we shall cover here is the simplest one, which can be performed by the industrial plant maintenance operator. There is another test which is sometimes necessary, requiring more sophisticated equipment and procedures, and which should be performed by factory trained service men or other specialists in this type of work. This latter test is the Verification Field Test. The testing procedure is fully described in General Electric's publication GET-2779G.

The need for preventive maintenance on molded case circuit breakers will vary depending on operating conditions. Where heavy dust conditions exist, for instance, and the circuit breakers are not called upon to operate, an accumulation of dust on the latch surfaces may affect the operation of the breaker. Dust accumulation can usually be cleared from the latch by occasionally manually turning the breaker "OFF" and "ON," thus exercising the circuit breaker.

The following constitutes a guide to tests which might be performed during routine maintenance.

The tests recommended are based on proven standard maintenance practices and are aimed at assuring that the breaker is functionally operable.

INSULATION RESISTANCE TEST

Extreme atmospheres and conditions may reduce the dielectric withstandability of an insulating material—including those of which molded case breakers are made—therefore, the first routine check recommended on installed breakers is a resistance measurement test. An instrument commonly known as a "megger" is used to perform this test. See Chapter 10.

The voltage recommended for this test should be at least 50 percent greater than the breaker rating, however, a minimum of 500 volts is permissible. Tests should be made between phases of opposite polarity as well as from current carrying parts of the circuit breaker to ground. Also, a test should be made between the line-and-load terminals with the breaker in the "OFF" position.

Resistance values below one megohm are considered unsafe and should be investigated for possible contamination on the surfaces of the molded case of the circuit breaker.

NOTE: For individual breaker resistance reading, load and line conductors should be disconnected. If not disconnected, the test measurements will also include the characteristics of the attached circuit.

CONNECTIONS TEST

Connections to the circuit breaker should be inspected to determine that a proper electrical joint is present. If overheating in these connections is evident by discoloration or signs of arcing, the connections should be removed and the connecting surfaces cleaned before the breaker is re-installed. It is essential that electrical connections be made properly to prevent and reduce overheating. Aluminum connectors (lugs) are plated and should not be abrasively cleaned. If damage is evident, the lugs should be replaced. In making connections with aluminum conductors, use a joint compound made for the purpose.

CONTACT RESISTANCE TEST

Extensive operations of the circuit breaker under load conditions beyond that for which the circuit breaker was intended may cause deterioration of the contacts. A simple way to test for such deterioration is by measuring resistance across each pole of a breaker. This may be done by use of a resistance bridge or by measuring the voltage drop across the circuit breaker while a current is flowing through the breaker. A milli-voltmeter will be required. Any convenient current value less than the rating of the breaker can be used as long as it is sufficient to obtain a voltage reading. A comparison between the poles of the breaker or similar breakers can be made. A difference of as much as 2:1 may indicate that the breaker's contacts should be cleaned. Excessive millivolt drops across a complete breaker can be an indication of several abnormal conditions within the circuit breaker such as eroded or contaminated contacts or loose connections. This test is an important indicator of the acceptability for continued use of the circuit breaker.

OVERLOAD TRIPPING TEST

A general indication of the proper action of the overload tripping components of the circuit breaker can be verified by selecting certain percentages of the breaker rating, such as 300 percent, and applying this separately to each pole of the circuit breaker to determine if it will open automatically. The significant part of this test is that the circuit breaker will operate, and since ambient conditions and types of connections used for the tests greatly affect the results of tripping times of the circuit breakers, they become of little significance in these tests. When verification of tripping characteristics, other than to determine if the circuit breaker is functional, is required, refer to VERIFICATION FIELD TESTING in General Electric GET-2779G Publication.

In routine tests, it is more important to determine that the magnetic feature is operating and will trip the breaker, rather than the exact value at which the instantaneous magnetic feature operates.

MECHANICAL OPERATION

During routine tests, mechanical operation of the breaker should be checked by turning the breaker "ON" and "OFF" several times.

If the circuit breaker is equipped with a Veri-fier, the circuit breaker should be tripped, reset, and turned "ON" several times. This will remove any dust accumulation on the mechanism and latch surfaces.

SUMMARY

Please note once again that common maintenance practices for electrical equipment should be adhered to in field testing both new and installed molded case circuit breakers. Usually the standard routine operating checks listed will be sufficient to assure proper functioning protective devices. Where molded case circuit breakers are factory calibrated and sealed, the seal should not be broken and the breaker itself should not be tampered with. Circuit breakers with removable covers may be checked for contact cleanliness, connections, and latch cleanliness by making careful visual inspection.

That ends our discussion on circuit protection for now. We shall cover this subject more in later chapters, as the occasion arises in the appropriate place.

We hope that we have not frightened you unduly, but we do wish to impress upon your mind, both conscious and subconscious, how important it is to know the operation of these protective measures, and why they should be maintained properly.

C H A P T E R 12
PLANT LIGHTING

A. Introduction

In today's industrial plant, the lighting equipment and systems are usually the responsibility of the electrical crew. This is true only in the case of an existing installation, of course. When a new plant or an addition to a plant is designed and built, the lighting is determined and designed by specialists in the office of the architect and engineer who are designing the entire plant or addition. In fact, the subject of lighting has expanded considerably in recent years, sufficient to develop a completely new field of engineering.

In this chapter we shall touch only on the basics of the subject, in order to give the electrical maintenance worker some basic knowledge of the principles behind the lighting in the plant. We do not for a minute expect that from this discussion you will be able to design the lighting for any new addition of appreciable size. We do feel that, in keeping with the philosophy of this entire book, the effective maintenance man is one who understands the basic principles behind the equipment and systems he is expected to maintain.

What are the uses and objectives of industrial plant lighting? Briefly, the following list will cover most of these objectives.

1. To ensure that those working in the plant will be able to perform their assigned tasks with the least amount of eye strain.

2. To spot those areas where the worker should tread with care, lest he becomes a victim of an accident.

3. To produce a pleasing atmosphere for the workers, and thus maintain a high degree of morale among them.

4. On exterior applications, lighting is often used for advertising the plant by judicious use of spotlighting and area lighting.

5. To maintain the plant security system.

Before we can embark on a meaningful discussion, we must cover the language which will be encountered. Like many other fields of work, a unique system of terms has been adopted, and we shall give here a few of the more common ones.

Most of the material following in this chapter has been adapted from material graciously contributed by the General Electric Company, The Phillips Lighting Company, and the educational material published by the Illuminating Engineering Society of North America. The specific references are listed in the Bibliography.

Candela: A standard unit of luminous intensity emitted from a point source. It is very nearly the same as the old International Candle, which produced one candlepower.

Footcandle: The intensity of light cast on the inside surface of a sphere one foot in radius, the center of which is a candela.

Lumen: The amount of light cast over one square foot of the inside surface of the same sphere producing one footcandle of intensity.

Luminaire: A general term for a light fixture, consisting of the lamp bulb, the holder, the reflector, and the diffuser. Also may include any ballast, or such device, if it is normally furnished with the fixtures.

Lux: The amount of light which falls on the inside surface of a sphere one meter in radius, the center of which is one candela.

Some equivalents involved in the above definitions are:

1 Candela = 12.57 lumens

1 Footcandle = 1 lumen/sq. ft

1 Footcandle = 10.76 Lux (use 10 for calculations)

1 Lux = 1 lumen/sq. meter

1 Lux = .0929 Footcandles (use .10 for calculations).

B. How Much Light Do We Need?

Before that question can be answered, there must be a method of measuring light intensity on any surface or at any designated spot. First we shall describe the basic method, then proceed to answer the above question, in a general manner, in keeping with our statements earlier concerning this being a manual of basic principles only.

Illumination is readily measured in the field by portable instruments which can be moved about to "read" or indicate the footcandle level at any point of interest. Since the meter indicates only the level of illumination where the light-sensitive element is placed, a number of readings are usually taken at different locations, from which the average illumination of an area can be derived. Footcandles are usually calculated and measured in either the horizontal or vertical planes. However, they can be measured in any plane for special purposes. Inexpensive hand-held meters are available as well as precision lightmeters that can measure

from a fraction of a footcandle to several thousand footcandles.

In addition to being able to accurately measure the light intensity on a surface or work area, there are a few fundamental principles which must be considered. The problem of providing sufficient light to permit a worker to perform his given task without damaging his eyesight is not a simple one of light intensity alone, but it is the starting point to the solution.

Some of the elements which have to be considered are:

age and seeing ability of those who are to perform the task

type of work to be performed; close, arm's length, walking, sitting, etc.

length of time worker is expected to perform without a rest

is the work expected to be done quickly and accurately, or moderately slow with a fair degree of accuracy

is the work in a plane parallel, or sloping, or vertical in respect to the worker's eyes

what are the surrounding factors which may affect the worker's sight, health, concentration ability, etc.

The last item in the preceding list includes such elements as

color and texture of surfaces at and around the work site.

Noise levels and movements surrounding the work site.

Atmospheric conditions such as temperature, stray odors, and drafts.

Activities of other workers in the area.

Obviously, some of the preceding elements are beyond the scope of this book, and outside professional help is required if there is a production or employee health problem.

Under good lighting—diffused, uniform, with no glare or reflections—the average human eye has efficiencies of vision approximately according to that shown in Table 12-1.

From Table 12-1, we deduce that above 50 footcandles, there is only a slight increase in eye efficiency for higher lighting intensities. Also, it is obvious that for most of the average tasks to be performed in an industrial plant, 50, or perhaps 60, footcandles is all that may be necessary. Some specialized tasks, such as very accurate assembly work, may have to be provided with spot lighting of higher intensities. If this is the case, then we recommend that a lighting expert be consulted, because to attempt to solve the problem without adequate knowledge of all factors that are involved, could result in serious damage to the worker's eyes.

The Illuminating Engineering Society recommendations for illuminance values, given in the *IES Lighting Handbook*, 6th Edition, cover a broad range of seeing tasks. A table of selected tasks (extracted from the *IES Handbook*) is shown in Table 12-2. Note that the illuminance values are given as a range.

It is recommended that the lower end of the range be used if the visual task is not critical and speed and accuracy are not important, the background is light and reflectances high, and the workers are young. Conversely, the older worker, with the more critical task and with poorer contrast, requires the upper end of the scale. For average conditions of people, tasks, and visual performance, choose the center of

the range. These values are for the task itself, not necessarily for the entire space.

The illuminance ranges given are typical for seeing tasks found in most factories. For a more detailed method of determining the values, the *IES Handbook* should be consulted.

C. Types and Evaluation of Lamps

There are so many different types, sizes, and styles of lamps and lamp fixtures that the first approach to the subject is confusing. Perhaps this brief discussion will assist the plant electrical maintenance man in making a little sense out of the subject.

To produce light from electrical energy, we find the same situation as we note in other segments of our industrial developments. The first attempts were startling to behold, but very inefficient. Over the years since the first electric bulb was successfully produced, presumably by Thomas Edison, the common electric light bulb has evolved into forms, shapes, and uses not envisioned by the inventor.

The first commercial lamps were simply carbon filaments, or wires, fed an electric current sufficient to cause it to glow in the absence of air or oxygen. The glow emitted light, as well as heat. In fact, the light was only a small by-product, most of the energy went into the production of heat, as we shall see a little later.

From this simple beginning has developed an amazing array of lamps, but they all fall into four general classes: incandescent, fluorescent, sodium lamps, and mercury lamps. There are other methods of breaking them down into classes, but this will do nicely for our discussion.

Table 12-3 lists the approximate comparative light outputs, in lumens per watt, for each type. Keep in mind that this table is only approximate, as developments may make it obsolete very quickly. Some other reference points are included, for orientation and comparison.

It is obvious from Table 12-3 that if energy efficiency is the only basis for selection, the low

Table 12-1.

Lighting Intensity, Footcandles	Efficiency, Percent
5	40
10	55
15	65
20	72
25	75
30	80
35	83
40	86
45	88
50	90
60	92
70	95
80	96

Table 12-2. Typical Industrial Lighting Intensities

	Footcandles Maintained on the Task
INDOORS	
Garages—Service	
□ repair	50–100fc
□ active traffic areas	10–20fc
Loading Platform	20fc
Machine Shops and Assembly Areas	
□ rough bench/machine work, simple assembly	20–50fc
□ medium bench/machine work, moderately difficult assembly	50–100fc
□ difficult machine work, assembly	100–200fc
□ fine bench/machine work, assembly	200–500fc[a]
Receiving & Shipping	20–50fc
Warehouses, Storage Rooms	
□ active—large items/small items, labels	15fc/30fc
□ inactive	5fc
OUTDOORS	
Storage Yards	
□ active/inactive	20fc/1fc
Parking Areas	
□ open—high activity/medium activity	2fc/1fc
□ covered—parking, pedestrian areas	5fc
□ entrances—day/night	50fc/5fc

[a]Higher illuminance values may be achieved through a combination of supplementary and general lighting.

pressure sodium lamp is the only one to choose. However, there are other elements to take into consideration, which will be covered later. We have already touched on the subject in a general manner in the introduction to this chapter.

The ordinary incandescent lamp bulb contains a filament of tungsten, chosen because of its relatively high output of white light, combined with other characteristics which make it

Table 12-3. Light Outputs In Lumens Per Watt

Approximate maximum	620
White Light	220
Low-pressure sodium lamps	100–180
High-pressure sodium lamps	80–140
Metal halide lamps	70–125
Fluorescent lamps	50–100
Mercury vapor lamps	30–160
Tungsten–halogen lamps	17–35
Tungsten-filament incandescent lamps	10–30
Tantalum-filament incandescent lamps	6
Carbon-filament incandescent lamps	3

highly favorable to manufacture on a high production basis. The white light consists of radiation of 10–20% of the energy input, while the remaining energy is used in the production of heat (mostly infrared radiation).

Incandescent lamps come in many different shapes and sizes, including various patterns of filament to produce different effects in the lighting. They are used as much for decoration as for utility.

The low wattage lamps are usually evacuated, while the larger and more prevalent bulbs are filled with a mixture of argon and nitrogen. As these bulbs age, the tungsten evaporates and is deposited on the inside of the glass bulb, giving the black appearance which you may have noticed. When this appears, it is coming close to the burn-out time for the bulb, as the light output has dropped considerably. All lamps deteriorate with time, but incandescent lamps are very pronounced in that respect.

The average incandescent lamp deteriorates to about 80% rating at its expected burn-out point. Also, the average lamp life is based on the number of hours of continuous service which 50% of a test lot attains. Thus, out of 1000 lamps, each rated at 1000 hours, 500 of them will last for 1000 hours. Of the remaining 500, some will fail at less than 500 hours, and some will burn longer than 500 hours.

The incandescent lamp is good for illumination levels up to about 20 or 25 footcandles. From that level and up, it is best to consider the use of the fluorescent lamp. There are many new types, shapes, and sizes on the market recently, making it much easier to find the right one for the particular job required. The best use of fluorescent lamps is for indoor lighting of work areas where a great deal of concentration and fine work is performed. They are not too popular for outdoor use, as the type of light given off often gives an eerie effect. New developments may change this situation, however.

The fluorescent lamp consists of a glass tube, lined with phosphor crystals, filled with argon gas at a very low pressure, and with a few drops of mercury.

By using different types of phosphors, the color of the lighting can be altered. The major ones available are:

Cool White	Approaches outdoor effect
Deluxe Cool White	Good color rendition
Standard Warm White	Warm tint, more natural
Deluxe Warm White	Warm, flattering effect to people
White	General lighting uses
Daylight	Slight blue tint

Fluorescent lights may also be classified by the starting characteristics, as follows:

Preheat Lamps	Slow start, 3 to 5 seconds to produce light
Instant Start	Instant light upon applying electricity
Rapid Start	Combination of above two, 2 to 3 second start up

The glass tubes may be coated or covered to produce various results. For instance, all-weather lamps have a glass jacket around the outside as an insulator. The latest safety feature is a plastic coating, which prevents the glass tube from shattering when it is broken. This is important, as the phosphor lining can cause a bad infection if the jagged glass cuts the skin.

The fluorescent lamp has an energy efficiency of about 22%, which means that about 22% of the applied electrical energy is used to produce light. About 36% of the energy produces infrared rays or energy, and the remaining 42% is used to produce heat. These figures are average for a typical cool white lamp, with some variations to be expected.

The light output depreciates from the beginning, with the greatest depreciation coming in the first 100 hours of operation. From that time until burn-out, the rate of depreciation is fairly steady. As an example, a standard cool white lamp will drop to about 80% of its output rating after about 10,000 hours of on-time. This varies considerably with the types and sizes, as well as the brand of lamp.

There is a table of typical performance data for fluorescent lamps supplied by Phillips Lighting Company in Section D of the Appendix.

One of the typical features of the fluorescent lamp is the requirement for what is known as a "ballast." It is a heavy piece of equipment, filled with asphalt insulation. Just what is its purpose? It serves several functions, as follows.

1. It is a combination auto-transformer and current limiting device, to supply proper voltage and current to the lamp and prevent it from overheating.

2. It supplies the proper amount of current to heat the electrodes, if required, for the particular type of lamp.

3. It provides a method of suppressing radio interference, often associated with fluorescent and neon lights.

Maintenance for fluorescent lighting mostly consists of the following.

1. Changing burned-out bulbs, or bulbs which are depreciated to the extent that the light is objectionable.

2. Changing the ballasts, or checking them periodically for signs of overheating, frayed wires, etc.

3. Cleaning the bulbs, reflectors, and refracting surfaces.

4. Checking the wiring circuits for overheating effects and aging.

5. Changing the starters on those lamps requiring them, such as the Preheat models.

Replacement ballasts should be chosen with care, since they must match the requirements of the lamp. There may be several styles of ballast available for any one lamp, and the wattage consumption may vary with each style. As there may be a large number of ballasts in the average industrial plant, the total wattage consumption may be considerable. So discuss all the characteristics of the lamp with the supplier for his recommendations.

In Section D of the Appendix there is a table of recommended ballast selections for the fluorescent lamps supplied by the Phillips Lighting Company.

Similar considerations should be given to the selection of replacement fluorescent lamps, as well as the HID lamps. There can be quite a substantial difference in light output per watt between the available styles, sizes, and brands. Therefore, it pays to look into all factors involved when making your selection. The nominal size designation is just the starting point for your search.

WARNING! ALL FLOURESCENT AND HID LAMPS ARE POTENTUALY DANGEROUS. BE SURE TO READ AND HEED THE WARNINGS INCLUDED WITH THE LAMPS BY THE MANUFACTURER!

There are service companies in operation which regularly make calls on industrial and commercial establishments, changing the bulbs, and performing the above maintenance functions. Your plant may have such a service, if not the above will serve as a rough guide for your use. A call to the major suppliers of fluorescent lighting equipment will probably get you a fairly complete maintenance instruction manual.

Of more recent design are the next higher types of artificial light producing sources—the High Intensity Discharge (HID) lamp. These are all similar in construction, but utilize different basic chemicals for the source of light.

The heart of this lamp is an arc tube, consisting of a glass tube from which all air has been removed, and which contains electrodes at the ends, with fillings containing either mercury, sodium, or one of several metal halides. We shall start the discussion with the mercury lamps.

The mercury—a few drops is all it takes—inside the arc tube is vaporized when the arc is struck across the ends of the tube. The mercury vapor, with some inert gas added, glows and gives off a blue-white light of very high intensity, approximately 30 to 60 lumens per watt.

The arc tube is itself enclosed in another sealed globe, and the space between the arc tube and the globe is filled with nitrogen, or other inert gas. This protects the inner tube, and the inert gas prolongs the life of the metal wires serving the arc tubes. The outer globe also screens out ultra-violet radiation, which is emitted by the arc tube. Treatment of the outer globe may also result in other effects desired, such as altering the color of the light.

The mercury vapor lamp requires a ballast to obtain the proper voltage and current to start and maintain the arc. When electricity is applied, it takes several minutes for the lamp to reach maximum output. If, after the lamp is in operation, power is interrupted, it will again take another four to seven minutes to reestablish the full brilliance of the lamp. This should be borne in mind when considering placing them on emergency circuits in critical locations.

Mercury vapor lamps are on the market at from 40 to 2000 watts, and they may be expected to last from 12,000 to 24,000 hours, and some even longer. The light output drops fairly rapidly for the first 3000 hours, then the drop-off in output falls at a steady rate to the end. After about 16,000 hours, the average lamp will still have about 85% of its rating available.

The metal halide lamps are very similar to the mercury vapor lamp, but in addition to the mercury and inert gas filling in the arc tube, a metal iodide is added. The resultant light output is much higher than for the mercury vapor lamp, being in the neighborhood of 70–125 lumens per watt. They may be expected to last about 7500 to 15,000 hours. The drop-off in light output is fairly rapid for the first 8,000 hours, at which point, about 85% of rating, the drop-off is less rapid for the additional life of the lamp. The metal halide lamps require a special type of ballast, similar to that used for the mercury vapor lamps.

High Pressure Sodium lamps are the next development in the lamp field. The construction is similar to the mercury vapor lamp, except that the arc tube glass is resistant to the action of the sodium filling. The filling contains mercury, along with inert gas, but the sodium vapor is the main producer of light.

High pressure sodium lamps are conspicuous by their yellow color. They are popular for street, freeway, and driveway lighting. The light output per watt varies from 80 to 140 lumens, and the life is from 12,000 to 24,000 hours. The drop-off with age is steady from installation to about 70% at 20,000 hours.

They require several minutes to reach full light output, and if the electricity is interrupted after full operation is attained, then reestablished immediately, they will relight in one to three minutes.

Ballasts must be of a type made especially for the sodium lamp. These have been covered elsewhere in this chapter.

D. Electrical Lighting Circuits

Now that you have a rudimentary knowledge of the lighting methods and requirements, as well as the different types of lamps available to supply those needs, there only remains the matter of supplying electricity to them. And, of course, it is important that the proper amount of effort and study be applied to see that this is done safely and adequately. That is the purpose of this section.

We are going to assume that a new area has been properly designed with a new lighting system by outside professionals, and the plant's personnel are to provide the electrical service to the new system. You may never have to do this, but it will serve to acquaint you with the basic problems involved.

The first question to be answered, when the problem arises concerning the addition of more lighting, is: in which portion of the existing power wiring should the new circuits be placed?

Assuming that the amount of power required for the new lighting has been determined, the next most important step is to run load tests on the present system to see where the new load can be applied. The lighting, being a single phase load, will have to be wired according to one of the schematics pictured in Fig. 12-1. In Chapter 16 there are wiring diagrams covering the power distribution circuits available for an industrial plant, and some of them show the location of the lighting circuits in those systems.

The most important thing to keep in mind when deciding where to place the new load is that the resultant total load on the system must be equalized across the phases of any three phase system you select. In other words, the current drawn through any two wires of a three phase circuit must be as nearly equal as possible, so that the total current draw on all three phases is nearly equal. It will not be possible for the loads to be equalized exactly, but a close approximation should be attempted. The loads across the various circuits in an industrial plant are changing constantly throughout the working shifts, so this balancing of the loads may take some considerable effort to achieve. But remember, the plant's efficient use of the available electric power is at stake, and consequently, the power bill will suffer if the loads are not properly balanced.

Now that the location of the load within the plant's power system has been decided, Figs. 12-1 and 16-2 will show the various methods of connecting the loads into the switching schematics. Notice that in some cases shown, there is a contact labelled "If Used." This must be checked with the local electrical codes, and in most cases you will probably find that the extra contact is required. The general rule is that the

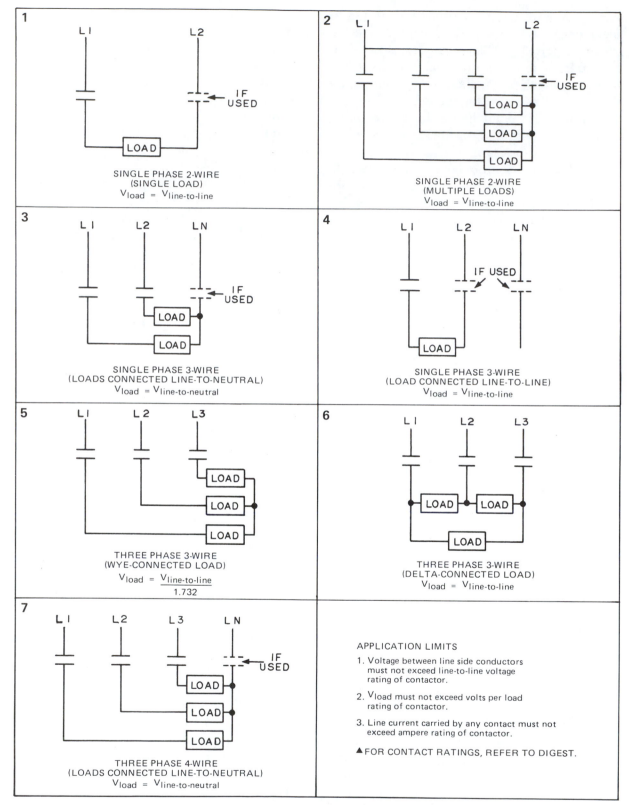

Fig. 12-1. AC Lighting load connections. (Courtesy of Square D Co.)

master circuit controlling any single phase circuit must be a two pole switch, breaking both paths for electricity to reach the load. This is to provide complete isolation when work must be performed on the load side of the circuit. This rule does not always apply to local control switches, such as thermostats, pressure switches, and other automatic switches.

The reason for this rule is that if only one side of the circuit is broken, there is always the possibility of stray or unwanted current coming in from a neutral or ground line which is supposed to be "dead" but isn't. So the life-saving rule for the maintenance man to follow when working on single phase loads is: never work on one unless both lines supplying the load are disconnected and tagged.

Figure 16-8 shows a typical lighting panel board as found in the average industrial plant, in a general purpose enclosure. Figures 12-2 through 12-4 show typical wiring diagrams for normal plant lighting systems—incandescent, fluorescent, and mercury vapor, as supplied by the Phillips Lighting Co. and the General Electric Co.

Unlike a filament lamp, a fluorescent lamp

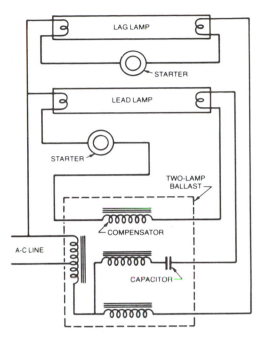

Fig. 12-3. Typical two-lamp high power factor, lead-lag ballast and circuit.

cannot be operated directly from the electric lighting circuit.

In most cases, the lamp would not light up and if it would, once the arc discharge was established between the electrodes of the lamp, the current would rise until the lamp was destroyed.

All fluorescent lamps must have an auxiliary, commonly known as a ballast, to limit the current and to provide the necessary starting voltages. Each lamp requires a ballast specifically designed for its characteristics and for the ser-

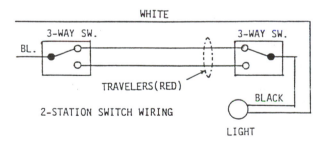

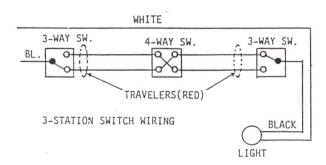

Fig. 12-2. Two and three station switch wiring.

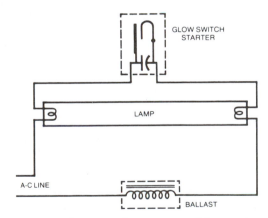

Fig. 12-4. Glow-switch starter and lamp circuit.

vice voltage on which it is to be operated. Two-lamp ballasts are either lead-lag or series-sequence design. In the lead-lag variety, each lamp operates on its own independent circuit—one lead and one lag. Such a ballast provides high power factor (between 90 and 100%) and practically eliminates stroboscopic effect.

In the popular series-sequence design, the lamps start in sequence and, when fully lighted, operate in series. These ballasts also operate at high power factor but do not correct for stroboscopic effect.

All Two-Lamp Preheat ballasts are lead-lag; Slimline and Instant Start are either lead-lag or series and most Two-Lamp Rapid Start ballasts are of the series type.

THE PREHEAT LAMPS

The original fluorescent lamps, developed for the New York World's Fair in 1939 and offered to the public in that year, were of the "preheat" type. The lamps are started by means of a starter and are made with bipin bases to permit the flow of current through the electrode filaments during the preheating cycle before lamp starting (Fig. 12-3).

There are two basic types of starters in use today for starting preheat lamps. The Glow Switch starter shown in Fig. 12-4 is the most popular starter.

A small glowlamp is the heart of a glow-switch starter. In one type, one electrode is a stiff wire and the other is a bimetal strip, both enclosed in a small glass bottle filled with an inert gas such as argon or neon. When a voltage is applied across the lamp, the same voltage is impressed across the starter. This causes a glow discharge and a small current flow between the electrodes. The heating effect of the current causes the bimetal strip to expand and to make contact with the other electrode. This allows the preheating current to flow through the lamp cathodes. For a short time there is enough residual heat in the switch to keep it closed. As the bimetal strip cools, it bends in the other direction, opening the contacts with a resultant high-voltage pulse that should start the lamp.

The thermal-switch starter shown in Fig. 12-5 consists of (1) a small heating element; (2) a bimetal strip which can be held in contact with either part 3 or part 4. The bimetal strip will move, either when a current passes through it or when it is affected by the heater. The heat moves the thermal switch to the open position and the lamp starts. With the lamp operating normally, a small amount of current continues to flow through the heater, but the power consumed is only about 1 watt. Thermal-switch starters are recommended for DC operation and for low temperature starting.

THE SLIMLINE AND INSTANT START LAMPS

These lamps are designed for starterless operation. The ballast provides sufficient voltage to strike the arc instantly. This requires cathodes that will withstand instant starting without preheat. With such a system, a single pin base can be used on the Slimline lamps.

The single pin base design simplifies lamp insertion. To eliminate electrical shock hazard, the pin also acts as a switch breaking the circuit to the ballast when the lamp is removed (Fig. 12-6). The push–pull spring pressure type of lamp holder supports the lamps firmly in place.

Instant Start lamps are similar to Slimline lamps in design and performance except that they are equipped with bipin bases. The two

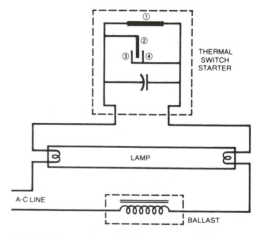

Fig. 12-5. Thermal-switch starter and lamp circuit.

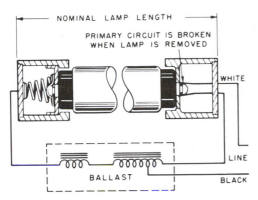

Fig. 12-6. Slimline lamp, holders, and circuit.

pins of each base are connected together electrically so that the primary current to the ballast may pass through the end of the lamp without damaging the electrode. As with Slimline lamps, the lamp pins act as a switch, breaking the circuit to the ballast when the lamp is removed.

Series-sequence type ballasts are used for Slimline and Instant Start lamps (Fig. 12-7). In this ballast circuit, the lamps start in sequence and, when fully lighted, operate in series. In the circuit diagram it will be noted that the auxiliary winding supplies the voltage to start Lamp No. 1. (approx. 680 volts).

Before the first lamp lights, the voltage of the auxiliary winding subtracts from the primary and secondary voltages, thus resulting in insufficient starting voltage for lamp No. 2. However, when lamp No. 1 lights, the current flow through the capacitor shifts the phase relationship between the auxiliary winding and the

secondary winding, causing the two voltages to add. This causes the voltage to be sufficient to start lamp No. 2. The lamps then operate in series, with the auxiliary winding contributing nothing to the circuit.

THE RAPID START LAMPS

Ballasts for Rapid Start circuits have separate windings to provide continuous heating voltage for the lamp cathodes as shown in Fig. 12-8. Unlike the preheat lamp that has no cathode heating circuit after the arc strikes, the Rapid Start lamp is provided with a small heating current even when the lamp is burning. Under normal conditions, the Rapid Start ballast will start the lamps in less than one second.

Two-Lamp Rapid Start ballasts start lamps in sequence and then operate them in series. After the circuit is turned on, the first operation is the heating of the cathodes to aid in starting the lamp by reducing the starting voltage requirements. The capacitor shunted across lamp No. 2 aids in starting lamp No. 1 first by momentarily applying nearly all of the ballast secondary voltage across lamp No. 1. Since the voltage drop across this lamp after starting is very low, practically all of the ballast voltage is available to start lamp No. 2. The two lamps then run in series with rapidly increasing current until stable operation at rated current is achieved. It is essential that proper cathode heat be maintained during lamp operation to ensure normal lamp life.

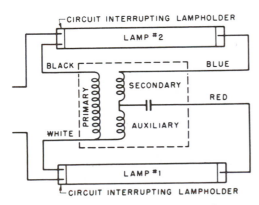

Fig. 12-7. Typical Two-Lamp Series Sequence instant Start circuit.

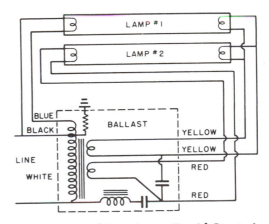

Fig. 12-8. Typical Two-Lamp Rapid Start circuit.

All lamps that operate on Rapid Start type ballasts must be mounted within 0.5 in. of a grounded metal strip at least one inch wide extending the full length of the lamp.

THE TRIGGER START CIRCUIT

The trigger start circuit is sometimes used for operating preheat fluorescent lamps up to 32 watts in size. This circuit was developed prior to the rapid start circuit and is quite similar in that it provides continuous heating of the cathodes and does not require a starter.

THE ELECTRONIC CIRCUIT

The electronic circuit converts the normal 60 hertz circuit voltage to high frequency (approximately 25 kilohertz). The high frequency increases the overall system efficiency by increasing the lamp efficacy and by reducing the watts lost in the ballast. The circuit can be designed to operate the lamps in the Rapid Start or Instant Start mode. Note Fig. 12-9.

DIMMING AND FLASHING

Standard Rapid Start lamps can be dimmed readily and flashed, but only when operated on ballasts or circuits specifically designed for these applications. Econ-o-watt lamps are not recommended. Dimming ballasts have continuous cathode heating circuits which supply full voltage to the lamp electrodes at all times. Special dimmer controls change the lamp arc current by means of a thryatron, adjustable voltage transformer, adjustable reactor or electronic circuit. Note Fig. 12-10.

Some dimming controls operate with standard ballasts and can be retrofitted into existing installations. They are installed in the branch circuit feeding a group of lamps.

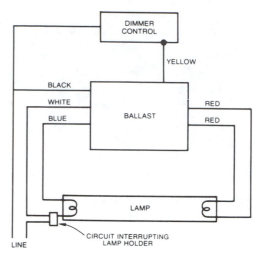

Fig. 12-10. Typical dimming circuit for fluorescent lamps.

Sensors are available that will continuously monitor the light level in a space. They operate with special controls that adjust the power to the lamps and thereby maintain a preset light level. These circuits permit a wide range of light intensity, depending on the circuit, and do not adversely affect lamp life with normal dimming service. For all dimming ballasts and controls, refer to the manufacturer for circuitry and performance data.

The basic problem in installing any of the fluorescent, mercury vapor, or sodium vapor lamps is the proper treatment of the ballast. These are important pieces of apparatus, and should be very carefully installed. The most important thing to remember is that they produce heat, and thus require an atmosphere which will handle that heat, and get rid of it before it builds up to such an extent to cause it to burn up. The first indication that it is running too hot is usually an appearance of black insulation material dripping down from it. When this happens, it is probably too late, and it must be replaced immediately, and the cause of overheating remedied.

Because ballasts produce heat, they consume electricity in doing it. They require an additional 20% of power over and above the power required for the lamps. Remember this when adding up the total load on a new circuit.

Most of the lamps described in this chapter are available for installation in areas where

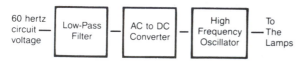

Fig. 12-9. Block diagram of a high-frequency electronic ballast.

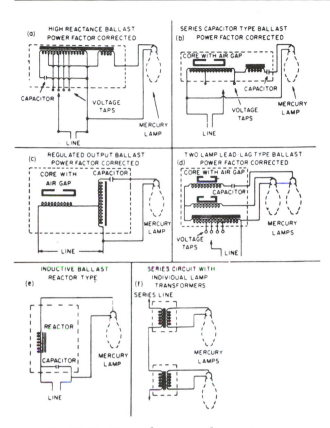

Fig. 12-11. Typical mercury lamp circuits.

there is a danger of fire or explosion. To assist the plant engineer, and others, in determining the correct fixture to use, as well as the proper wiring and electrical gear in general, the **National Electric Code**® was devised. This code classifies the various atmospheres and conditions under which electrical power is expected to operate with safety. Following this code will not definitely assure freedom from danger, but it will give an exceedingly high degree of freedom from worry. It is still up to the plant maintenance engineer to properly apply the code, and to maintain a reasonable degree of proper maintenance of the equipment being used in hazardous locations, as well as all other normal applications of electrical power in the plant.

Articles 500 through 517 of the **National Electric Code**® cover the definition, and provide the basic ground rules for the application and installation of electrical equipment and lighting fixtures in hazardous locations. The insuring agency for the plant may have already established the areas in which special attention

must be given to prevent fires and explosions. There may be in the engineering files one or more drawings on which the plant has been divided into areas of hazardous and nonhazardous atmospheres. If so, then it may be badly outdated, and should be brought up to current codes and plant conditions. If it does not exist, then now is the time for one to be produced, with the help of the **National Electric Code**®.

To assist the plant electrical department, we give here brief excerpts from Article 500 of the **Code**, with the admonition that it is only a guide, and the latest copy of the **Code** should be followed completely.

Class I locations are those in which flammable gases or vapors are or may be present in the air in quantities sufficient to produce explosive or ignitible mixtures. In general, this includes the following hazardous materials:

Group A: Atmospheres containing acetylene.

Group B: Atmospheres containing, fuel, and combustible process gases containing more than 30% hydrogen by volume, or gases or vapors of equivalent hazard such as butadiene, ethylene oxide, propylene oxide, and acrolein.

Group D: Atmospheres such as acetone, ammonia, benzene, butane, ethanol, gasoline, hexane, methenol, methane, natural gas, naphtha, propane, or gases or vapors of equivalent hazard.

Group C: Atmospheres such as cyclopropane, ethyl ether, ethylene, or gases or vapors of equivalent hazard.

Class II locations are those that are hazardous because of the presence of combustible dust. In general, this includes the following hazardous materials:

Group E: Atmospheres containing combustible metal dusts, or other combustible dusts of similarly hazardous characteristics having resistivity of less than 100 ohm-centimeter.

Group F: Atmospheres containing carbon black, charcoal, coal, or coke dusts that have more than 8% total volatile material or atmospheres containing these dusts sensitized by other materials so that they present an explosion hazard, and having resistivity greater than 100 ohm-centimeter but equal to or less than 10^8 ohm-centimeter.

Group G: Atmospheres containing combustible dusts having resistivity of 10^8 ohm-centimeter or greater.

Class III locations are those that are hazardous because of the presence of easily ignitible fibers or flyings, but in which such fibers or flyings are not likely to be in suspension in the air in quantities sufficient to produce ignitible mixtures.

All three classes listed previously are further divided into two divisions, generally as follows:

Division 1: An environment in which, as a normal operating procedure the hazardous materials listed in the appropriate groups are constantly present and constitute a danger.

Division 2: An environment in which the hazardous materials listed in Groups A through G are not normally present in the atmosphere, but which may under certain conditions contaminate the area, and produce a Division 1 condition.

In actual practice, the establishment of the hazardous areas in the plant is done by an electrical engineer in conjunction with the insuring agency. Application of the proper equipment, both electrical and illumination, is the responsibility of the plant facility staff, on a routine daily basis, after the plant is in operation. Chapter five of the **National Electric Code**® covers the matter of hazardous areas in complete detail, with the required practices and materials.

E. Maintenance of Lighting Systems

By now, most of the requirements of maintenance of the plant's lighting equipment should be apparent to the average reader. However,

we shall list the major items which should not be overlooked, and these will be of a general nature covering all lighting types. We recommend that the plant maintenance schedule be designed around the specific types and brands of lamp fixtures in the plant. The fixture supplier will provide adequate recommendations when asked to do so.

In replacing fluorescent lamps, remember to be very careful when handling them. The coating inside the bulb is highly toxic and can cause blood poisoning if the lamp is broken and the skin is scratched with the broken edges. Fluorescent lamps are now available with plastic coatings on the outside to prevent shattering when broken, and these are much safer and well worth the extra cost.

When cleaning or servicing lamps, be sure the lamps have been dark long enough to cool down before touching. And above all, do not use a wet or even a damp rag to clean the lamp while it is still hot. The result may be a violent explosion, with glass splinters flying into your face.

One of the frequent accidents which damages lamp fixtures is the moving of scaffolding, ladders, or other long, awkward apparatus around the shop. If this is unavoidable, then the lamp fixtures should be protected with wire cages, or such methods.

Section D in the Appendix contains complete trouble shooting and maintenance instructions for fluorescent lamps published by Phillips Lighting Company in their bulletin, *Guide To Fluorescent Lamps.*

A complete set of servicing instructions for fluorescent and high intensity discharge lighting systems may also be obtained from the local Sylvania representative or servicing center. Ask for their Engineering Bulletin 0-330 and 0-345.

Fluorescent lamps generally require only changing of the lamps when they become too dim to be effective, or when they start to flicker. The flicker may be caused by the lamps becoming loose in their sockets, dirt on the terminals, or by a failing starter. In the latter case, the starter will have to be replaced by one with the same wattage rating as the lamps.

Very seldom will the ballast cause problems, unless the fixture is in the wrong location for the style, or the wiring is faulty. Replacement ballasts are available for all the lamps requiring them, whether fluorescent, mercury, or sodium vapor type.

One thing that is very seldom done often enough is to clean the reflector of the lamp. Often that is all that is needed to restore most of the lamp's original light output. The maintenance engineer should make it a standing rule that all lamp fixtures receive a good cleaning each time servicing of any kind is performed upon them. In dirty locations, such as in a foundry or polishing room, the lamps will require cleaning much oftener, of course.

Turning a lamp on causes a flow of air past the ballast, if the fixture is of the open type. This air carries dirt and dust, which can cause problems in the attached wiring circuit. Lamps for this type of service may be obtained with filters which clean the air before it enters the ballast chamber. These filters often are of the activated carbon type, and require checking and replacing, much as the air filter on the automobile engine.

There are commercial service firms available who will perform the above maintenance functions on the plant's lighting equipment. This is a very good thing to consider, especially for the small plant with limited maintenance personnel available. These service firms can be used when a new plant is being put into operation, while the maintenance crews are still new and inexperienced. After the plant is in full and normal operation, the service can be discontinued, after the plant's personnel have learned the duties and functions required.

Such service firms usually follow a well-engineered method of replacing the lamps in the plant at the proper lamp life to assure maximum lighting intensity for the power consumption. They are very efficient in their methods, and can be relied upon to work out the best program for the plant.

Aside from the items mentioned above, the remainder of the maintenance procedures follow those regularly followed for the normal electric circuit. These items include such things as:

a) cleaning contact points of switches and breakers

b) replacing fuses

c) Checking wiring and insulation

d) Recording current draw on all circuits to check condition of wiring

e) Replacing worn switches and breakers, or any other equipment which has received much wear and tear.

The latest tool for testing fluorescent and High Intensity Discharge lamps is shown in Fig. 12-12, known as the LT-277 Lamp Tester, made by the BEHA Division of Greenlee Tool Company, 4455 Boeing Dr., Rockford IL 61109. We include it here as we feel it is one of the best maintenance tools available for the electrical maintenance worker. Like all electrical instruments, it must be used and cared for in a proper manner.

LAMP TESTER

The LT-277 system provides a very simple method of troubleshooting "Gas Lamp" lighting fixtures. This diagnostic, troubleshooting

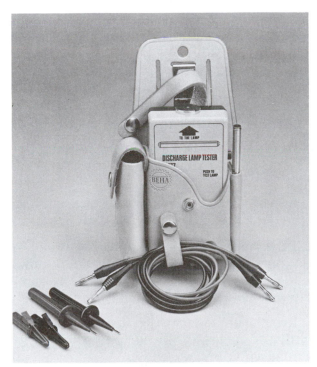

Fig. 12-12. Fluorescent and HID Lamp Tester.

tool enables the user to first test the bulb for electrical integrity, then, depending on the results of the bulb test, it is possible to test:

1. Fluorescent Fixture—ballast, input voltage, output voltage and wiring continuity

2. H.I.D. Fixture—capacitor, igniter, ballast, input voltage, and wiring continuity.

TEST PROCEDURES UTILIZING THE LT-277 TEST LEADS— OVERVIEW

To utilize the LT-277 leads, simply insert the shrouded banana probes in their respective receptacles in the bottom of the LT-277 and decide which of the accessories you need: the push-on probes or the push-on alligator clips depending upon the application. The LT-277 is a unique voltage and continuity testing tool. The sound frequency is directly proportional to the resistance being tested. Thus, the higher the tone produced, the lower the resistance of the circuit being tested. Conversely, the lower the tone, the higher the resistance. The LT-277 is capable of checking a range of resistance from 0 to 250,000 ohms! The human ear is capable of detecting a difference in resistance as small as 30 ohms. In most cases, using the LT-277 is faster than using a meter . . . with the added advantage of allowing the user to free the eyes to do other important tasks while letting the ears determine the results of the testing.

Without the need to flip switches or push buttons, the LT-277 will detect AC or DC voltages up to 500 VOLTS by sound and by L.E.D. on the face of the unit.

AC Voltage Indication—strong buzz or intermittent sound. L.E.D. glows strongly.

DC Voltage Indication—strong steady tone in reverse polarity position only. L.E.D. glows strongly. Correct polarity will, in turn, provide no sound.

NOTE—THE LT-277 WILL NOT BE DAMAGED WHEN CONTACTING THE LEADS TO VOLTAGES UNDER 500 VOLTS!!!

BULB TESTING PROCEDURES

Fluorescent Bulbs—When diagnosing any system, it is necessary to visually assess the situation. Look for obvious abnormalities such as no light, slow starting, blinking on and off, low light output, "raccoon tails," flickering, ends only lighted, and end blackening.

USING THE LT-277 LAMP TESTER ON FLUORESCENT TUBES

1. Fluorescent Tube in Fixture—With the power OFF, touch the emitter antenna to the tube and press the button on the lamp tester. (In the case of a power saving type tube, it is necessary to touch the metal end of the tube.) If the tube does not light or lights only at the point the emitter is touching the tube, the tube has definitely failed! If the tube lights, turn the power on and repeat the test. If the tube now lights, and continues to remain on after the LT-277 has been removed, the tube is bad because one or more ends have failed. The LT-277 is, in effect, "jump starting" the tube. (The elements are "open.") This may be verified by testing the continuity of the resistance elements at each end with the LT-277 leads as outlined later in this manual. Check to be certain the tube base pins are seated properly in their respective fixture sockets!

2. Fluorescent Tube Out of Fixture—Touch the LT-277 emitter antenna to the tube or the metal end of the tube and press the button. If the tube does not light, it is faulty and must be replaced. (Remember, a power saving tube can only be tested by contacting the metal end.) If the tube glows, follow up by testing the continuity of the ends of the tube with the LT-277 test leads. NOTE: If the tube lights only in an area of close proximity to the contact point, the tube has lost the properties of the gas in the tube and should be replaced.

If excessive raccoon tails are present, oxygen has entered the tube and destroyed the integrity of the gas in the tube.

TESTING FLUORESCENT BALLASTS

Rapid Start—2 Tube Ballast

Using the LT-277 leads and the push-on alligator clips, contact either one of the test leads to the black wire and the other to the white wire. Listen for the sound of perfect continuity.

Any other sound is an indication of failure of the primary coil. Remove either one of the leads and attach it to either one of the yellow secondary wires. Any sound at all is an indication of insulation failure.

Remove the lead from the yellow wire and attach it to one of the blue wires. (The blue wires are tapped off the primary coil.) There should be an indication of perfect continuity!

Remove the lead from the blue wire and attach it to one of the red wires. Listen for the "chirp" of the capacitor. If it does not charge, or if a continuous sound is heard, the ballast has failed!

There are various other types of ballasts that may be checked in a similar manner as just discussed. Use the principles as outlined in previous paragraphs to develop a test procedure.

NEON AND LOW PRESSURE SODIUM LAMPS

A neon lamp may be tested by simply touching the LT-277 emitter to the bulb or base and pressing the button. If the bulb fails to glow, it is bad.

MERCURY VAPOR AND METAL HALIDE LAMPS

Touch the LT-277 emitter to the bulb base and press the button. (Be certain not to touch the metal base with your hands since doing so will diminish the results of the test.) The arc tube should glow steadily. A bad bulb will not glow at all!

When diagnosing the situation with the lamp in the fixture and power on, and before applying the LT-277, note whether or not the bulb is cycling on and off or appears unstable. A bulb exhibiting these characteristics should be replaced. Be certain the fixture is not being subjected to unusual or excessive heat which could cause the thermostatic "switch" inside the bulb to open and close repeatedly, thus effecting the same results as a defective bulb.

HIGH PRESSURE SODIUM LAMPS

Before proceeding with this test, please check visually for obvious defects such as a brown coating on the inside wall of the arc tube, white flaky particles loose inside the bulb, oxidized internal mount parts, or a white glass area just above the base. Visual inspection will make the troubleshooting effort much simpler.

Touch the LT-277 emitter to the bulb base and press the button. (Be certain not to touch the metal base with your hands since doing so will diminish the results of the test.) Should the arc tube light appear as a clean blue "line" the full length of the tube, with little or no "yellow" at either end, the bulb may be considered as good!

If the arc tube does not light at all, be certain the contact point is clean and try the test again. A no-light situation indicates a bad bulb. (Be certain the batteries are good!)

TESTING HID FIXTURES

Depending on the results of the lamp test, it may be necessary to test the componentry of the fixture. A simple fixture test that works on the fixtures utilizing "regulator ballasts" is to use the LT-277 probes in the following manner.

UNDER NO CIRCUMSTANCES SHOULD THE LT-277 BE USED TO TEST THE HIGH VOLTAGE OUTPUT SIDE OF AN HID FIXTURE.

1. Touch either of the probes to the center contact in the empty bulb socket.

2. Touch the other probe to the socket threads.

3. Listen for a rapidly decreasing sound dwindling to nothing. A constant sound indicates a faulty fixture. No sound at all also indicates a faulty fixture.

NOTE: This test should be monitored for each fixture brand in use to be certain of test validity!!!

HID FIXTURE—COMPONENT TESTING

Capacitor—Use the leads across the capacitor terminals and listen for the sound of "charging." The sound will start at a high pitch and rapidly die off to nothing if the capacitor has charged. Reverse the leads and test again. If charging is indicated in both positions, the capacitor should be considered as good. Should a continuous sound be heard, the capacitor is shorted. If no sound is present, the capacitor is "open" and should also be considered as defective.

Igniter (Starter)—Use the alligator clips across the white and red leads. A very quick capacitor charge should be heard or seen! Any other sound across any other combination of wires should indicate a defective igniter. Note: Certain types of specialized electronic starter boards cannot be tested in this manner. In case of encountering this type of starter (a relatively unusual case), use the process of elimination to determine the faulty component!

Ballast—The ballast is actually a specialized transformer capable of producing extremely high voltages. Heat is it's enemy! Visually inspect the ballast for obvious signs of overheating. A quick check can be made on a de-energized ballast with the LT-277. Use the alligator clamps and check each pair of leads in succession. Virtually perfect continuity should be heard—the same sound as one would hear when contacting the LT-277 leads directly together. Should all other components in the fixture check out as good and the ballast be suspected, there is little or no need for further

testing. However, if further verification is required, a "true RMS" multimeter may be used to check the open circuit voltage and short circuit current values according the manufacturer's specifications.

PHOTOCELL TESTING PROCEDURE FOR FIXTURES CONTROLLED BY LIGHT

To test for a defective photocell, cover the "eye" completely with a piece of electrical tape allowing absolutely no light to enter. Allow time for the contacts to close. Use the LT-277 leads (with alligator clips attached) across the photocell contacts to determine if the contacts have actually closed. Be certain the power to the photocell is turned off! Next, remove the tape and listen for the "sound" to cease. If the sound continues, the photocell is defective!

BATTERY REPLACEMENT

The batteries must be tested regularly since the lamp tester will not light the arc tubes brightly enough for a good test (or not at all) if the batteries are weak! The batteries may be tested by touching the leads together. A strong tone and light are an indication that the batteries are functioning properly.

The LT-277 is powered by 6 AA alkaline batteries. It is recommended that they be replaced by alkaline batteries—type AA. Life expectancy is approximately one hour of continuous usage of the bulb tester. To replace the batteries, remove the four phillips head screws in the back of the unit and very carefully remove the cover.

BE CAREFUL NOT TO BREAK THE WIRES LEADING TO THE SWITCH AND THE INDICATOR LIGHT MOUNTED IN THE COVER!

Remove the battery case by unsnapping the connector. Note position of the batteries! Polarity is important! Replacement is the reverse of removal.

The case of the LT-277 is a very durable polymer material; however, take care to avoid dropping it to prevent damage!

We have left out much available material on the selection and application of the vast number of lamps and fixtures available, as the average plant electrician will not be involved in all phases of plant lighting. However, should more data and information be desired, then we refer the reader to the various equipment suppliers, and to the listings in the Bibliography for the Illuminating Engineering Society of North America.

CHAPTER 13
DIRECT CURRENT MACHINES

A. Introduction

In Chapter 7 we explained the general principles of generating an emf with a rotating machine which produced an alternating current.

In this chapter we shall cover the generation of a direct current and then describe the motors which utilize that direct current, or DC. As you shall see later, the major differences are in the method of taking the power off of the generator, and in the resultant wave form of the emf produced. The AC generator produces a wave form consisting of alternating pulses of emf traveling in opposite directions through the circuit. The DC wave form will be an undulating, smooth pulse in one direction only, resembling the waves on the surface of water.

There are also differences in the methods of controlling the input to the motors utilizing the two wave forms of electric power, and in this chapter we shall describe those used for controlling DC motors.

PRINCIPLES OF OPERATION

A generator is a machine which converts mechanical energy into electrical energy. This is done by rotating an armature, which contains and moves conductors, through a magnetic field, thus inducing an emf in the moving conductors. In any generator, a relative motion between the conductors and the magnetic field must always exist by the application of a constant mechanical force or twist on the shaft. In the direct current generator, the magnetic lines of force (called the field) are stationary. The armature, which contains the conductors, rotates through the stationary field.

The voltage generated by the basic DC generator varies from zero to its maximum value twice for each revolution of the loop. This variation of DC voltage is called ripple, and may be reduced by using more loops, or coils, as shown in Fig. 13-1. As the number of loops is increased, the variation between maximum and minimum values is decreased, and the output voltage of the generator approaches a steady DC value.

B. Major Components of DC Generators

The principal components of a DC generator are the armature, the commutator, the field poles, the brushes and brush rigging, the yoke or frame, and the end bells or end frames.

a. ARMATURE: The armature (Fig. 13-2) is the structure upon which are mounted the coils which cut the magnetic lines of force. It is fixed on a shaft which is suspended at each end of the machine by bearings set in the end bells or end frames. The armature core is circular in cross section and is built

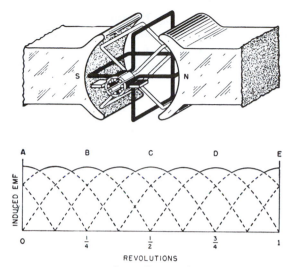

Fig. 13-1. Output voltage from a four-coil armature.

up from sheets of soft iron. The circumferential edge of the laminated core is slotted in order to receive the coil windings. The windings are held in place and in their slots by wooden or fiber wedges. Steel bands are sometimes wrapped around the completed armature and, with the wedges, hold the coil windings in place.

b. COMMUTATOR: The commutator is that component of the generator which rectifies the generated alternating current to provide direct current output and connects the stationary output terminals to the rotating armature. A typical commutator consists of commutator bars, which are wedged-shaped segments of hard-drawn copper, insulated from each other by thin strips of mica.

c. BRUSHES: Brushes make contact with the commutator bars, collect the current generated by the armature coils, and, through the brush holders, pass the current to the main terminals. As the commutator bars are insulated from each other, each set of brushes, as it makes contact with the commutator bars, collects current of the same polarity, resulting in a continuous flow of direct current. The finer the division of the commutator bars, the less ripples will be present, and the smoother will be the flow of the DC output. See Fig. 13-3.

The brushes of a generator are the points of contact between the external circuit conductors and the commutator. These points "brush" the commutator in such a way as to take off the generated emf. Several types of brushes may be used, depending on the application. It is of utmost importance that replacement brushes are always of the same type as the original brushes on the equipment. The types of brushes are as follows.

Brushes are usually made of high grade carbon. These brushes are primarily restricted in use to low-speed machines with low current densities (amperes per unit area of brush face) and where economy in brush cost is a major factor.

Electrographite brushes are made from carbon which has been processed at high temperatures in an electric graphitizing furnace. This treatment lowers its hardness and increases its electrical and thermal conductivity and toughness. This results in a brush that has low fric-

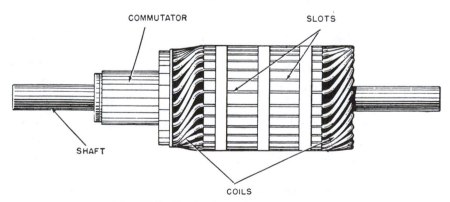

Fig. 13-2. Basic drum type armature.

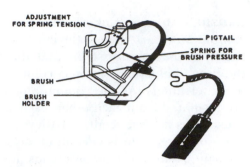

Fig. 13-3. Commutator brush assembly.

tion, great resilience (hugs commutator without bouncing), and high current capacity, and that is nonabrasive and cool running.

Natural graphite brushes can be identified by their silvery appearance and soft flaky structure. They cannot be made as mechanically strong as the carbon or the electrographite brush. They are also more prone to selective action and shunt burning at high current densities. Their application is therefore rather restricted; but with proper allowance, these brushes can give long brush life with minimum maintenance.

Copper-graphite brushes are made from a mixture of powdered copper and graphite pressed together and baked at relatively low temperatures. They are primarily used on rings and commutators where high current densities are being carried.

A flexible braided-copper conductor, commonly called a pigtail, connects each brush to the external circuit. The brush rigging (Fig. 13-3) consists of brushes set in brush holders fastened to a rocker arm which in turn is connected to the yoke or frame of the generator. The brush holders hold the brushes in place as the brushes ride over the surface of the commutator. Each brush is free to slide up and down in its holder so that it may follow irregularities in the surface of the commutator. Each brush is also insulated from its holder. A spring on each brush holder forces each brush to bear on the commutator with from $1\frac{1}{2}$ to 2 pounds of pressure for every square inch of brush surface riding on the commutator. These springs are usually mounted so that the brush pressure is adjustable. The rocker arm to which the brush holders are fastened permits shifting the brush

positions about the commutator without changing the relative position of the brushes.

    d. FIELD POLES: Field poles are required to produce the magnetic field, or flux, which passes through the conductors of the armature. The minimum number of field poles required to complete the magnetic circuit is two—a north pole and a south pole. Most commercial generators are made with four or more field poles, depending on the speed of the generator. The slower the generator, the more poles are needed to produce the same output. See Fig. 13-4.

C. Types of DC Generators

DC generators are classified by the method of supplying excitation current to the field coils. The two major classifications are separately excited and self-excited generators. Self-excited generators are further classified by the method of connecting the field coils, as series-connected, shunt-connected, and compound-connected generators.

A DC generator which has its field supplied by another generator, batteries, or some other outside source is referred to as a separately excited generator (Fig. 13-5). When operated at

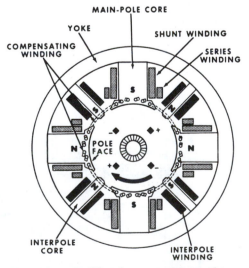

Fig. 13-4. Field poles on a DC machine.

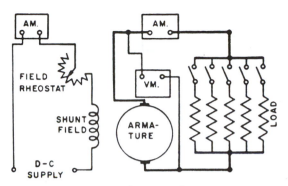

Fig. 13-5. Separately excited DC generator.

constant speed with constant field excitation but not supplying current, the terminal voltage of this type of generator will equal the generated voltage. When the unit is delivering current, the terminal voltage will be less than the generated voltage.

When all of the windings are connected directly in series with the armature, the generator is series-connected (Fig. 13-6). At no load, the only voltage present is that due to the cutting of the flux established by residual magnetism. (Residual magnetism is that which is retained by the poles of a generator when it is not in operation.) However, as the load is applied, the current through the field coil increases the flux and, therefore, the generated voltage. Since the terminal voltage of series generators varies under changing load conditions, they are generally connected in a circuit that demands constant current. When so used, they are sometimes referred to as constant-current generators, despite the fact they do not in themselves tend to maintain a constant current.

When the field windings are connected in parallel with the armature, the generator is

shunt-connected (Fig. 13-7). It is advantageous to use a shunt generator in place of a separately excited or a series generator, where a constant voltage at varying load is required. Shunt generators are readily adaptable to applications where the speed of the prime mover cannot be held constant, as in the aircraft and automotive fields. When so used, it is not desirable to have a constantly fluctuating terminal voltage. It is necessary, therefore, to control field current by varying the shunt field resistance to compensate for changes in speed of the prime mover.

Terminal voltages associated with series and shunt-connected generators vary in opposite directions, with load-series connections increasing and shunt connections decreasing slightly with an increase in load. Thus, as shown in Fig. 13-8, if both a series and a shunt field were included in the same unit, it would be possible, by proper design of the respective fields, to obtain a generator with a voltage-load characteristic somewhere between that of either previous type.

If the turns of the series field are more in number than is necessary to give approximately the same voltage at all loads, the generator is overcompounded. Thus, the terminal voltage at full load will be higher than the no-load voltage. This is desirable where the power must be transmitted some distance. The rise in generated voltage compensates for the drop in the transmission line.

If the relationship between the turns of the series and shunt fields is such that the terminal voltage is approximately the same over the entire load range, the unit is flat-compounded.

If the series field is wound with so few turns

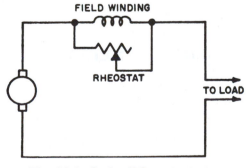

Fig. 13-6. Series DC generator.

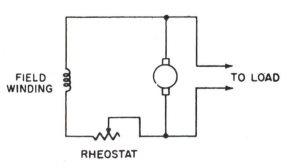

Fig. 13-7. Shunt DC generator.

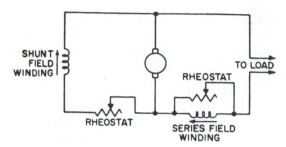

① CUMULATIVE-COMPOUNDED CONNECTIONS.

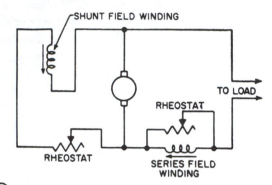

② DIFFERENTIALLY COMPOUNDED CONNECTIONS.

Fig. 13-8. Compound-wound DC generator.

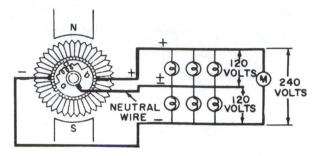

Fig. 13-9. Three-wire generator.

unbalanced current in the two circuits. The chief advantage of the three-wire system is a saving in copper, because the neutral wire carries less than 25 percent of the rated current output of the generator. Therefore, it can be much smaller than either of the two outer wires.

This is the Edison DC system used from about 1890 to 1940 in many public buildings and local compact distribution grids.

D. Control of DC Generators

Generally, a DC generator is controlled by a variable resistance called a rheostat, after the generator is brought up to proper speed by the prime mover. The rheostat may be manually or automatically operated. The adjustment of the rheostat controls the amount of exciter current fed to the field coils. Metering requires the use of a DC voltmeter and ammeter of appropriate ranges in the generator output circuit. Matched sets of shunt-wound or compound-wound generators with series-field equalizer connections are used for parallel operation. Precautions must be observed when connecting the machines to generator buses.

SERIES GENERATOR

The series generator may be classified as a constant-current generator and, as such, may be used to supply series motors, series arc-lighting systems, and voltage boosting on long DC feeders. The voltage increases with load since the load current provides the necessary

that it does not compensate entirely for the voltage drop associated with the shunt field, the generator is undercompounded. In this type, the voltage at full load is less than the no-load voltage.

With this type of generator, the terminal voltage decreases rapidly as the load increases. Undercompounded generators are used in applications where a short might occur, such as in welding machines.

The three-wire generator develops 240 volts across the armature terminals. The generator is arranged so a third wire, or neutral wire, is brought out from a point midway in potential between the positive and negative terminals. This provides for a lead at half generator voltage.

The midpoint ("c," Fig. 13-9) has a potential midway between the potential of the brushes connected to the outside wires. If the load taken from one side of the circuit is equal to the load taken from the other, no direct current will flow in the neutral wire nor through the coil. But if the loads are unbalanced, the neutral wire and coil will carry a direct current equal to the difference in currents, or the amount of

SHUNT MOTORS

In a shunt motor, the field is across the line or in parallel with the armature (Fig. 13-13). The rheostat is used for speed control. The field current stays the same, regardless of changes in armature current. Therefore, when the armature current is doubled, the torque is doubled. The speed of a shunt motor changes very little with change of load, the speed increasing when the load decreases. The characteristic of the shunt motor is an almost constant speed for all reasonable loads.

COMPOUND MOTORS

Compound motors differ from the stabilized shunt types by having a more predominant series field. Like compound generators, compound motors can be divided into two classes, differential and cumulative, depending on the connection of the series field in relation to the shunt field.

Differential-Compound Motors: A diagram of a differential-compound motor is shown in Fig. 13-14. In this type, the series field opposes the

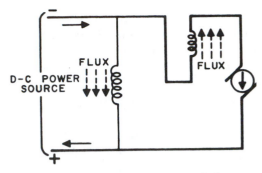

Fig. 13-14. Differential-compounded motor connections.

connected shunt field. Therefore, this motor operates at practically a constant speed. As the load increases, the armature current increases to provide more torque.

Cumulative-Compound Motor: The cumulative-compound motor diagrammed in Fig. 13-15 is connected so that its series and shunt fields aid each other. From this comes the name—cumulative-compound motor. A motor thus connected will have a very strong starting torque, but poor speed regulation. Motors of this type are used for machinery where speed regulation is not necessary, but where great torque is desired to overcome sudden application of heavy loads.

APPLICATION DATA

The several different types of DC motors that have just been discussed have varied particular

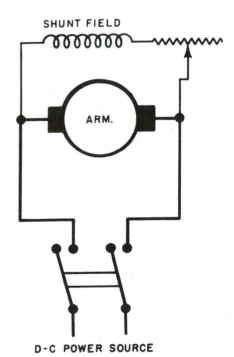

Fig. 13-13. Shunt DC motor connections.

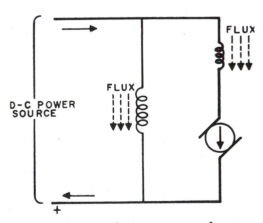

Fig. 13-15. Cumulative-compound motor connections.

uses depending on their construction. Continued use of any DC motor depends on the kind of load to be carried, the speed, and the torque. DC motors not adapted to the load or of improper rating should not be used even for short periods due to possible damage by overheating, inefficient operation, or improper speed control. The location in which the motor is to be operated is also a factor to consider when selecting a machine for a particular job. A dusty atmosphere requires an enclosed motor; high temperatures necessitate a motor with special insulation, and so on.

See Section C in the Appendix for a convenient selection table for DC motors.

F. DC Motor Controllers

A controller is a device for regulating the operation of electrical equipment. A controller for electric motors is simply a mechanism which conveniently and safely performs several or all of the following functions: connection to the power line, limitation of starting current, control of acceleration, control of speed, and disconnection from the power line. Because of this functional variety in controllers, they can be classified in several ways.

MANUAL

The manual type controller is one having all of its basic functions performed by hand. The basic functions are usually line closing, acceleration, retardation, and reversing. Manual control permits regulation of machines from only one position and is limited in the size and capacity of the equipment that can be so controlled, up to about 2 hp.

SEMIMAGNETIC

This controller has part of its basic function performed by electromagnets, and part by other means (Fig. 13-16).

FULL MAGNETIC

The full magnetic controller performs all of its basic functions by electromagnets. The power circuits to the motor are closed and opened by magnetic contactors (Fig. 17-1). The contactors are controlled by a pilot device which has small current capacity. The pilot device may be manually operated, by push-button or master switch, or it may be automatically operated by a float switch or thermostat. Magnetic controllers make it possible to control motors automatically. This has advantages over the manual types. For example, the operator may accelerate the motor too rapidly with the manual type controller, with the result that the motor may take excessive current. This causes fuses to blow or circuit breakers to open. Furthermore, the starting resistance or resistances may burn out. With automatic controllers, the starting resistances may be cut out at the maximum safe rate by magnetically operated contacts. The operator need only press a

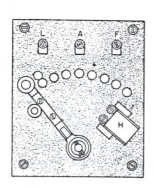
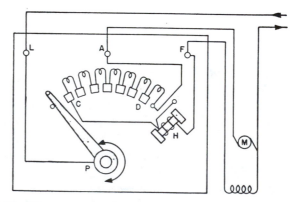

Fig. 13-16. Motor starting box.

button and the electromagnetic relays start the motor and bring it up to speed automatically in the proper time sequence. The motor is stopped by merely pressing the stop button. Magnetic controllers are used principally for:

a) smooth acceleration, retardation, and reversing of motors without damage

b) control of a motor from one or more stations

c) automatic control where an attendant is not present

d) operation of high-voltage equipment

e) conservation of space, by locating the controller in an out-of-the-way place and operating it by a pilot device.

These types of controllers may be further divided into general classes; starters and speed regulators. The starter is designed for accelerating a motor to normal speed in one direction of rotation. (If it is designed for starting a motor in either direction of rotation, it is usually called a reversing controller.) The speed regulator is designed for operating a motor at a speed either below or above normal. There is a more complete description of the magnetic controller in Chapter 17.

FACEPLATE CONTROLLERS

The difference in appearance between a faceplate starter (Fig. 13-16) and faceplate controller (Fig. 13-17) is the arrangement of contact segments on the face of the control. On a starter, the contact segments are mounted in a complete outer circle and another set (fewer in number) in a complete inner circle. On a faceplate controller, the segments are contacted by brushes attached to (but insulated from) the horizontal arm, which is actuated by a handle. The two sets of brushes at each end of the arm are connected together, thus forming a circuit between the inner contact segments and outer contact segments. When the controller is in the OFF position, the brushes rest on insulation pieces. When the handle is moved in the FOR-

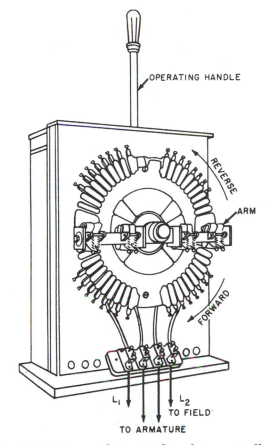

Fig. 13-17. Forward-reverse faceplate controller.

WARD direction, sections of starting resistances are cut out in steps, thus causing the motor to accelerate. When the handle is moved in the REVERSE direction, the motor reverses.

DRUM CONTROLLERS

A drum controller consists essentially of a drum cylinder insulated from a control shaft to which an operating handle is keyed, Fig. 13-18. When the controller handle is moved forward one notch, the motor starts, with resistance in its armature circuit. As the handle is turned further (notch by notch), the resistance is cut out of the armature circuit in steps and is inserted in the field circuit. When the handle reaches its limit, all the resistance is cut out of the armature circuit and the motor is operating at its maximum speed. By moving the handle from OFF position in the opposite direction, the motor will be caused to run in the reverse direction. The lower portion of Fig. 13-18 is a diagrammatic representation of the same drum controller with contact fingers rolled out flat.

STARTING AND ACCELERATING

There are two methods of starting DC motors—on full voltage and reduced voltage. The full-voltage controller connects the motor directly to the power lines. The reduced-voltage controller impresses, at first, less than full line voltage on the armature terminals, and then, by one or more steps, increases the voltage at the armature to full-line voltage. Good practice dictates that a DC motor of over $\frac{1}{3}$ hp be started by a reduced-voltage controller either manually or magnetically operated. Acceleration of DC motors is obtained by manual control, with

a faceplate type starter, Fig. 13-16, or drum controllers and resistors, Fig. 13-18, and by magnetic control. Starters of this type usually have not more than three points of acceleration, and are satisfactory where the motor load and line voltage are fairly constant.

Current-limit acceleration requires current relays which are adjusted to close the contacts at a value of current higher than that drawn by the load to be accelerated. The contactor, in turn, cuts out one step of resistance, so that the motor accelerates more.

In definite-time acceleration, time-element starters cut out the accelerating resistors within a fixed time limit, regardless of the load. If the motor is heavily overloaded and stalled on the first point, the starting and acceleration are automatically forced on a later point. The overload relay will protect the motor from very high currents as well as normal overloads.

SPEED REGULATION

Faceplate type starters are made and used for various services, such as starting duty with speed regulation by armature resistance, and starting duty with speed regulation by both armature and field resistance, Fig. 13-19.

Drum controllers, standard, listed types, and resistors can be used for speed regulation by either armature resistance or field resistance, but not a combination of both.

Other methods for regulating speed are by the variable-voltage or Ward–Leonard system, by multivoltage control, and by shunting the armature with resistance.

RETARDATION

Deceleration is obtained by the following means.

1) The resistance in the armature circuit is increased, or in the case of the adjustable speed DC motor, resistance is reduced in the field circuit.

2) Dynamic braking is applied for quick-stopping of shunt and compound motors, and sometimes as a step in reversing service.

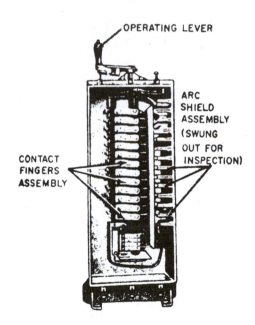

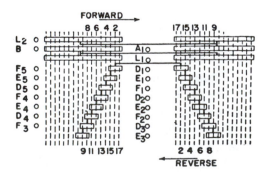

Fig. 13-18. Drum type controller.

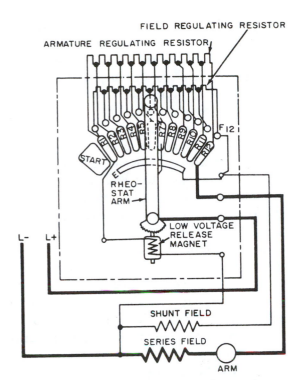

Fig. 13-19. Starting and speed-regulating rheostat.

When the motor circuit is opened, a dynamic-braking resistor is instantly connected in parallel with the armature. The shunt field remains energized. The motor, running by inertia, now acts as a generator, and the braking resistor acts as a load on the generator. Dynamic braking is seldom applied to series motors because the series field must be reversed to be effective. This complicates the control.

3) Another method of bringing motors to a quick stop is plugging. This is done by reversing the motor connections to the power line. It is used most often with series motors or heavily compound motors.

APPLICATION

Full-voltage controllers, either manual or magnetic, can be used to connect small DC motors of up to about 2 hp directly to the line.

Reduced-voltage controllers, DC either manual or magnetic, are used to operate DC motors of $\frac{1}{2}$ hp (rated) and larger.

Manual reduced-voltage starters and speed regulators of the faceplate type, with self-contained resistors, are nonreversing and are used for motors of up to about 25 hp. Separate devices are necessary for reversing service.

Drum controllers with separately mounted resistors for starting or speed regulating duty may be used with motors of up to about 50 hp.

Magnetic controllers are used for any size of motor and can be obtained for nonreversing or reversing service and other basic functions.

G. Modern DC Developments

For years, the use of DC equipment was relegated to a very few special applications, with the majority of industrial uses being taken over by the AC approach. The reasons for this have been given in a previous chapter. With modern designs of electronic controllers, coupled with the demand for close control of electric drive motors, the picture is changing. We now see more and more automatic and high-tech equipment being operated by DC motors, whose speed is under very precise control by combination AC/DC power and control modules. These consist of solid state, neatly packaged units which take in standard three phase AC power and convert it to DC power, at voltages varying to meet the requirements of the DC motor drives on the plant equipment.

See Section C in the Appendix for proper selection data of DC motor controllers.

H. Maintenance and Repair

The major problems which are unique to DC machines, as well as with some of the AC machines, center around the points of current passing between brushes and commutator surfaces. This is a constant source of friction, and the troubles ordinarily associated with friction. Added to that is the possibility of sparking as the current jumps across the face of the stationary brushes and the moving commutator surface. They all add up to trouble, and the maintenance worker will spend a good share of his maintenance hour budget on this portion of the generators and motors under his care.

The following procedures are an excellent guide to follow, but they require the availability of the manufacturer's literature, including at least the maintenance manual containing the clearances, angles, pressures, and other pertinent data in setting the brushes, switches, etc., involved.

BRUSH FAILURE

Brush failure is usually the result of one or several of the following conditions: a rough or eccentric commutator; high mica; vibration due to imbalance or loose bearings; or the improper adjustment of spring tension, brush position, brush angle, brush spacing, or brush staggering. For a description of the various types of brushes and their functions, see Section B.

BRUSH SETTING

Brush setting is the fixed angle of contact between the brush and the commutator, depending upon the direction of rotation of the commutator. The three brush settings are the trailing, the leading, and the radial, (Fig. 13-20). The trailing setting forms an angle between brush and commutator of less than 90 degrees from periphery of the commutator, be-

hind or trailing the direction of rotation. The leading setting forms an angle between brush and commutator of less than 90 degrees from periphery of the commutator in advance of, or leading, the direction of rotation. A radial setting is formed by the brushes meeting the periphery of the commutator at right angles. Where a machine has reverse rotation, the radial setting is usually considered best, although such motors are sometimes set with one brush leading and its counterpart trailing. The leading setting is recommended because it causes less brush-in-holder friction and more readily follows commutator irregularities. With the leading setting, all the brush-spring tension is used in holding the brush against the commutator, whereas with radial and trailing settings, the brushes exert a greater side thrust on the brush holder when the machine is in operation.

The correct setting of the brush angle is determined by the speed and direction of rotation of the commutator, and by the service requirements of rotation in either direction. For low commutator-surface speeds (approximately 5,500 feet per minute), the brushes should be set leading at an angle of approximately 20 degrees. For high commutator-surface speeds (9,000 feet per minute) the brushes should be set leading at an angle of about 30 degrees to 40 degrees from vertical. The formula for determining surface speed is surface speed in fpm = diameter in inches × pi × rpm divided by 12.

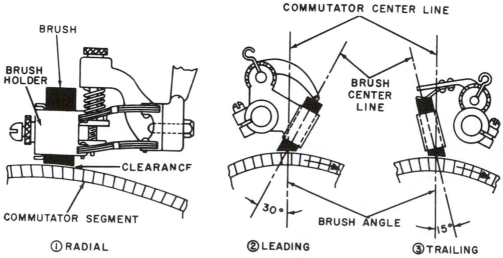

Fig. 13-20. Brush settings and measurement of brush angle.

As a general rule, the brushes have been carefully set in the neutral position by the manufacturer. The brushes are either permanently fixed or the neutral position is marked. Various methods of marking are used by manufacturers. For example, painted marks are made on the armature slots and the ends of the commutator bars, or a straight line is drawn across the yoke and the end shield with an arrow indicating the direction of rotation. Figure 13-21 illustrates use of letters to designate the direction of rotation. Where these markings are not given, neutral must be located by flashing the field. Normally, mechanical neutral is found at the midpoint between two main pole center lines. Quite often, the brushes of commutating pole machines are set slightly off mechanical neutral in order to obtain best commutation. The procedures below should be followed when making adjustments for either kind of marking.

1) Disconnect the machine from its power source.

Warning: Make certain that the power source is disconnected and cannot be connected while working on the machine.

2) Loosen the brush-yoke adjustment screws.

3) Use a bar or pipe as a lever to aid in making small incremental adjustments.

4) Insert the bar into the brush yoke at a point between the brush studs, where it will not interfere with the brushes or commutator.

5) Move the lever slightly in the direction of rotation for the forward lead and counter to the direction of rotation for the backward lead. These brush positions are the same, respectively, as the leading setting and the trailing setting described above. Adjustments are very critical, and a slight imperceptible movement may be sufficient for improving commutation.

Warning: Do not attempt to shift the brushes when the machine is connected to a source of power.

6) Tighten the brush-yoke adjustment screws.

7) Apply power source.

8) Load the machine to full-rated load.

9) Notice if commutation is improved to the point of sparkless commutation.

10) If sparking is worse, shut down the machine, disconnect from power source, and carry out steps 1) to 8) above with the exception that in 5) above shift the brushes in the opposite direction to the direction in which they were originally set.

SPRING-PRESSURE ADJUSTMENT

Spring pressure should be kept in proper adjustment, because neglect of spring tension causes much unnecessary maintenance. Uneven tension causes selective action. Although it might be assumed that low tension will reduce brush wear, in reality low tension often causes streaking, arcing, and grooving of the commutator. Such weak brush spring tension normally results from wear of the brush and/or the commutator segments. It might also seem reasonable that brush chatter can be reduced by increasing the spring tension, but if the chatter is caused by high mica or flat bars, increased tension will result in chipped or broken brushes. Maintenance of proper tension on all of the brushes is highly important.

Brush-spring pressure should be from $1\frac{3}{4}$ to $2\frac{1}{2}$ psi (pounds per square inch) for the contact area between the brush and the commutator

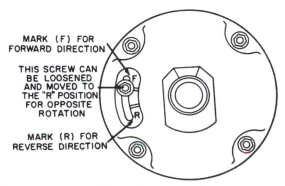

MARK (F) FOR FORWARD DIRECTION

THIS SCREW CAN BE LOOSENED AND MOVED TO THE "R" POSITION FOR OPPOSITE ROTATION

MARK (R) FOR REVERSE DIRECTION

Fig. 13-21. Factory markings on the brush yoke.

for light-metalized, carbon, or graphite brushes, and from 3 to 5 psi for heavy-metalized brushes or collector rings. In making adjustments for brush pressure, allowance should be made for the weight of the brush. This is particularly important in the case of heavy-metalized brushes on collector rings. A heavy spring pressure may give less brush wear than a light spring pressure.

ADJUSTING BRUSH PRESSURE
(Figure 13-22)

1) Insert a strip of paper (newspaper will do) between the brush and the commutator.

2) Provide a means for adapting the attachment of the spring-balance hook to the contact end of a brush-spring arm.

3) Attach the spring-balance hook.

4) Raise the spring balance until the paper can be withdrawn from between the brush and the commutator.

5) Notice the reading; if it does not meet specifications, repeat steps 3) and 4) after adjusting the spring tension.

6) Repeat 1) through 5) above for remaining brushes.

7) On some types of brush holders, an extension of the helical spring wire is moved along a notched holder, in line with the supporting shaft of the helical spring, for the purpose of adjusting spring pressure.

INSTALLATION AND SPACING ON THE COMMUTATOR

To ensure maximum commutation, the brushes must be carefully spaced around the commutator and set at the proper angle. Faulty adjustments in the brush rigging can cause considerable error in stud alinement with the commutator. Brushes must fit correctly in the brush boxes and be snugly supported while still being free to move the entire length of the brush box. The contact surfaces of the brushes must conform to the surface of the commutator and be properly spaced and staggered to ensure even minimum wear on the commutator.

SETTING THE BRUSH HOLDERS

The setting of brush holders depends somewhat on the type of brush holder, but in all cases, the holders must be set an equal distance from the commutator.

1) Determine the approximate distance of the studs from the commutator or the correct angle of the brush holders when the holders are set about $\frac{1}{8}$-inch maximum from the commutator. The studs supporting the brush holders must be carefully set and not just tightened. If they are not set properly, correct commutation cannot be obtained.

2) Loosen the setscrews or clamping bolts of the brush holder.

3) Use a fiber strip of the proper thickness as a gage, and insert it between the bottom edge of the brush holder and the commutator.

4) Move the holder down until it rests firmly on the fiber gage.

5) Tighten the setscrews or clamping bolts and fix the holder securely in place.

6) Remove the fiber gage.

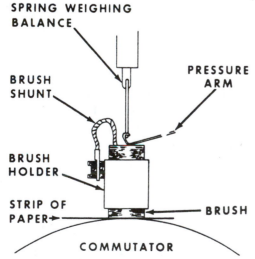

SPRING WEIGHING BALANCE

BRUSH SHUNT

PRESSURE ARM

BRUSH HOLDER

STRIP OF PAPER

BRUSH

COMMUTATOR

Fig. 13-22. Checking brush-spring pressure.

FITTING OF BRUSHES IN THE HOLDERS

1) Check the brush holders or boxes for cleanliness and freedom from obstructions.

2) Clean the brush box with a clean cloth moistened with solvent if it appears to be dirty and greasy.

3) Use sandpaper on the inside of the brush box to remove small burrs and stubborn dirt.

4) Insert the brush and check for sideplay. If there is appreciable sideplay, the brush should be discarded and replaced.

5) Sand the sides of tight-fitting brushes with a medium-coarse grade of sandpaper until the sides make a close but free fit.

FITTING THE BRUSHES TO THE COMMUTATOR

After the brushes are placed in the holders, fit them to the commutator by sanding them with a medium-coarse grade of sandpaper. Sand in the direction that tends to push the brush upward in the holder.

1) Use a strip of sandpaper (medium-coarse grade) wide enough to engage the entire contact surface of the brush.

2) Place the sandpaper strip between the brush-contact surface and the commutator with the rough side of the sandpaper toward the brush-contact surface.

3) Apply pressure to the brush with the thumb of the left hand and pull the sandpaper through with the right hand, in the direction which tends to push the brush upward in the holder.

4) Lift the brush and continue, following 2) and 3) above, until the brush face makes a good fit with the commutator curvature.

5) Repeat 1), 2), 3), and 4) above for all brushes.

6) After the brushes have been sanded to a good fit and the machine put into operation, brush fit can be improved by applying a brush-seating stone to the commutator. Once the brushes have been fitted, blow out the machine; wherever possible, blow from the back of the commutator toward the outside of the machine.

SPACING OF BRUSHES AROUND THE COMMUTATOR

It is not safe to assume that, after the studs and holders are accurately set, the brushes are correctly spaced once they are placed in the holders. Slight inaccuracies in the brush rigging can be accumulative, causing errors in brush spacing. The following procedure will aid in obtaining accurate spacing.

1) Count the number of commutator bars and divide them into as many equal groups as there are brush holder studs, for the purpose of using the result as a guide to check brush spacing. The commutator bars and the insulation between them vary in thickness.

2) By means of a strip of paper wound tightly around the commutator, mark off the exact circumference of the commutator.

3) Remove the paper, and accurately divide the circumference length into as many divisions as there are brush groups.

4) Draw a line squarely across the width of the strip at each division, to guide in locating the brushes.

5) Remove the outer brush in each group.

6) Replace the marked strip, aligning a mark with the toe of the next brush of a group, and firmly affix the strip with cellophane or masking tape.

7) Replace the outer brushes and, if the toe of the outer brush of a brush group does not meet the mark, slowly turn the armature until it does so.

8) Notice if the toes of the other brush groups meet the equidistant marks.

9) Adjust any brush groups which do not meet the marks.

I. Commutator Adjustment

Most of the difficulties of DC motors and generators are due to faulty commutation. Adjustments on small DC motors are not possible since the area or range of most efficient commutation is inherent in the design of the motor. On large DC motors, the air gaps of field poles in relation to armature poles is a design feature, and only the brushes need adjusting after being set on neutral. Adjustment on variable-speed motors that give difficulty in obtaining satisfactory commutation are made after observation. This is done by starting at top speed with rated load and gradually working down to low speed. Reversing constant-speed DC motors may require an adjustment only on the average neutral setting. If further adjustment is required so that the motor will run in either direction at the same speed and load and with the same shunt-field current, the most reliable adjustment is flashing the field. Before attempting to set electrical neutral, an accurate brush fit and a smooth commutator surface are necessary. Make no commutation adjustment until the brushes have operated satisfactorily at full fit, and until the commutator has taken on a polish.

1) Open the main switch, disconnecting the motor from the power source.

Caution: Take measures to be certain that the main switch will not be closed while adjusting the machine.

2) Connect a low-range voltmeter between two adjacent brush studs.

3) Adjust the field rheostat for maximum resistance.

4) Apply current to the field only.

5) Adjust the field rheostat until the voltmeter gives an appreciable reading.

6) Adjust the field rheostat in the opposite direction until the voltmeter reading nearly approaches 0.

7) Momentarily open and close the field switch.

8) Note the kick of the voltmeter pointer.

9) Open the field switch.

10) Adjust the brushes.

11) Retain the original field-rheostat adjustment and repeat 7), 8), 9), and 10) above until the voltmeter reading is minimum or 0 for either direction of movement of the brushes.

12) Run the motor, noticing the field current value at rated load and speed.

13) Reverse the rotation, maintaining the same field current value as for the rated load, and notice the speed.

14) Notice if the motor comes up to the same speed as for the speed of the opposite direction.

15) Adjust the brushes if the motor does not come up to the speed of the opposite direction.

16) Repeat 13), 14), and 15) above until the brushes are adjusted so the the motor runs at the same speed in either direction.

17) Adjustments are very critical, and minute adjustments of small increments of movement may be all that is necessary.

That is as far as we shall go in this matter, as any further repair work is best left up to those who specialize in motor repair. We strongly suggest that should the maintenance worker feel inadequate at this stage, it is an excellent idea to call in more professional help.

J. Troubleshooting DC Equipment

DC GENERATORS

1. Failure to Build Up Voltage

Probable Cause	Remedy
Voltmeter not operating	Check output voltage with separate voltmeter. Replace voltmeter.
Open field resistor	Repair or replace resistor.
Open field circuit	Check coils for open and loose connections. Replace the defective coil or coils. Tighten or solder loose connections.
Absence of residual magnetism in a self-excited generator	Flash the field. Accomplished by touching leads from an outside DC source to the field winding terminals, $(+)$ to $(-)$, and $(-)$ to $(+)$, to reinforce the excitation and bring the generator up to capacity.
Dirty commutator	Clean or dress commutator.
High mica	Undercut mica.
Brushes not making proper contact	Free if binding in holders. Replace and reseat if worn.
Newly seated brushes not contacting sufficient area on the commutator	Run in by reducing load and use a brush-seating stone.
Armature shorted internally, or to ground	Remove, test, and repair or replace.
Grounded or shorted field coil	Test, and repair or replace.
Shorted filtering capacitor	Replace.
Open filter choke	Replace.
Open ammeter shunt	Replace ammeter and shunt.
Broken brush shunts or pigtails	Replace brushes.

2. Output Voltage Too Low

Probable Cause	Remedy
Prime mover speed too low	Check speed with tachometer. Adjust governor on prime mover.
Brushes not seated properly	Run in with partial load, use brush-seating stone.
Commutator is dirty	Clean, or if film is too heavy, replace brushes with complete set of proper grade.
Field resistor not properly adjusted	Adjust field strength. Tighten all connections. Make shim adjustment.
Reversed field coil or armature connection	Check and connect properly.

3. Output Voltage Too High

Probable Cause	Remedy
Prime mover speed too high	Check speed with tachometer. Adjust governor on prime mover.
Faulty voltage regulator	Adjust or replace.

4. Armature Overheats

Probable Cause	Remedy
Overloaded	Check meter readings against nameplate ratings. Reduce load.
Excessive brush pressure	Adjust pressure or replace tension springs.
Couplings not aligned	Align units properly.
End bells improperly positioned	Assemble correctly.
Bent shaft	Straighten or replace.
Armature coil shorted	Repair or replace armature.
Armature rubbing or striking poles	Check for bent shaft, loose or worn bearings. Straighten and realign shaft. Replace bearings, tighten pole pieces, or replace armature.
Clogged air passages (poor ventilation)	Clean equipment.
Repeated changes in load of great magnitude (improper design for the application)	Generator should be used with a steady load application.
Unequal brush tension	Equalize brush tension.
Broken shunts or pigtails	Replace brushes.
Open in field rheostat	Repair or replace rheostat.

5. Field Coils Overheat

Probable Cause	Remedy
Shorted or grounded coils	Repair or replace.

5. Field Coils Overheat

Probable Cause	Remedy
Clogged air passages (poor ventilation)	Clean equipment. Remove obstructions.
Overload (compound generator)	Check meter reading against nameplate rating. Reduce load.

6. Sparking at Brushes

Probable Cause	Remedy
Overload	Check meter readings against nameplate ratings. Reduce load.
Brushes off neutral plane	Adjust brush rigging.
Dirty brushes and commutator	Clean brushes and commutator.
High mica	Undercut mica.
Rough or eccentric commutator	Resurface commutator.
Open circuit in the armature	Repair or replace armature.
Grounded, open- or short-circuited field winding	Repair or replace defective coil or coils.
Insufficient brush pressure	Adjust or replace tension springs.
Brushes sticking in the holders	Clean holders. Sand brushes.

DC MOTORS

1. Failure to Start

Probable Cause	Remedy
Open circuit in the control	Check for open. Replace open resistor or fuse.
Low supply voltage	Check with voltmeter and apply proper voltage.
Frozen bearing	Replace bearing and recondition shaft.
Overload	Reduce load or use larger motor.
Excessive friction	Check for air gap, bent shaft, loose or worn bearings, misaligned end bells. Straighten shaft, replace bearings, tighten pole pieces, align end bells.

2. Stops After Running a Short Time

Probable Cause	Remedy
Failure of supply voltage	Apply proper voltage, replace fuses, or reset overload relay.

	Check meter readings against nameplate ratings. Reduce load.
Overload	Check meter readings against nameplate ratings. Reduce load.
Ambient temperature too high	Ventilate space to reduce ambient temperature.
Overload relays set too low for application	Adjust relays for the application.

3. Attempts to Start, But Overload Relays Trip Out

Probable Cause	Remedy
Motor field weak or nonexistent	Check field circuit. Repair or replace defective field coils.
Overload	Check meter readings against nameplate ratings. Replace motor with one suitable to the application.
Relays adjusted too low for the application	Adjust relays for the application.

4. Runs Too Slow

Probable Cause	Remedy
Line voltage low	Apply proper voltage.
Brushes ahead of neutral plane	Adjust brush rigging.
Overload	Check meter reading against nameplate readings. Reduce load.

5. Runs Too Fast Under Load

Probable Cause	Remedy
Weak field	Check field circuit. Replace open coils or open starter resistors.
Line voltage too high	Reduce line voltage.
Brushes off adjustment with neutral plane	Adjust brush rigging.

6. Sparking at Brushes

Probable Cause	Remedy
Same as DC generator (paragraph 6)	Same as DC generator (paragraph 6).

7. Overheating

Probable Cause	Remedy
Same as DC generator (paragraphs 4 and 5)	Same as DC generator (paragraphs 4 and 5).

C H A P T E R 14
SINGLE-PHASE MOTORS

A. General Introduction

In Chapter 7 we explained how alternating current is generated, and some of the principles involved were covered, also. In this chapter we shall explain how those same principles, and others explained in earlier chapters, are combined to produce rotary power.

First, we must start with a general description of the action inside the electric motor, and then we shall proceed to cover the various types and their characteristics. Electric motors are like any other type of tool to be found in the shop; each has its own particualr application where it gives the best service.

You are going to see the term "synchronous speed" used in this section for the first time. This simply means that the motor is turning at the same speed, in revolutions per minute, as the AC supply system, as stated in cycles per minute.

Single-phase motors, as their name implies, operate on a single-phase power supply. These motors are used extensively in fractional horsepower sizes in commercial and domestic applications. The advantages of using single-phase motors in small sizes are that they are less expensive to manufacture than other types, and they eliminate the need for 3-phase AC lines. Single-phase motors are used in interior communications equipment, fans, refrigerators, portable drills, grinders, and so forth.

A single-phase induction motor with only one stator winding and a cage rotor is like a three-phase induction motor with a cage rotor except that the single-phase motor has no magnetic revolving field at start and hence no starting torque. However, if the rotor is brought up to speed by external means, the induced current in the rotor will cooperate with the stator currents to produce a revolving field, which causes the rotor to continue to run in the direction in which it was started.

Several methods are used to provide the single-phase induction motor with starting torque. These methods identify the motor as split phase, capacitor, shaded pole, repulsion, and so forth.

Another class of single-phase motor is the AC series (universal) type. Only the more commonly used types of single-phase motors are described. These include the 1) split-phase motor, 2) capacitor motor, 3) shaded-pole motor, 4) repulsion-start motor, and 5) AC series motor.

B. Split-Phase Motors

The split-phase motor (Fig. 14-1) has a stator composed of slotted laminations that contain an auxiliary (starting) winding and a running (main) winding. At start, these two windings

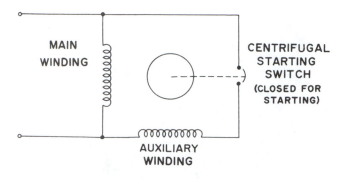

Fig. 14-1. Split-phase motor.

produce a magnetic revolving field that rotates around the stator airgap at synchronous speed. As the rotating field moves around the airgap, it cuts across the rotor conductors and induces a voltage in them, which is in phase with the stator field. The interaction of the rotor currents and the stator field causes the rotor to accelerate in the direction in which the stator field is rotating.

When the rotor has come up to about 75 percent of synchronous speed, a centrifugally operated switch disconnects the starting winding from the line supply, and the motor continues to run on the main winding alone. The motor has the constant-speed variable-torque characteristics of the shunt motor. Many of these motors are designed to operate on either 120 volts or 240 volts. For the lower voltage, the stator coils are divided into two equal groups and these are connected in parallel. For the higher voltage, the groups are connected in series. The starting torque is 150 to 200 percent of the full-

load torque, and the starting current is 6 to 8 times the full-load current.

In Fig. 14-2 we show the wiring diagram and terminal connections for a dual voltage split-phase motor. When wired for low voltage operation, all windings receive the same voltage, as shown by the diagram for 115 volts. When wired for the high voltage, the running windings are in series, and the starting winding is in parallel with one of the running windings. This provides proper voltage distribution.

The method of accomplishing this may vary with the motor manufacturer, and the wiring diagram given on the motor nameplate must be followed.

Fractional-horsepower split-phase motors are used in a variety of equipment, such as washers, oil burners, and ventilating fans. The direction of rotation of the split-phase motor can be reversed by interchanging the starting winding leads.

C. Capacitor Motors

The capacitor motor is a modified form of split-phase motor, having a capacitor in series with the starting winding. An external view is shown in Fig. 14-3, with the capacitor located on top of the motor. The capacitor produces a greater phase displacement of currents in the

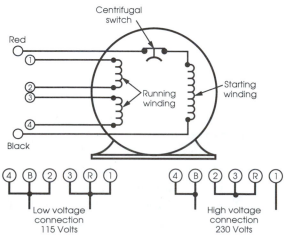

Fig. 14-2. Dual voltage split-phase motor wiring diagram.

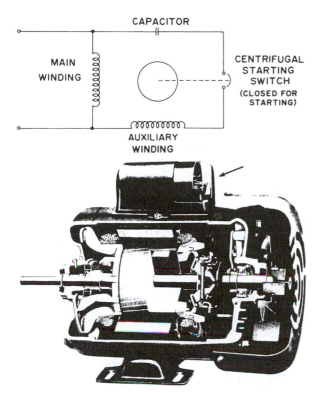

Fig. 14-3. Capacitor-start motor.

starting and running windings than is produced in the split-phase motor. The starting winding current is displaced approximately 90 degrees from the running winding current. These conditions produce a higher starting torque than that of the split-phase motor. The starting torque of the capacitor motor may be as much as 350 percent of the full-load torque.

If the starting winding is cut out after the motor has increased in speed, the motor is called a CAPACITOR-START MOTOR (see Fig. 14-3). If the starting winding and capacitor are designed to be left in the circuit continuously, the motor is called a CAPACITOR-RUN MOTOR (see Fig. 14-4). Capacitor motors of both types are made in sizes ranging from small fractional horsepower motors up to about 10 horsepower. They are used to drive grinders, drill presses, refrigerator compressors, and other loads requiring relatively high starting torque. The direction of rotation of the capacitor motor may be reversed by interchanging the starting winding leads.

Permanent-split capacitor (PSC) motors, Fig. 14-5, are similar to capacitor-start motors ex-

cept that the same value of capacitance is used for both starting and running conditions. Starting torque is much lower than that for capacitor-start motors, and the breakdown torque is suitable for loads that require load peaks no greater than normal load torque, such as fans and blowers.

No starting mechanism is used on PSC motors, therefore, they are adaptable to variable speed control and can be operated at reduced speeds (below design speed) by lowering the effective supply voltage. A PSC motor should not be operated at a speed less than that at which torque breakdown occurs. With a standard low-slip motor, torque breakdown occurs at about 75 percent of the motor's synchronous speed. With a high-slip design, torque breakdown occurs at less than 75 percent of the synchronous speed.

Motor currents much higher than normal

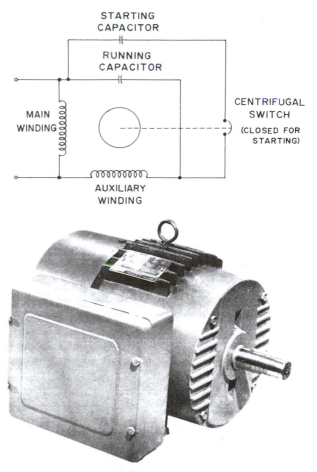

Fig. 14-4. Two-value, Capacitor-run motor.

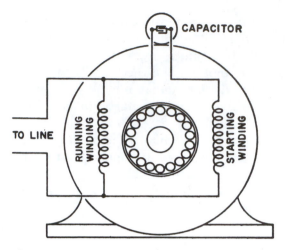

Fig. 14-5. Permanent-split capacitor motor.

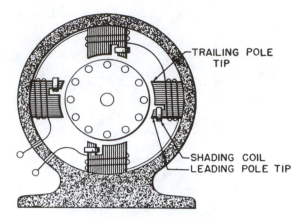

Fig. 14-6. Shaded-pole motor.

will be drawn if the PSC motor is operated at a speed lower than that at which torque breakdown occurs.

D. Shaded-Pole Motors

The shaded-pole motor employs a salient-pole stator and a cage rotor. The projecting poles on the stator resemble those of DC machines except that the entire magnetic circuit is laminated and a portion of each pole is split to accommodate a short-circuited copper strap called a SHADING COIL (Fig. 14-6). This motor is generally manufactured in very small sizes, up to 1/20 hp.

Most shaded-pole motors have only one edge of the pole split, and therefore the direction of rotation is not reversible. However, some shaded-pole motors have both leading and trailing pole tips split to accommodate shading coils. The leading pole tip shading coils form one series group, and the trailing pole tip shading coils form another series group. Only the shading coils in one group are simultaneously active, while those in the other group are on open circuit.

The shaded-pole motor is similar in operating characteristics to the split-phase motor. It has the advantages of simple construction and low cost. It has no sliding electrical contacts and is reliable in operation. However, it has low starting torque, and high noise level. It is used to operate small fans. The shading coil and split pole are used in clock motors to make them self-starting.

E. Repulsion-Start Motors

The repulsion-start motor, Figure 14-7, has a form-wound rotor with commutator and brushes. The stator is laminated and contains a distributed single-phase winding. The motor has a centrifugal device which removes the brushes from the commutator and places a short-circuiting ring around the commutator. This action occurs at about 75 percent of synchronous speed. Thereafter, the motor operates with the characteristics of the single-phase induction motor.

The starting torque of the repulsion-start induction motor is developed through the interaction of the rotor currents and the single-phase stator field. The rotor currents are induced through transformer action.

The function of the commutator and brushes is to divide the rotor currents along an axis that is displaced from the axis of the stator field in a counterclockwise direction. The motor derives its name from the repulsion of like poles between the rotor and stator.

The starting torque is 250 to 450 percent of the full-load torque, and the starting current is 375 percent of the full-load current. This motor is made in fractional horsepower sizes and in

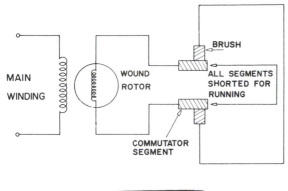

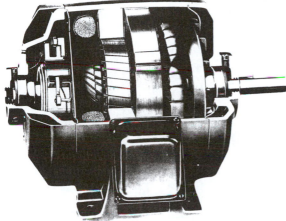

Fig. 14-7. Repulsion-start induction motor.

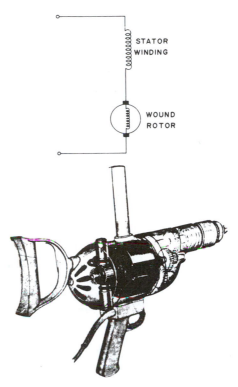

Fig. 14-8. Universal motor.

larger sizes up to 15 horsepower, but has been replaced in large part by the cheaper and more rugged capacitor motor. The repulsion-start motor has higher pull-out torque (torque at which the motor stalls) than the capacitor-start motor, but the capacitor-start motor can bring up to full speed loads that the repulsion motor can start but cannot accelerate.

F. Universal or Series Motors

The universal or series motor, Fig. 14-8, is a high-speed motor that will operate on either alternating or direct current. It is usually a special-purpose motor, often used on drills, grinders, sanders, sprayers, vacuum cleaners, and food mixers. The advantages of this type of motor include high starting torque, high power-to-size ratio, and rapid acceleration of the load to speed.

The operating speed of these motors depends on the load. They do not operate at a constant speed, but run as fast as the load permits. If not loaded, they will overspeed, which may damage the motor.

G. Characteristics

Table 14-1 summarizes the characteristics of the single-phase AC motors just covered. The majority of those with which you will come in contact are included.

H. Single-Phase AC Motor Troubleshooting

Modern design of motors, with the usually excellent quality control in their manufacture engendered by a highly competitive market, has produced motors which will last for years in service, if properly used and maintained. Experience has shown that the major sources of

Table 14-1. Single-Phase Motor Characteristics

Type	Horsepower ranges	Load-starting ability	Starting current	Characteristics	Electrically reversible	Typical uses
Split-phase	$\frac{1}{20}$ to $\frac{1}{2}$	Easy starting loads. Develops 150 percent of full-load torque.	High; five to seven times full-load current.	Inexpensive, simple construction. Small for a given motor power. Nearly constant speed with a varying load.	Yes.	Fans, centrifugal pumps; loads that increase as speed increases.
Capacitor-start	$\frac{1}{8}$ to 10	Hard starting loads. Develops 350 to 400 percent of full-load torque.	Medium, three to six times full-load current.	Simple construction, long service. Good general-purpose motor suitable for most jobs. Nearly constant speed with a varying load.	Yes.	Compressors, grain augers, conveyors, pumps. Specifically designed capacitor motors are suitable for silo unloaders and barn cleaners.
Two-value capacitor	2 to 20	Hard starting loads. Develops 350 to 450 percent of full-load torque.	Medium, three to five times full-load current.	Simple construction, long service, with minimum maintenance. Requires more space to accommodate larger capacitor. Low line current. Nearly constant speed with a varying load.	Yes.	Conveyors, barn cleaners, elevators, silo unloaders.
Permanent-split capacitor	$\frac{1}{20}$ to 1	Easy starting loads. Develops 150 percent of full-load torque.	Low, two to four times full-load current.	Inexpensive, simple construction. Has no start winding switch. Speed can be reduced by lowering the voltage for fans and similar units.	Yes.	Fans and blowers.
Shaded-pole	$\frac{1}{250}$ to $\frac{1}{2}$	Easy starting loads.	Medium.	Inexpensive, moderate efficiency, for light duty.	No.	Small blowers, fans, small appliances.
Wound-rotor (Repulsion)	$\frac{1}{6}$ to 10	Very hard starting loads. Develops 350 to 400 percent of full-load torque.	Low, two to four times full-load current.	Larger than equivalent size split-phase or capacitor motor. Running current varies only slightly with load.	No. Reversed by brush ring readjustment	Conveyors, drag burr mills, deep-well pumps, hoists, silo unloaders, bucket elevators.
Universal or series	$\frac{1}{150}$ to 2	Hard starting loads. Develops 350 to 450 percent of full-load torque.	High.	High speed, small size for a given horsepower. Usually directly connected to load. Speed changes with load variations.	Yes, some types.	Portable tools, kitchen appliances.

trouble with motors are overlubrication, over-loading, or dirty operating conditions. In the case of single-phase motors, to this may be added the failure of capacitors and the centrifugal starting switches. Chapter 10 explains how to check capacitors and centrifugal switches.

A warning here is in order, concerning the cleaning of all electrical equipment.

1. Do not use any solution containing carbon tetrachloride, or any other liquid which the label declares toxic and dangerous.

2. Do not use the plant air hose to blow out the dirt. The air hose may contain water or oil, and the blast may imbed foreign particles inside the mechanism. Use a shop vacuum cleaner instead, along with a brush for loosening the caked dust and dirt.

Chapter 10, Section D, contains instructions for checking capacitor and split-phase motors, using an ohmeter. Figure 14-9 and 14-10 contain instructions for using a test lamp for testing single-phase motors. Instructions for using the test lamp are given in Chapter 15, Section C. The test lamp is a cheaper way to do it, but it takes a fair amount of practice to learn how to read the results from the glowing lamp. The ohmeter gives an exact readout, with very little

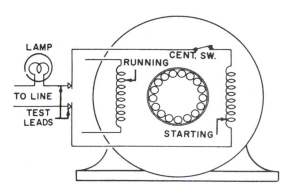

Fig. 14-10. Testing for open starting winding circuit.

chance for errors, if the instrument is properly used.

We give below a condensed troubleshooting list, with a brief method of solving each problem.

REPULSION - INDUCTION MOTORS

1. Failure to Start

Probable Cause	Remedy
Open fuse	Replace fuse.
Overloaded	Check meter readings against nameplate ratings. Reduce load or install larger motor.
Low supply voltage. Lead wires have insufficient current capacity	Apply correct voltage. Install larger lead wires.
Stator coil open	Check and replace coil or coils.
Stator coil shorted	Check and replace shorted coil or coils.
Stator coil grounded	Check and replace defective coil or coils.
Centrifugal mechanism not operating properly	Disassemble, clean, inspect, adjust, repair, or replace.
Incorrect brush setting	Locate neutral plane by shifting brushes until there is no rotation when current is applied. Shift brushes in the direction of the desired rotation, $1\frac{1}{3}$ bars from neutral on 4-pole motors of $\frac{1}{2}$ hp and smaller, and $1\frac{3}{4}$ bars on larger 4-pole motors. On 2-pole motors, set $\frac{1}{3}$ bar farther than setting given above.

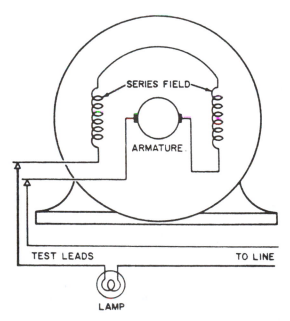

Fig. 14-9. Testing universal motor for an open circuit.

| Bearings | Check for bent shaft or worn, loose, dry, or overlubricated bearing. Straighten and realign bent shaft. Replace worn and loose bearings, lubricate dry bearings, relieve overlubrication. |

2. Runs Slow

Probable Cause	Remedy
Overloaded	Check meter readings against nameplate rating.
Centrifugal mechanism not operating properly	Disassemble and clean.
Bearings binding	Clean and lubricate bearings.

3. Overheating

Probable Cause	Remedy
Overloaded	Check meter readings against nameplate ratings. Reduce load or install larger motor.
Incorrect supply voltage	Apply correct voltage.
Centrifugal mechanism not operating properly	Disassemble, clean, inspect. Repair, adjust, or replace.
Bearings	Check for bent shaft, loose, dry, or overlubricated bearings. Straighten and realign bent shaft. Replace worn or loose bearings, relieve overlubrication.

4. Noisy Operation

Probable Cause	Remedy
Bearings	Check for bent shaft, or worn, loose, dry, or overlubricated bearings. Straighten and realign bent shaft. Replace worn or loose bearings, lubricate dry bearings, relieve overlubrication.
Excessive end play	Adjust end-play takeup screw, or add thrust washers to shaft.
Motor not aligned with driven machine	Realign.
Loose motor mounting and accessories	Tighten all loose components.

5. Motor Produces Shock when Touched

Probable Cause	Remedy
Grounded stator coil	Replace defective coil or coils. Check motor-frame connection or connections to ground. Clean and tighten.
Static charge	Check motor-frame connection or connections to ground. Clean and tighten.

SPLIT-PHASE AND CAPACITOR MOTORS

1. Failure to Start

Probable Cause	Remedy
Open fuse	Replace fuse.
Low supply voltage	Apply correct voltage.
Stator coil open	Replace open coil or coils.
Centrifugal mechanism not operating properly	Disassemble, clean, inspect. Adjust, repair, or replace.
Defective capacitor	Replace capacitor
Stator coil grounded	Check and replace grounded coil or coils.
Bearings	Check for bent shaft, or worn, loose, dry, or overlubricated bearings. Straighten and realign bent shaft. Replace worn or loose bearings, relieve overlubrication.
Overloaded	Check meter readings against nameplate ratings. Reduce load or install larger motor.

2. Overheating

Probable Cause	Remedy
Shorted coil	Replace shorted coil or coils.
Centrifugal mechanism not operating properly	Disassemble, clean, inspect. Adjust, repair, or replace.
Incorrect voltage	Apply correct voltage.
Overloaded	Check meter readings against nameplate ratings. Reduce load or install larger motor.
Bearings	Check for bent shaft, or worn, or overlubricated bearings. Straighten and realign bent shaft, replace worn or loose bearings, lubricate dry bearings, relieve overlubrication.

3. Noisy Operation

Probable Cause	Remedy
Worn bearings	Replace. Realign.
Shaft bent	Straighten shaft. Realign or replace rotor.
Excessive end play	Adjust screw of end-play takeup device, or put shim washers on shaft between end bells and rotor.
Loose motor mounts or accessories	Tighten all loose components.

C H A P T E R 15
THREE-PHASE MOTORS

A. Induction Motors

Of all AC motors, the induction motor is the most widely used. Its design is simple and its construction rugged. The induction motor is particularly well adapted for constant speed applications, and because it does not use a commutator, most of the troubles encountered in the operation of DC motors are eliminated. An induction motor can be either a single-phase or a polyphase machine. The operating principle is the same in either case, and depends on a revolving, or rotating, magnetic field to produce torque. The key to understanding the induction motor is a thorough comprehension of the rotating magnetic field.

Consider the field structure of A of Fig. 15-1 where the poles have windings which are energized by three AC voltages; a, b, and c. These voltages have equal magnitude but differ in phase, as shown in B.

At the instant of time shown as 0, the resultant magnetic field produced by the application of the three voltages has its greatest intensity in a direction extending from pole 1 to pole 4. Under this condition, pole 1 can be considered as a north pole and pole 4 as a south pole.

At the instant of time shown as 1, the resultant magnetic field will have its greatest intensity in the direction extending from pole 2 to pole 5 and, in this case, pole 2 can be considered as a north pole and pole 5 as a south pole.

Thus, between instant 0 and instant 1, the magnetic field has rotated clockwise.

At time 2, the resultant magnetic field has its greatest intensity in the direction from pole 3 to pole 6, and it is apparent that the resultant magnetic field has continued to rotate clockwise.

At instant 3, poles 4 and 1 can be considered as north and south poles, respectively, and the field has rotated still farther.

At later instants of time, the resultant magnetic field rotates to other positions while traveling in a clockwise direction, a single revolution of the field occurring in 1 cycle. If the exciting voltages have a frequency of 60 cycles per second, the magnetic field makes 60 revolutions per second, or 3,600 revolutions per minute. This speed is known as the synchronous speed of the rotating field.

When the rotor of an induction motor is subjected to the revolving magnetic field produced by the stator windings, a voltage is induced in the longitudinal bars. The induced voltage causes a current flow through the bars. This current, in turn, produces its own magnetic field which combines with the revolving field in such a way as to cause the rotor to assume a position in which the induced voltage is minimized. As a result, the rotor revolves at very nearly the synchronous speed of the stator field, the difference in speed just being sufficient to induce current in the rotor to overcome

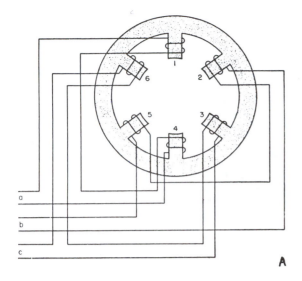

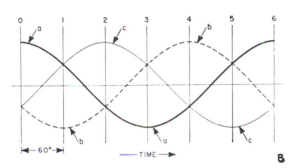

Fig. 15-1. Rotating magnetic field caused by three-phase voltages.

the mechanical and electrical losses in the rotor. If the rotor were to turn at the same speed as the rotating field, the rotor conductors would not be cut by any magnetic lines of force, no voltage would be induced in them, no current could flow, and there would be no torque. The rotor would then slow down. For this reason, there must always be a difference in speed between the rotor and the rotating field. This difference in speed is called slip and is expressed as a percentage of the synchronous speed. For example, if the rotor turns at 1,750 rpm and the synchronous speed is 1,800 rpm, the difference in speed is 50 rpm. The slip is then equal to 50/1,800 or 2.78 percent.

The direction of rotation of a three-phase induction motor can be changed by simply re-

versing two of the leads to the motor. The same effect can be obtained in a two-phase motor by reversing connections to one phase (see Figs. 15-2 and 15-3). In a single-phase motor, reversing connections to the starting winding will reverse the direction of rotation. Most single-phase motors designed for general application have provision for readily reversing connections to the starting winding. Nothing can be done to a shaded-pole motor to reverse the direction of rotation because the direction is determined by the physical location of the copper shading ring.

If, after starting, one connection to a three-phase motor is broken, the motor will continue to run but will deliver only one-third the rated power. Also, a two-phase motor will run at one-half its rated power if one phase is disconnected. Neither motor will start under these abnormal conditions.

The STATOR of a polyphase induction motor consists of a laminated steel ring with slots on the inside circumference. The motor stator winding is similar to the AC generator stator winding. Stator phase windings are symmetrically placed on the stator and may be either Y or delta connected.

There are two types of ROTORS—the CAGE ROTOR and the FORM-WOUND ROTOR. Both types have a laminated cylindrical core with parallel slots in the outside circumference to hold the windings in place. The cage rotor has an insulated bar winding, whereas the form-wound rotor has performed coils like those on a DC motor armature.

CAGE ROTORS

A cage rotor is shown in Fig. 15-2(A). The rotor bars are of copper, aluminum, or suitable alloy placed in the slots of the rotor core. These bars are connected together at each end by rings of similar material. The conductor bars carry relatively large currents at low voltage. Hence, it is not necessary to insulate these bars from the core because the currents follow the path of least resistance and are confined to the cage winding.

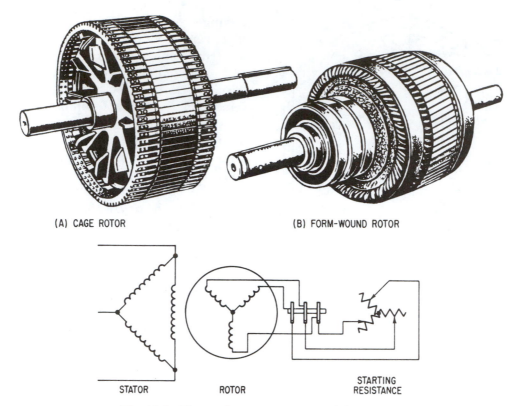

(A) CAGE ROTOR   (B) FORM-WOUND ROTOR

STATOR   ROTOR   STARTING RESISTANCE

Fig. 15-2. Three-Phase motor rotors. (A) Cage rotor. (B) Form-wound rotor. (C) External variable resistance.

FORM-WOUND ROTOR

A form-wound rotor (Fig. 15-2 (B)) has a winding similar to three-phase stator windings. Rotor windings are usually wye connected with the free ends of the winding connected to three sliprings mounted on the rotor shaft. An external variable wye-connected resistance (Fig. 15-2 (C)) is connected to the rotor circuit through the sliprings. The variable resistance provides a means of increasing the rotor-circuit resistance during the starting period to produce a high starting torque. As the motor accelerates, the rheostat is cut out. When the motor reaches full speed, the sliprings are short circuited and the operation is similar to that of the cage motor.

LOSSES AND EFFICIENCY

The losses of an induction motor include (1) stator copper loss and rotor copper loss; (2) stator and rotor core loss; and (3) friction and windage loss. The power output may be measured on a mechanical brake, or calculated from a knowledge of the input and the losses. The efficiency is equal to the ratio of the output to input power; and at full load, it varies from about 85 percent for small motors to more than 90 percent for large motors.

CHARACTERISTICS OF THE CAGE-ROTOR MOTOR

The cage-rotor induction motor is comparable to a transformer with a rotating secondary. The power factor of the motor with no load is very poor. It may be as much as 30 percent lagging. Because there is no drag on the rotor, it runs at almost synchronous speed.

When load is added to the motor, the rotor slows down slightly. The motor torque increases more than the decrease in speed and the power output increases. A small reduction in speed and counter emf in the primary may be accompanied by large increases in motor

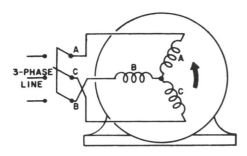

① TO REVERSE, INTERCHANGE ANY TWO MOTOR LEADS.

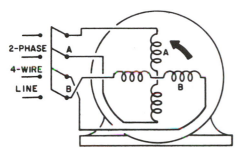

② TO REVERSE, INTERCHANGE THE LEADS OF ONE PHASE.

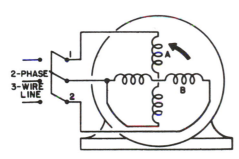

③ TO REVERSE, INTERCHANGE THE OUTER TWO MOTOR LEADS, 1 AND 2.

Fig. 15-3. Reversing rotation of AC motors.

current, torque, and power output. Thus, the cage-rotor motor has especially constant-speed variable torque characteristics.

When the rotor of an induction motor is locked (stalled), the voltage applied to the primary winding should not exceed 50 percent of its rated voltage.

At standstill, stator current is nearly 5 times normal; hence, constant-potential motor circuits like the one supplying the motor are equipped with time-delay automatic-overload protective devices. Sustained overload causes a

circuit breaker to open and thus protects both the motor and the circuit from damage.

The cage-rotor induction motor has a fixed rotor circuit. The resistance and inductance of the windings are determined when the motor is designed and cannot be changed after it is built. The standard cage-rotor is a general purpose motor. It is used to drive loads that require a variable torque at approximately constant speed with high full-load efficiency—such as blowers, centrifugal pumps, motor-generator sets, and various machine tools. If the load requires special operating characteristics, such as high-starting torque, the cage rotor is designed to have high-resistance. The starting current of a motor with a high-resistance rotor is less than that of a motor with a low-resistance rotor. The high-resistance rotor motor, like the cumulative compounded DC motor, has wider speed variations than the low-resistance rotor motor. These motors are used to drive cranes and elevators when high-staring torque and moderate-starting current are required, and when it is desired to slow down the motor without drawing excessive currents.

CHARACTERISTICS OF WOUND-ROTOR MOTORS

The wound-rotor, or slipring, induction motor is used when it is necessary to vary the rotor resistance in order to limit the starting current or to vary the motor speed. Maximum torque at start can be obtained with a wound-rotor motor with about 1.15 times full-load current, whereas a cage-rotor may require 5 times full-load current to produce maximum torque at start. The wound-rotor motor is desirable for an application which requires frequent starts.

The advantages of the wound-rotor induction motor over the cage-rotor induction motor are: 1) high-starting torque with moderate starting current 2) smooth acceleration under heavy loads, 3) no excessive heating during starting, 4) good running characteristics, and 5) adjustable speed. The chief disadvantage of the wound-rotor motor is that the initial and maintenance costs are greater than those of the cage-rotor motor.

DUAL-VOLTAGE MOTORS

Many motors are available for service on two alternate voltages of the ratio of 1 to 2 between the two choices, such as 120/240, 230/460, etc. This makes the selection easier from a reserve stocking and replacement parts aspect. The dual voltage motor costs slightly more, but it is usually well worth the extra cost. There is a decided cost advantage if the higher of the two voltages is selected when wiring the motor, as the size of the wiring required is a smaller circular mil value. This not only means cheaper wiring, but there may be a reduction in the conduit size, with more flexibility in routing possible. There is no difference in power output or performance at the higher voltage.

Figures 15-4 and 15-5 illustrate the wiring diagrams for the connections of typical three-phase motors shipped from the factory in dual-voltage arrangement. Remember, these are only typical, the motor nameplate must be consulted for the proper wiring sequence. Notice that all three-phase dual voltage motors are provided with nine leads, and the wiring diagrams merely show how to connect the six extra leads to adapt to the three incoming power leads. The motor shipped from the factory may have the leads already arranged for one of the two voltages, which is another reason why the nameplate must be read very carefully.

From the wiring diagrams for single-speed three-phase motors shown in this chapter, it will be noticed that there are three phase coils, wired in either the delta or the Y configuration internally. In the case of dual-voltage three-phase motors, there are a total of six coils, with each of the three phase coils being split into two separate coils. The wiring diagrams in Figs. 15-4 and 15-5 merely provide for connecting each pair of coils on a phase in series for the high voltage service, and in parallel for the low voltage service.

CHOOSING A MOTOR

In Chapter 3 we started the reader on a typical problem situation involving installing a

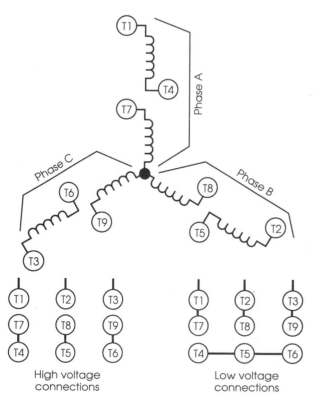

Fig. 15-4. Dual-voltage three-phase motor, Y connections.

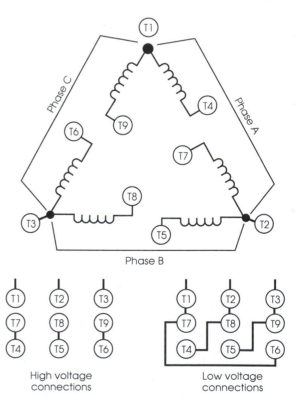

Fig. 15-5. Dual-voltage three-phase motor, delta connections.

new 50 hp motor in the plant, with about 225 feet of cable. As another step in completion of that project, we are now ready to investigate some of the steps in selecting the motor for the load. We stated at the time that the motor would be a standard duty type, with normal starting torque. The starting current for this type of motor will be very high, around five or six times the normal running full-load current. This will place a very heavy draw on the plant's electrical supply system, and could cause trouble in the entire plant by dropping the voltage for the several seconds it takes the motor to reach full speed.

Motors of such design as this are satisfactory when they start up on very little load. Pumps, blowers, and such which build up the load as the speed increases are typical applications for the normal starting torque motor.

Let us assume that the load being applied is one that starts right out at a high torque demand, such as a conveyor belt, direct drive power pump, gear driven mixer, or other equipment of like characteristic. The motor then would have to be a type which could handle this high starting torque without a corresponding high current demand. The wound-rotor motor is the proper type to use on this load, whereas the cage rotor motor would be the one offered by the supplier for the standard duty motor.

When ordering a new motor for your application, always remember that the supplier will have to know the type of load it is going to handle, and as much other information you can give him which will assist in choosing the correct motor. For assistance, please refer to page 315 in the Appendix.

The starting equipment for the two different types of motors will also be different. We will get into that problem in a later chapter.

B. Synchronous Motors

The synchronous motor is one of the principal types of AC motors. Like the induction motor, the synchronous motor makes use of a ro-

tating magnetic field. Unlike the induction motor, however, the torque developed does not depend on the induction of currents in the rotor. Briefly, the principle of operation of the synchronous motor is as follows. A multiphase source of AC is applied to the stator windings and a rotating magnetic field is produced. A direct current is applied to the rotor windings and another magnetic field is produced. The synchronous motor is so designed and constructed that these two fields react upon each other in such a manner that the rotor is dragged along and rotates at the same speed as the rotating magnetic field produced by the stator windings.

A rheostat placed in series with the DC source provides the operator of the machine with a means of varying the strength of the rotor poles, thus placing the motor under control for varying loads (see Fig. 15-6).

If a synchronous motor is driven by an external power source, and the excitation, or voltage applied to the rotor, is adjusted to a certain value called 100 percent excitation, no current will flow from or to the stator winding. In this case, the voltage generated in the stator windings by the rotor, or cemf, exactly balances the applied voltage. However, if the excitation is reduced below the 100-percent value, the difference between the cemf and the applied voltage produces a reactive component of current which lags the applied voltage. The machine then acts as an inductance. Similarly, if the excitation is increased above the 100-percent value, the reactive component leads the applied voltage, and the machine acts as a capacitor. This feature of the synchronous motor permits use of the machine as a power-factor correction device. When so used, it is called a synchronous capacitor.

When used as a synchronous capacitor, the motor is connected on the AC line in parallel with the other motors on the line, and run either without load or with a very light load. The rotor field is overexcited just enough to produce a leading current which offsets the lagging current of the line. Unity power factor results.

The synchronous motor can be made to pro-

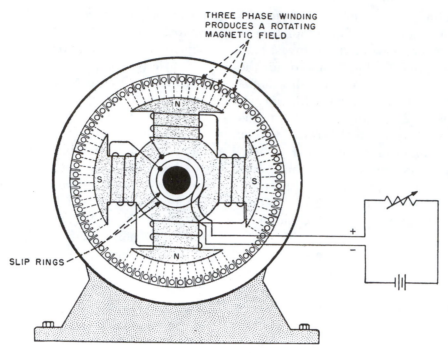

Fig. 15-6. Synchronous motor and control.

duce up to 80-percent leading power factor. However, because leading power factor on a line is just as detrimental as a lagging power factor, the synchronous motor is regulated to produce just enough leading current to compensate for lagging current in the line.

The synchronous motor is not basically a self-starting motor. The rotor is heavy and, from a dead stop, it is impossible to bring the rotor into magnetic lock with the rotating magnetic field. For this reason, all synchronous motors have some kind of starting device. A simple starter is another motor, either AC or DC, which brings the rotor up to approximately 90 percent of its synchronous speed. The starting motor is then disconnected and the rotor locks in step with the rotating field. Another starting method is a second winding of the squirrel-cage type on the rotor. This induction winding brings the rotor almost to synchronous speed and, when the DC is connected to the rotor windings, the rotor pulls into step with the field. The latter method is the more commonly used.

C. Checking a Motor for Shorts

The troubleshooting chart in Section D refers to situations where it is necessary to check a stator or field coil for shorts or grounds. In this section we shall give you the simplest methods available for doing that. These are taken from the U.S. Army Training Manual TM-5-764, as listed in the Bibliography. For more complete instructions, it is best to contact the Government Printing Office for their price list of current literature covering electrical maintenance, or contact the supplier of the motors in your shop.

Most failures of insulated windings are not immediately detectable; therefore, certain tests must be made before the reason for failure can be determined. The simplest method for detecting failures of insulated windings is visual inspection. A burned-out winding, for example, is revealed by charred and blackened insulation. Data about the condition of the winding, collected over a period of time, can provide clues as to the nature of failure so that time is not wasted in unnecessary testing to locate and isolate the defect. Before disconnecting any of the field, stator, or armature windings, test the winding circuit for grounds, shorts, or open circuits. If any of these are apparent, disconnect the separate coil connections and test each coil winding separately to locate the defective one.

ARMATURES

Place one lead of the ohmmeter on the shaft or core of the armature and, with the other lead, contact each successive bar around the commutator. A deflection of the meter pointer, when the ohmmeter lead is touched to a commutator bar, indicates that a coil is grounded.

WOUND ROTORS

All of the methods described above can be applied to checking and testing wound rotors. The only difference is that the collector rings are contacted or connected one to the other instead of the commutator bars. If a ground is indicated at any one of the collector rings, disconnect the coil connections and test each coil separately to locate the ground.

STATORS AND FIELDS

The most likely locations of grounds of field coils are shown in Fig. 15-7. The following method of checking and testing for grounds is the simplest one.

OHMMETER METHOD

Connect one lead of the ohmmeter to the stator or field frame and the other lead to the winding being checked. A deflection of the meter pointer indicates that ground exists. When it is determined that a ground is in the winding circuit, disconnect each phase and check it to ground. When the grounded phase is located, disconnect each coil group of the phase (Fig. 15-8) and check them separately until the ground is located. DC field circuits are first tested as circuits (shunt, series, commutating), then the grounded circuit coils are disconnected and checked separately until the grounded coil or coils are located.

THE TEST LAMP

A very useful tool for testing for shorts, open circuits, and grounds is the test lamp, illustrated in diagram form in Fig. 15-9. It is a simple device, and one every electrical repair shop should have. The voltage source is the 1.5 volt dry cell, and the entire package can be made in

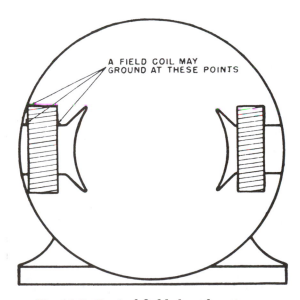

Fig. 15-7. Typical field short locations.

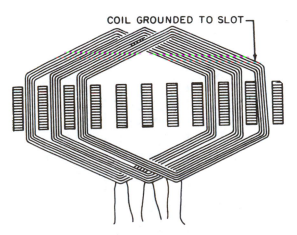

Fig. 15-8. Testing each coil separately.

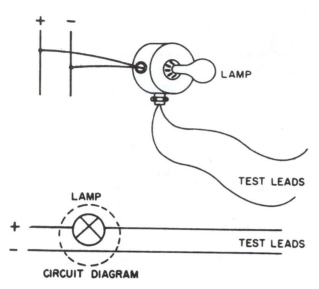

Fig. 15-9. Diagram of a test lamp.

one assembly, easily portable. It tests for continuity of a circuit, as a glowing lamp indicates the presence of a short in the windings.

Figures 15-10 and 15-11 illustrate how the test lamp is used to check for phase winding troubles in star-connected stators and delta-connected stators.

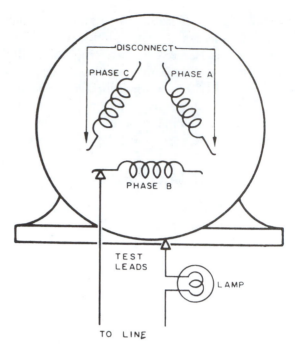

Fig. 15-11. Locating trouble in a delta-connected stator.

SHORT CIRCUITS, OPEN CIRCUITS, AND JOINTS

Methods similar to those described above for grounds will be applicable for these tests. Figure 15-12 shows how to check field pole windings for open circuits.

ARMATURES

Connect one side of a low-voltage, DC power source (1.5 volt dry cell) to the commutator bar and the other side, as shown in Fig. 15-13 in series with a variable resistance.

Use a strap made of webbing, leather, or some strong, pliable insulating material to hold the series-circuit connections to the commutator bars, so that the hand will be free to work the millivoltmeter leads. The electrode connecting points must be spaced as follows: for a two-pole machine, the electrodes must span $\frac{1}{2}$ the total number of commutator bars; for a four-pole machine, $\frac{1}{4}$; a six-pole, $\frac{1}{6}$; and so on, for the number of poles on the machine. Touch the millivoltmeter leads to a pair of ad-

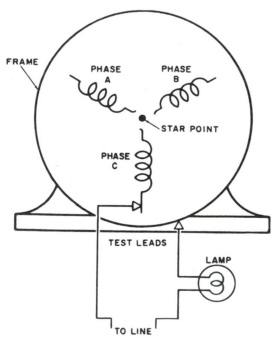

Fig. 15-10. Locating trouble in a star-connected stator.

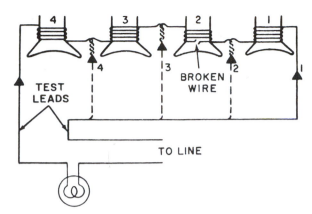

Fig. 15-12. Determining which pole winding is open circuited.

jacent bars within the area spanned by the electrodes. Adjust the variable resistance so that a $\frac{3}{4}$ scale deflection of the meter pointer is obtained (Fig. 15-13). Retain this resistance adjustment throughout the entire test procedure. With the series circuit so adjusted, proceed to test each pair of adjacent bars with the meter. Start at the first electrode on the left and, as the meter leads are moved to the right, be certain that the left meter lead contacts the bar on which the right meter lead was previously applied (Fig. 15-13) until the electrode on the right is reached. Move the left electrode of the strap to the bar which the right electrode occupied and proceed to test the next pole section of the armature. Continue in this manner until every bar of commutator has been tested. Any material increase in the meter reading indi-

cates an open circuit or a high-resistance joint, and any decrease indicates a short-circuited coil.

WOUND ROTORS

All of the methods described above for armature testing are suitable for testing wound rotors, provided the coil ends are disconnected from the collector rings.

STATORS AND FIELDS

Figure 15-14 diagrams the typical markings on the leads of a compound motor to aid in testing.

OHMMETER METHOD

Check the resistance of an armature coil with the ohmmeter (Figs. 15-15 and 15-16). Refer to armature-connection diagrams as furnished by the motor supplier for aid in determining the location of an armature coil. Knowing the resistance of one armature coil, it will be possible to determine, by comparative reading checks, whether a coil is short-circuited on itself. A short-circuited coil is indicated by a lower value of resistance on the ohmmeter scale. To

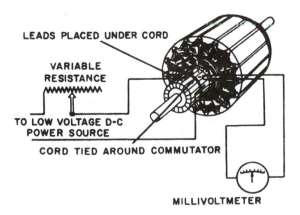

Fig. 15-13. Testing for shorts in an armature.

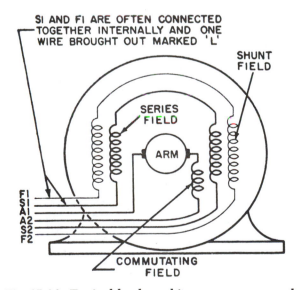

Fig. 15-14. Typical lead markings on a compound motor.

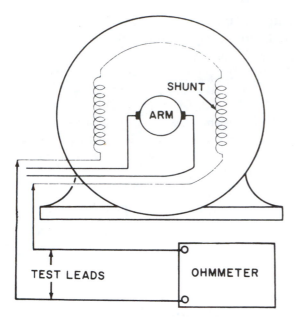

Fig. 15-15. Testing a shunt motor for open circuits.

than the resistance value of a single coil, when checking a single coil, indicates a high-resistance connection. For similar ohmmeter tests on open field coils, see Figs. 15-15 and 15-16.

TEST LAMP METHOD (FIGURES 15-17 TO 15-19)

It is practical to apply this method only in checking for short circuits of one coil side to another. The procedure will be the same as that given for an ohmmeter, as outlined above. By application of the knowledge gained of a particular light intensity for a given coil resistance, it is possible, in some cases, to use this visual

determine the short circuit of one coil to the next, connections to the commutator must be opened and the ohmmeter applied to the coil sides of two separate coils. A meter-pointer deflection indicates one coil side of a coil is short-circuited to the coil side of another coil. Absence of a deflection of the meter pointer indicates an open circuit, and a reading higher

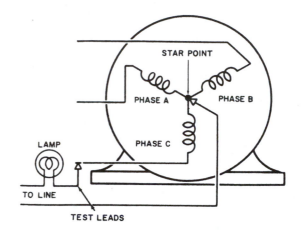

Fig. 15-17. Testing for open phase, star-connected stator.

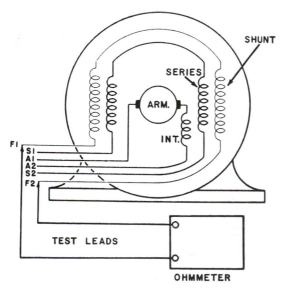

Fig. 15-16. Testing compound motor for open circuits.

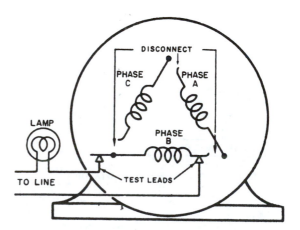

Fig. 15-18. Testing for open phase, delta-connected stator.

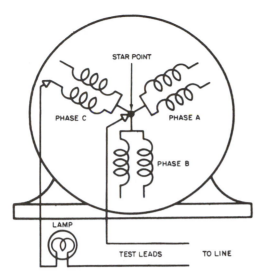

Fig. 15-19. Locating open coil in a two-parallel stator.

means for determining a coil shorted on itself or a high resistance joint. A coil shorted on itself will give a lamp glow of a bright intensity, while a lessening of light intensity will indicate a high resistance.

MEGOHMMETER OR MEGGER

This method is practical only if a dual or multiple scale megohmmeter is available. However, a single high-scale-reading megger may be used for checking shorts between coils. A zero reading indicates a short circuit. The procedure outlined above for an ohmmeter is applicable for these tests.

IMPROPER CONNECTIONS

Improper connections of insulated windings in motors and generators can generally be located in:

1) connections of coils in groups

2) connections of these groups

3) connections of external leads.

Improper connection in any one of above categories may lead to serious trouble.

D. Troubleshooting Three-Phase Motors

AC INDUCTION MOTORS

Failure to Start

Probable Cause	Remedy
Circuit breaker or fuse open	Check for grounds. Close breaker or replace fuse.
Overload relay open	Wait until motor cools and relay closes.
Low supply voltage	Apply correct voltage.
Stator or rotor windings open or shorted	Check and replace shorted coil or coils.
Winding grounded	Check and replace grounded coil or coils.
Overload	Check meter readings against nameplate ratings. Reduce load or install larger motor.

Noisy Operation

Probable Cause	Remedy
Unbalanced load or coupling misalignment	Balance load and check alignment.
Air gap not uniform	Center rotor by replacing bearing.
Lamination loose	Tighten bolts. Dip in Varnish and bake. Repeat several times.
Coupling loose	Tighten.

Overheating

Probable Cause	Remedy
Overloaded	Check meter readings against nameplate ratings. Reduce load.
Electrical unbalance	Balance supply voltage.
Open fuse	Replace line fuse.
Restricted ventilation	Clean. Remove obstruction.
Rotor winding shorted, open, or grounded	Check and replace defective coil or coils.
Stator winding shorted, open, or grounded	Check and replace defective coil or coils.
Bearings	Check for worn, loose, dry, or overlubricated bearings. Replace worn or loose bearings, lubricate dry bearings, relieve overlubrication.

AC WOUND ROTOR MOTORS

Runs Slow with External Resistance Cutout

Probable Cause	Remedy
Cables to control box have insufficient current-carrying capacity	Replace with larger cables.
Open circuits in rotor, cables, or controls	Clean, remake connections, and repair.
Excessive brush sparking	Clean sliprings and reseat brushes.

AC SYNCHRONOUS MOTORS

Failure to Start

Probable Cause	Remedy
Open fuse	Replace fuse.
Faulty starter	Check and repair or replace contacts or contactor coils.
Low supply voltage	Apply correct voltage.
Bearings	Check for bent shaft or worn, loose, dry, or overlubricated bearings. Replace and realign bent shaft. Replace worn and loose bearings, lubricate dry bearings, relieve overlubrication.
Overloaded	Check meter readings against nameplate ratings. Reduce load or install larger motor.
Stator coil open or shorted	Repair or replace coil or coils.
Field exciter current is being applied	Make sure that field contactors are open, and that field-discharge resistors are connected.

Runs Slow

Probable Cause	Remedy
Overloaded	Check meter readings against nameplate. Reduce load or install larger motor.
Low supply voltage	Adjust correct voltage.
Field excited too soon	Adjust time-delay relay so that exciter current will not be applied until rotor reaches synchronous speed.

Failure to Pull Into Step

Probable Cause	Remedy
No field excitation. Open rotor coils. Exciter inoperative. Faulty field contactor	Tighten or solder open or loose connections. Repair or replace defective rotor coils. Be sure field contactor is operating properly.
Overloaded	Check meter readings against nameplate ratings. Reduce load or install larger motor.

No Field Excitation

Probable Cause	Remedy
Grounded or open rotor coil	Repair or replace rotor coil or coils.
Grounded or short sliprings	Check and reinsulate.
No output from exciter	See DC generator, troubleshooting section.

Pulls Out of Step, or Trips Breakers

Probable Cause	Remedy
Low exciter voltage	Readjust voltage regulator on exciter to increase voltage.
Intermittently open or shorted cables	Check, and replace defective cables.
Reversed field coil	Check polarity. Change coil leads.
Low supply voltage	Increase voltage if possible. Raise excitation voltage.

Hunting

Probable Cause	Remedy
Fluctuating load	Increase or decrease size of flywheel on load or loads. Increase or decrease excitation current.
Uneven commutator	Recondition commutator.
Open phase coil	Check, and repair or replace faulty coil or coils.
Rotor not centered	Check for bent shaft, loose or worn bearings. Straighten and realign shaft. Replace bearings.
Unbalanced circuits	Repair loose connections, or correct wrong internal connections.
Shorted coil	Check, and replace faulty coil or coils.

Field Overheats

Probable Cause	Remedy
Shorted field coil	Check, and replace faulty coil or coils.
Excitation current too high	Reduce exciter current by adjusting DC voltage regulator.

Overheating

Probable Cause	Remedy
Overloaded	Check meter readings against nameplate ratings. Reduce load or install larger motor.
Underexcited rotor	Adjust to rated excitation.
Improper ventilation	Remove obstruction and clean air ducts.
Improper supply voltage	Adjust to rated voltage.
Reverse field coil	Check polarity. Change coil leads.

E. Dual-Voltage Motors

From Figs. 15-4 and 15-5, it can be found that dual-voltage three-phase motors have six coils instead of three as in the single-voltage motors. Also, this results in nine leads being available instead of three. From a servicing analysis, this increases the testing required to find a shorted coil, using the techniques already described for single-voltage three-phase motors. Also, it is very important that the leads be properly identified for accurate testing and location of shorted coils. Any markings on the leads must be maintained intact and in readable order, which is not always easy to attain. One or more tags often become detached and lost from the work of connecting or disconnecting the leads to the power lines.

Should this happen, with only one or a few of the tags missing, it is usually possible to identify the remaining leads by means of a continuity check with a test lamp or ammeter. This may tax the ingenuity of the serviceman if too many of the tags are missing, and will require a little analytical thinking to enable the serviceman to identify and tag each lead in turn, until they are all properly tagged.

When being required to install a used motor which has been in storage for some time, and is known to be a dual-voltage motor, we then offer the following suggestions.

1. Check the nameplate data to see how it should be wired and tagged, and to familiarize yourself with the motor characteristics.

2. Remove the terminal box cover very carefully, so as not to disturb the leads.

3. Before touching the leads, notice the condition of the terminals and the tags.

4. Check for loose tags in the box or in the cover. Attempt to determine where they belong without disturbing the leads.

5. Make a note of which tags were loose, and draw a diagram showing where you attached the loose tags.

6. Very carefully remove the remainder of the leads, one at a time, noting the tags on each, and tightening each loose tag.

7. When all of the leads are exposed, check your diagram with the one on the nameplate.

8. If the tagging is wrong for the voltage available, then check Fig. 15-4 and 15-5, and correct your diagram to show the lead changes required.

Before connecting the motor to the power, it should be checked, cleaned, and lubricated, using the instructions furnished by the manufacturer, if available.

If the motor is going to be used on the high voltage rating, then follow the instructions under "Preparing to Energize Circuit" below.

If the motor is going to be used on the low voltage rating, then omit items 1 and 7.

Any motor which has been in storage for some time should be in suspect condition. After putting it in service on a load, it should be watched very carefully for several days. The current draw in each leg, and the motor and bearing temperatures should be checked often.

Should none of the nine leads be tagged, then

the problem becomes rather complicated, as it is necessary to start from square one. To help solve this dilemma, the following procedure is offered.

Identifying Leads of Unmarked Nine-Lead Motor

1. Use ohmmeter to find Y and other three pairs.

2. Tag Y leads 7, 8, and 9.

3. With cheater cord energize 7 and 8 with 120 volts.

4. Read voltage across pairs to locate high and low pairs.

5. Tag the low pair 3 and 6.

6. Energize 7 and 9.

7. Tag the low pair 2 and 5, and the remaining pair 1 and 4.

8. Jumper 4 and 7 and energize 7 and 8 with the 120 volts.

9. Read voltage between 8 and 7, and 8 and 1.

10. If voltage is high between 8 and 1, numbers are correct.

11. If voltage is low between 8 and 1, change numbers 1 and 4.

12. Jumper 5 and 8, and energize between 7 and 8 with 120 volts.

13. Read voltage between 8 and 7, and 7 and 2.

14. If voltage is high between 7 and 2, numbers are correct.

15. If voltage is low between 7 and 2, change numbers 2 and 5.

16. Jumper 6 and 9, and energize between 7 and 9 with 120 volts.

17. Read the voltage between 7 and 9, and 7 and 3.

18. If voltage is high between 7 and 3, the numbers are correct.

19. If voltage is low between 7 and 3, change numbers 6 and 3.

Preparing to Energize Circuit

1. Connect the leads for the high voltage connection.

2. Install fuses for the low voltage ampere rating.

3. With the circuit de-energized, check each phase on the load side of the disconnect switch for short circuits. On the load side of the magnetic starter, test for resistance. Use a low reading ohmmeter and check across T1 & T2, T1 & T3, and T2 & T3. If there is some resistance, then the circuit is OK to energize.

If there is no resistance at all, then a short circuit exists and do not energize.

4. If resistance checks are OK, have a clamp-on ammeter ready and set on the highest scale. Energize with low voltage after reading steps 5 and 6.

5. If the motor makes a loud magnetic hum, de-energize immediately and recheck tags, connections, and line voltage.

6. If motor sound alright, make a quick check of the current in all three phase legs.

7. Reconnect to lower voltage connections and repeat steps 4–6.

C H A P T E R 16
PLANT ELECTRICAL DISTRIBUTION SYSTEMS

A. Main Transformer Station

The Public Utility supplies the electric power at high voltage, usually on poles beside the plant property, as a three-phase three-wire supply. This high voltage must be brought down into the plant premises, then reduced to the basic voltage which the plant equipment is designed to use, which is usually 460-volt, three-phase, three-wire. The various arrangements were covered in Chapter 7. See Fig. 16-1 for a picture of how this is accomplished for a typical residential area. For an industrial area, the last transformer would be wired on the secondary side to match the three-phase plant requirement.

Figure 16-2 is a one-line diagram of the basic plant distribution system starting at the utility poles outside the plant boundary. The high voltage power is brought down to the transformer station, then is fed through various protective devices, such as circuit breakers, past lightning grounding rods, through a meter, then into the transformer. This transformer may be fairly large, and must be cooled either by natural air circulation or by fans. It is often isolated from the plant to prevent fire danger to the plant, and must be protected against unauthorized access. This will mean a chain link fence with a locked gate, or if it is in a room inside the plant, it will have a

locked door. In either case, there will be a sign warning of "High Voltage." See Figs. 16-3 and 16-4 for typical plant transformer stations.

Fire is always a possibility at the transformer station. Therefore, if it is inside a building, there will probably be a fire smothering system installed, either carbon dioxide or Halon. They are usually set to smother the station upon detection of excess heat over the transformer, or in its vicinity. This equipment must be checked at regular intervals, and the storage bottles maintained at full pressure at all times.

If the station is outside, then an automatic Halon smothering system is not of much use, as the gas would disperse before it could smother the flames. In this case, amply sized carbon dioxide extinguishing equipment is kept nearby, all according to local codes and Fire Marshall regulations.

There may be a circuit breaker mounted on the utility pole, with a long rod extending down to the grade, arranged for closing the breaker again, once the trouble has been corrected. The rod will be locked to prevent unauthorized opening of the circuit to the plant. Its main purpose is to protect both the plant and the Utility's system in case of a dead short in the plant's system or at the transformer station. It trips on high current flow, which is the normal purpose of circuit breakers.

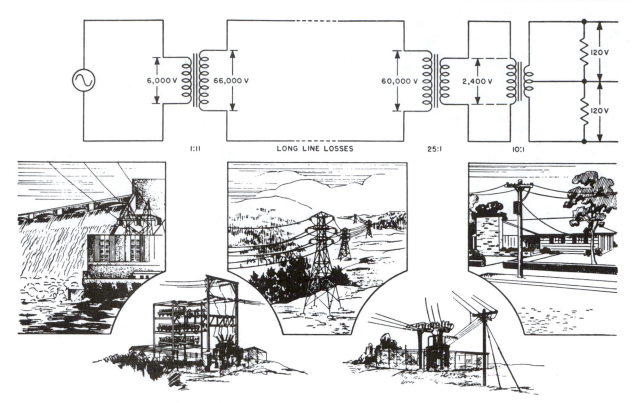

Fig. 16-1. Long distance power transmission.

B. 460-Volt Three-Phase Distribution Panels

The reduced voltage is fed from the transformer station into one or more 460-volt three-phase distribution panels by means of either heavy electrical cable, described in Chapter 3, or by heavy copper bars, known as busbars. Figure 16-5 is a composite drawing of a typical type of busbar, with its enclosure. For large current flows, this arrangement is much easier, cheaper, and safer to use than heavy stranded cable inside large conduit. One of its main attractions is the ease with which tie-ins may be made at various locations along its length. This is shown in use in Fig. 16-6.

There are many variations of cable and busbar distribution systems available on the market, and the styles and features are constantly being improved. We have shown only one; there are systems available from simple open cable trays, enclosed rectangular ducts, to the ones shown in Figs. 15-5 and 15-6.

The distribution panel may take one of several forms, from simple open front panelboards to large, enclosed walk-in style compartments. These latter arrangements must usually meet local building codes, and will have some type of forced ventilation to dissipate the heat from the electrical equipment.

The usual type of panelboard will be inside of the plant, will have a NEMA code enclosure suitable for the environment, and will have a locked, dead front door. Some, however, have the controls exposed, such as the switches, pilot lights, etc.

The main item inside of the distribution panel will be additional circuit protection in the form of a circuit breaker for each branch electrical service being fed out of the panel and into various portions of the plant. These circuit breakers are important, as they isolate each circuit or branch from the remainder of the plant, in case of a dead short in any one of the branch systems.

An important part of the protection system

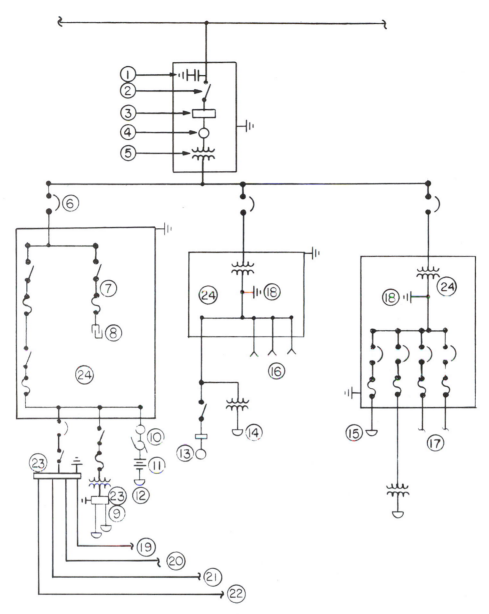

Fig. 16-2. Typical plant one-line distribution diagram.

1 - Lightning arrestor and grounding rod. **2** - Manual disconnect switch. **3** - Oil immersed circuit breaker. **4** - Plant metering station. **5** - Plant transformer, 460-volt output. **6** - Air circuit breakers, typical all branch feeders. **7** - Fused switches, typical as shown. **8** - Power factor correction capacitors. **9** - Miscellaneous 240-volt three-phase plant equipment. **10** - Motor generator set or rectifier battery charger. **11** - Battery bank. **12** - Plant emergency lighting. **13.** - Machine tools. **14** - Machine tool lighting systems. **15** - Plant lighting system. **16** - 230-volt convenience outlets. **17** - 115-volt convenience outlets, office circuits, etc. **18** - Circuit grounding connection. **19** - Cranes, hoists. **20** - Welders. **21** - Elevators. **22** - X-ray equipment, induction heaters. **23** - Motor control centers. **24** - Distribution panel.

here will be the grounding connection from the panel, either into a central grounding cable network, or grounded directly into a grounding rod at each panel.

Installed in the distribution panels will usually be found heavy, box-like capacitors, simi-

lar to those described earlier, for improvement of the plant's power factor. These must be checked regularly to guard against shorts developing.

All panelboards must be rigidly anchored to resist the thrust which develops if an exceed-

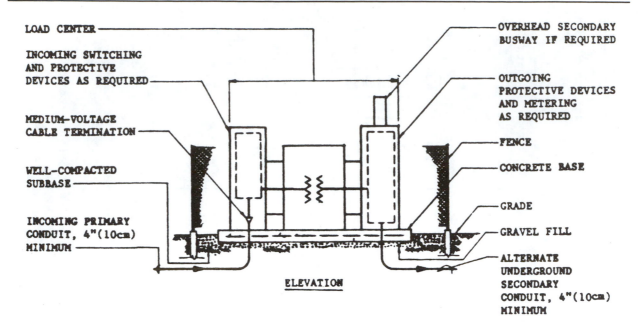

LOAD CENTER

INCOMING SWITCHING
AND PROTECTIVE
DEVICES AS REQUIRED

MEDIUM-VOLTAGE
CABLE TERMINATION

WELL-COMPACTED
SUBBASE

INCOMING PRIMARY
CONDUIT, 4"(10cm)
MINIMUM

OVERHEAD SECONDARY
BUSWAY IF REQUIRED

OUTGOING
PROTECTIVE DEVICES
AND METERING
AS REQUIRED

FENCE

CONCRETE BASE

GRADE

GRAVEL FILL

ALTERNATE
UNDERGROUND
SECONDARY
CONDUIT, 4"(10cm)
MINIMUM

ELEVATION

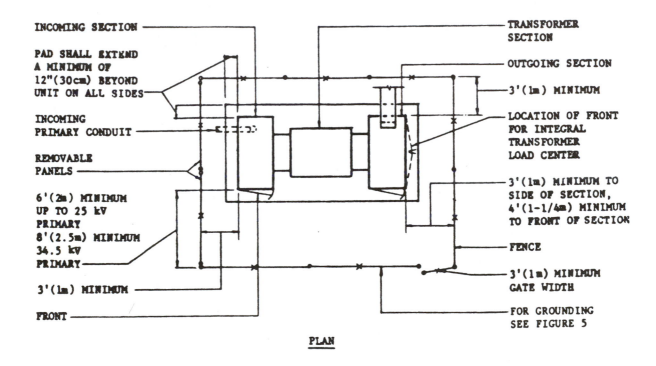

INCOMING SECTION

PAD SHALL EXTEND
A MINIMUM OF
12"(30cm) BEYOND
UNIT ON ALL SIDES

INCOMING
PRIMARY CONDUIT

REMOVABLE
PANELS

6'(2m) MINIMUM
UP TO 25 kV
PRIMARY
8'(2.5m) MINIMUM
34.5 kV
PRIMARY

3'(1m) MINIMUM

FRONT

TRANSFORMER
SECTION

OUTGOING SECTION

3'(1m) MINIMUM

LOCATION OF FRONT
FOR INTEGRAL
TRANSFORMER
LOAD CENTER

3'(1m) MINIMUM TO
SIDE OF SECTION,
4'(1-1/4m) MINIMUM
TO FRONT OF SECTION

FENCE

3'(1m) MINIMUM
GATE WIDTH

FOR GROUNDING
SEE FIGURE 5

PLAN

NOTE: FIGURE IS SHOWN FOR AN EXTERIOR INSTALLATION. SIMILAR FOR
INTERIOR INSTALLATION EXCEPT FENCE ENCLOSURE WILL BE WALL
OF BUILDING, CONCRETE PAD, ETC., WILL BE FLOOR SLAB AND
VENTILATION, LIGHTING, ETC., IS NECESSARY.

Fig. 16-3. Transformer substation installation.

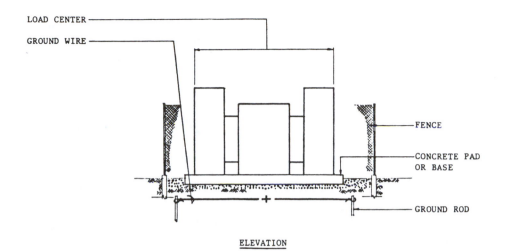

Fig. 16-4. Transformer substation grounding.

ingly large surge of current passes through the panel's circuiting. This surge, as from a dead short, can cause an induced current in the panelboard, producing a powerful thrust, which must be resisted by the panelboard itself.

Between the incoming 460-volt feeder and the distribution busbar to the various motor control centers downstream from the distribution panels will be found fused switches, in series with the circuit breakers. These give additional protection, as well as permitting isolation of the various motor control centers.

C. 240-Volt Three-Phase Distribution panels

As these are usually fed directly from the 460-volt busbar from the main transformer station, a transformer will be required within the 240-volt panel to provide the reduced voltage.

These panels will be similar in most other respects to the 460-volt panels described in Section B above. They will serve the smaller three-phase loads and motors in the plant, such as small compressors, pumps, and similar equipment.

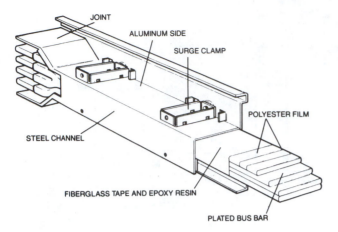

Fig. 16-5. Typical feeder busway construction. (Courtesy of Square D Co.)

The feeder from the main transformer will be by busbar or stranded cable, as for the 460-volt panel.

D. Three-Phase Motor Control Centers

Motor Control Centers often provide the most suitable method for grouping electric motor controllers, circuit breakers, and fusible switch disconnects, and other related devices in a compact, economical arrangement.

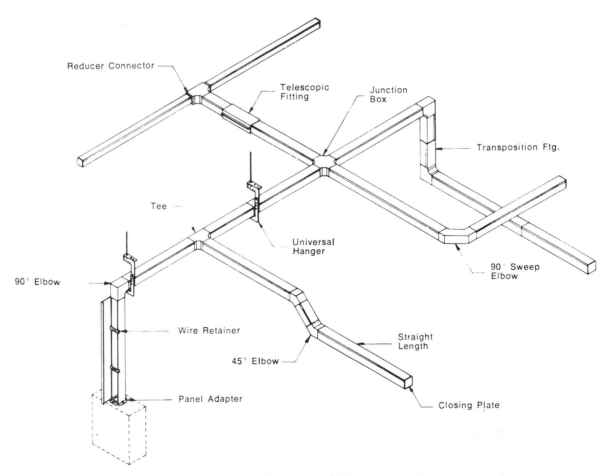

Fig. 16-6. Typical area distribution system. (Courtesy of Square D Co.)

MCC's are floor-mounted or wall-mounted assemblies of one or more enclosed vertical sections. Each vertical section in an MCC lineup normally contains a horizontal common power bus along with an uncommon vertical bus connected to the horizontal bus. This enables power to be distributed throughout the MCC lineup. The vertical sections principally contain combination motor control units. These units are mounted one above the other.

In some cases, the MCC may only handle 240-volt services, for such things as machine tools, heavy duty hand power tools, etc. Also, the circuits for the 240-volt convenience outlets will come off this panel.

The combination magnetic starters and fusible switch disconnects will be described in the next chapter.

MCC structures are totally enclosed, dead-front, free-standing assemblies. The enclosures for the MCC's are chosen for the application. Enclosures can be dust-tight, rainproof, depending upon the environmental conditions.

See Fig. 16-7 for a typical motor control center.

E. 240/120-Volt Single-Phase Distribution Panels

These panels will be similar to the 240-volt three-phase distribution panels, with a transformer to reduce the voltage to 240 volts, but with only single-phase leads taken off the transformer. See Chapter 9 for this arrangement.

This panel will contain the usual protection, such as grounding, circuit breaker, proper enclosure, etc. In addition, there will be fused switches, pilot lights, and ground fault current interruptors.

The 120-volt single-phase distribution panel may be combined with the 240-volt panel described above, or it may be a separate panel. In the latter case, it will be similar in most respects to the 240-volt single-phase panel. It will contain the lighting and 120-volt convenience outlet circuit switches, and other low voltage systems for specialized services. Emergency lighting must come from one of the other distribution panels, to ensure a more reliable backup. Often it will be provided from the 460-volt three-phase panel, along with the battery charging system. See Fig. 16-8 for a typical

Fig. 16-7. Motor control center. (Courtesy of Square D Co.)

Fig. 16-8. Typical single-phase distribution panel. (Courtesy of Square D Co.)

F. Local Control Stations

Near to, and in sight of, the motor or other load being supplied with electrical power from one of the MCC's will usually be found a local switching station, for the purpose of isolation when servicing the load. The size and complexity will depend upon the size of the load. The average station will contain a disconnect switch, either fused or non-fused, with a locking device on the switch handle. This disconnect must be visible from the motor being serviced, to ensure that the circuit is out of service.

Some of the equipment found in the MCC's may be duplicated here, or it may be more convenient to place it in the local panel. This is true of magnetic starters, which are often combined with the fused disconnect switches in one housing. This housing may also contain auxiliary switches and pilot lights, producing a complete local control station.

Smaller load centers may only contain on-and-off switches, fused or non-fused. Such convenience switches as "jogging" switches may be located on this local panel, bracket, post, or similar support.

These items will be covered in more detail in the next chapter.

G. Load Distribution Diagrams

One of the most useful items for maintenance in the plant is a one-line diagram of your entire electrical system similar to Fig. 16-2. If you do not have one, then we suggest you start to produce one, by walking the entire plant system, and sketching each item of equipment, with its location relative to the remainder of the equipment in the system. It will prove to be an excellent training exercise, as it will make you acquainted with the electrical system under your care.

Or perhaps your plant files contain such a diagram, but it is badly outdated, as so often happens in today's industrial plant. Then it is time

for you, the maintenance mechanic, to update it.

Figure 16-2 is a simplified one-line diagram of a typical industrial plant electrical distribution system, provided for instructional use only. We gave you, in Figs. 4-10, 4-11, and 4-12, typical electrical design and construction drawings for a new technical trades educational building. These three drawings are typical of what should be in the plant's files, as provided by the architect/engineering firm doing the original design of the plant, and furnished as part of the drawings of record. They could very well be out of date by now, but they are very useful as a starting point for updating by the electrical maintenance staff as part of a training program for new employees.

H. Underground Distribution Systems

So far in this chapter we have assumed that all wiring is readily available, being above ground. This is not always the case, as much of the distribution of power may be below ground, not only leading into the transformer substation, but also within the plant boundaries.

There are good reasons for doing this. Mainly, it keeps the plant premises clean and clear of dangerous overhead power lines between buildings and areas which are widely separated, thus minimizing possibility of accidents from truck and crane travel.

The underground systems are of four general types, and we will give a brief discussion of each.

The simplest, and possibly the cheapest, is to use directly buried cable, which is insulated and armored, and laid in trenches, encased in loose sand. Usually this cable, if properly installed, will be covered with either plain wood planking or a slab of red colored concrete as protection. The cable in this case is difficult to break into for installing branch power supplies at intermediate points. When this has been done, a steel or concrete box at grade level may usually be found. This system of direct buried cable is usually found where only one or two cables will supply the power required.

A variation of this system is to run the cable inside metal conduit. When this is done, the cable is of a cheaper grade of covering, and the conduit is usually galvanized steel. There will be pull boxes or junction boxes at least every 600 feet, and junction boxes at points of power takeoff.

The next method, and one quite often used, especially when more than two cables are re-quired, is the multiple duct system, shown by details in Fig. 16-19. When encased in concrete, as shown here, the conduit is usually plastic, and several cables are run in each conduit. The cables in this case may be of a cheaper grade due to the protection given by the conduit and the concrete. Both electric power and communication cables may be run inside separate conduits in the concrete encasement, with sufficient separation between them to prevent interference from the electrical field induced by the power cable, which could cause trouble with the communication signals.

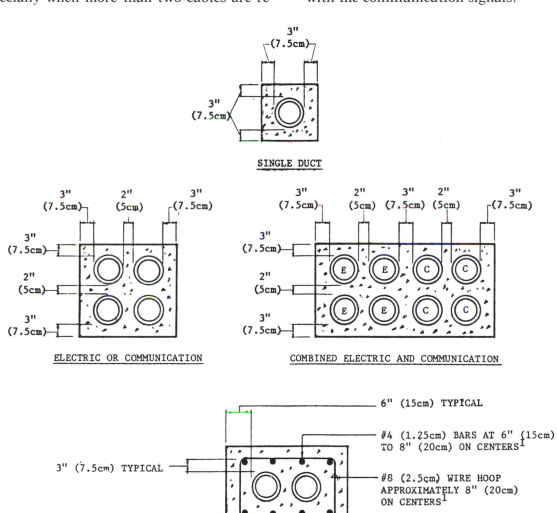

1WHERE REINFORCEMENT IS PROVIDED UNDER RAILROAD TRACKS IT SHALL EXTEND AT LEAST 12' (3.5m) BEYOND THE OUTER RAILS.

REINFORCED DUCT BANKS

Fig. 16-9. Concrete and duct power distribution.

The concrete encasement will be colored red or deep pink as a warning to any construction crew which may scrape against it. Also there will probably be colored plastic tape, sensitive to metal detection equipment, laid along the top of the concrete.

To take branch power or communication lines off from the concrete and conduit duct system, it is necessary that manholes or vaults be installed at junction points.

The most expensive method of running underground cabling between buildings and areas is inside tunnels. This is quite often done when there are a large number of high capacity cables being handled, or when other utilities are being run inside the same tunnel. This last arrangement can be dangerous, as any flooding of the tunnel or vaults can short out the electrical power, a fact to be kept in mind when maintenance work must be performed. For this reason, it is very essential that the tunnels and vaults be properly drained.

We have given only the bare facts regarding the methods of underground power transmission, as an orientation for the electrical maintenance worker. There are many variations of the above systems, and each will have its own maintenance problems.

I. Maintenance of the Distribution System

Much of the electrical gear installed in the plant distribution system has already been covered in previous chapters, and some will be covered in the next chapter. There are a few other general suggestions to be made on the complete system, however.

The outdoor plant transformer station is subject to troubles from any weather or climate difficulties prevalent in the area. Rain, hail, snow, and dust all take their toll on the equipment. Assuming that this has been considered when the station was built, it is the duty of the maintenance man to see that the weather-tight integrity is not breached in any manner. One of the main things to watch is the vulnerability of the main feeder from the high voltage line into the transformer station. Does it cross railroad lines or truck routes? Is it weighted down with sleet and ice in the winter? If there is a cooling tower or steam vent upwind of it, then there is the possibility of ice forming on the feeder or the transformer equipment in the winter. If this happens, it is best to consult with the Utility Company to see what can be done to correct the condition and the best way to clear away the ice and dirt endangering your plant.

The internal equipment for the most part has been covered, as we mentioned above. There is always the matter of maintaining adequate clearance in front of all of the electrical gear, as the Electrical Code requires about 3 feet clearance for servicing and any operation which has to be performed at the equipment. This is also an excellent idea for your own convenience and safety.

All electrical panels must be carefully serviced to maintain any seal around the doors and seams required to meet the NEMA enclosure class, and to prevent in-seepage of dust and moisture which could short out the system or cause a fire.

One helpful word of advice here: before opening any panel door in a high dust area, wipe all of the joints or seams which are going to be opened, and all portions which will discharge dirt into the panel after it is opened.

As an example of what should be done when servicing a motor control center, we will duplicate here the instructions furnished by the Allen-Bradley Company for their motor control centers, similar to that shown in Fig. 16-7. This is for your guidance, to be modified to suit the model in your shop.

PERIODIC SERVICING

The purchaser should establish a periodic maintenance program to avert unnecessary downtime caused by neglect of equipment. The frequency of the servicing required will vary with the extent of the equipment usage and the environment within which it operates. The following is a suggested checklist.

Warning: De-energize motor control center before servicing.

1. Inspect the motor control center a minimum of once per year.

2. Carefully inspect doors and enclosure sides for evidence of excessive heat. As a general rule, any temperature which the palm of the hand cannot stand for about 3 seconds may indicate trouble.

3. Check for any moisture or signs of dampness or drippings inside the motor control center. Condensation in conduits or dripping from an outside source is a common cause of motor control center failure. Seal off any conduits, cracks, or openings which have allowed moisture to enter the enclosure. Eliminate any source of dripping on the enclosure and any other source of moisture. Replace or dry and clean any insulating material which is damp or wet. Check devices such as contactors, circuit breakers, disconnect switches, relays, push buttons, etc., for wetness, contamination, or corrosion. Replace any damaged or malfunctioning parts.

(Be sure that the cause of any wetness or contamination has been identified and eliminated.)

4. Air Filters—Where blowers are used, air filters should be cleaned or changed periodically depending on specific environmental conditions encountered.

5. Operating Mechanisms—Check for proper functioning and freedom from sticking and binding. Replace any broken, deformed malfunctioning badly worn parts or assemblies.

6. Locking or Interlocking Devices—Check these devices for proper working condition and capability of performing their intended functions. If necessary, readjust, repair, or replace.

7. Contacts—Check contacts for excessive wear and dirt accumulations. Vacuum or wipe contacts with a soft cloth if necessary to remove dirt. Contacts are not harmed by discoloration and slight pitting. Contacts should never be filed, as it only shortens contact life. Contact spray cleaners should not be used as their residues on magnet pole faces or in operating mechanisms may cause sticking, and on contacts can interfere with electrical continuity. Contacts should only be replaced after silver has become badly worn. Always replace contacts in complete sets to avoid misalignment and uneven contact pressure.

8. Terminals—Check for loose connections in power and control circuits. Loose connections can cause overheating that could lead to equipment malfunction or failure. Replace any damaged parts or wiring.

9. Inspect main and vertical bus joints, and tighten if necessary. This will necessitate the removal of those units located in front of the horizontal bus.

10. If the motor control center horizontal bus has been spliced at the site, check the splices for tightness. Bus splices can be easily identified by the label on the interior of the vertical wireway doors or the interior right-hand side plate of frame-mounted units.

11. Coils—If a coil exhibits evidence of overheating (cracked, melted, or burned insulation), it must be replaced. In that event, check for and correct overvoltage or undervoltage conditions, which can cause coil failures. Be sure to clean any residues of melted coil insulation from other parts of the device, or replace such parts.

12. Pilot Lights—Replace any burned-out lamps or damaged lenses.

13. Fuses—If replacement is necessary, always install the same type and rated fuse that was originally furnished with your motor control center.

14. Inspect bus stab connections for wear or corrosion of the stab. Wear and corrosion could cause increased resistance causing increased temperature of the contact point which could lead to a failure. Re-

place stabs if wear is extensive. Lubricate bus stabs with NO-OX-ID grease before replacing units.

15. Remove accumulated dust and dirt by vacuuming. **Do not use compressed air as it may contain moisture and blow the debris elsewhere in the control equipment.**

16. **Note:** Also see **NFPA 70B. Electrical Equipment Maintenance.**®

MAINTENANCE AFTER A FAULT CONDITION

Opening of the short circuit protective device (such as fuses or circuit breakers) in a properly coordinated motor branch circuit is an indication of a fault condition in excess of operating overload. Such a condition can cause damage to control equipment. After a fault has occurred, repair the cause of the fault, inspect all equipment per **NEMA Standards Publicaton NO. ICS2.2-1983 Maintenance of Motor Controllers After a Fault Condition.** Make any necessary repairs or replacements prior to putting the equipment in service again. Be sure that all replacements, if any, are of the proper rating and are suitable for the application.

That about covers the specifics of the plant distribution equipment as far as servicing and maintenance are concerned.

J. Keep Your Eyes Open for Trouble

During the course of the daily shift, the plant maintenance man is always on the go, from one side of the plant to the other, above and below. While walking around the plant, the alert worker will be watching for any sign of developing trouble, especially regarding the portions of the plant under his care. This does not mean that extra time must be allotted to this function, but that a sharp glance at the equipment as he passes by will often catch trouble before it has gone too far. Some of the things to watch for will be listed here. This is only a general guide; the plant maintenance man will learn where the potential trouble spots are in his plant, and be forever on guard.

1. Appearance of smoke from any electrical gear or conductors.

2. Watch for water, oil, dirt or debris falling on the electrical gear or conductors.

3. Damaged panels, conduits, busbars, etc.

4. Stacked material or products infringing on the clearances around the equipment.

5. Workers from other departments or outside contractors working around the electrical equipment.

6. Any abnormality in the readings of the meters installed throughout the plant.

7. Unauthorized people loitering around sensitive areas.

8. Broken grounding cables.

All industrial plants have a definite rhythm or pattern of operation they follow from day to day, and often from week to week. Learn that pattern, and how it affects the electrical apparatus, and you will soon be a highly effective electrical maintenance worker.

C H A P T E R 17
ALTERNATING CURRENT CONTROL DEVICES

A. Purpose and Scope

In this chapter we shall cover those items of electrical control usually found fairly close to the load being served. These are unit controls, as only one electric power user or circuit is connected to them. They serve the same purpose as a valve in a pipeline, as they may be manual, automatically operated, or tied into other control systems.

Since over 90% of all motors are used on AC, DC motors and their control will not be discussed. The squirrel cage induction motor is the most widely used motor. Therefore, its control is the subject of this manual. The use of high voltages (2400, 4800, and higher) introduces requirements which are additional to those for 600-volt equipment; and although the basic principles are unchanged, these additional requirements are not covered here.

B. Selection

The motor, machine, and motor controller are interrelated and need to be considered as a package when choosing a specific device for a particular application. In general, several basic factors influence the selection of a controller.

ELECTRICAL SERVICE

Establish whether the service is DIRECT CURRENT (DC) or ALTERNATING CURRENT (AC). IF AC, determine the number of phases and frequency, in addition to the voltage.

MOTOR

The motor should be matched to the electrical service, and correctly sized for the machine load (horsepower rating). To select proper protection for the motor, its FULL-LOAD CURRENT (FLC) rating, service factor, and time rating must be known.

OPERATING CHARACTERISTICS OF A CONTROLLER

The fundamental job of a motor controller is to start and stop the motor, and to protect the motor, machine, and operator.

The controller might also be called on to provide supplementary functions, which could include reversing, jogging or inching, plugging, operating at several speeds or at reduced levels of current and motor torque.

ENVIRONMENT

Controller enclosures serve to provide protection for operating personnel by preventing accidental contact with live parts. In certain applications, the controller itself must be protected from a variety of environmental conditions which might include: water, rain, snow, or sleet; dirt or non-combustible dust; cutting oils, coolants, or lubricants.

Both personnel and property require protection in environments made hazardous by the presence of explosive gases or combustible dusts.

See Chapter 11, Section F, for a tabulation of the more common enclosures available.

FULL-LOAD CURRENT (FLC)

The current required to produce full-load torque at rated speed.

To find the approximate full-load current for three-phase motors:

Full-Load Current = (Horsepower × 2.5) + 2, for 240-volt motors

$$\text{Full-Load Current} = \frac{(\text{Horsepower} \times 2.5) + 2}{2},$$

for 480-volt motors.

Notice that we stipulated "approximate." They are to be used for estimating only. For accurate FLC requirements, consult the manufacturer's literature for the actual motor being used.

LOCKED ROTOR CURRENT (LRC)

During the acceleration period, at the moment a motor is started, it draws a high current called the "inrush" current. The inrush current, when the motor is connected directly to the line (so that full line voltage is applied to the motor), is called the locked rotor or stalled rotor current. The locked rotor current can be from 4 to 10 times the motor full-load current. The vast majority of motors have an LRC of about 6 times FLC, and therefore this figure is generally used. The "6 times" value is often expressed as 600% of FLC.

AMBIENT TEMPERATURE

The temperature of the air where a piece of equipment is situated is called the ambient temperature. Most controllers are of the enclosed type and the ambient temperature is the temperature of the air outside the enclosure, not inside. Similarly, if a motor is said to be in an ambient temperature of 30° C (88° F), this is the temperature of the air outside the motor, not inside. Per NEMA Standards, both controllers and motors are subject to a 40° C (104° F) ambient temperature limit.

TEMPERATURE RISE

Current passing through the windings of a motor results in an increase in the motor temperature. The difference between the winding temperature of the motor when running and the ambient temperature is called the temperature rise.

The temperature rise produced at full load is not harmful provided the motor ambient temperature does not exceed 40° C (104° F).

Higher temperature caused by increased current or higher ambient temperatures produces a deteriorating effect on motor insulation and lubrication. An old "rule of thumb" states that for each increase of 10° above the rated temperature, motor life is cut in half.

TIME (DUTY) RATING

Most motors have a continuous duty rating permitting indefinite operation at rated load.

Intermittent duty ratings are based on a fixed operating time (5, 15, 30, 60 minutes) after which the motor must be allowed to cool.

MOTOR SERVICE FACTOR

If the motor manufacturer has given a motor a "service factor," it means that the motor can be allowed to develop more than its rated or nameplate horsepower without causing undue deterioration of the insulation. The service factor is a margin of safety. If, for example, a 10 hp motor has a service factor of 1.15, the motor can be allowed to develop 11.5 hp. The service factor depends on the motor design.

JOGGING (INCHING)

Jogging describes the repeated starting and stopping of a motor at frequent intervals for short periods of time. A motor would be jogged when a piece of driven equipment has to be positioned fairly closely, e.g., when positioning the table of a horizontal boring mill during setup. If jogging is to occur more frequently than 5 times per minute, NEMA standards require that the starter be derated.

PLUGGING

When a motor running in one direction is momentarily reconnected to reverse the direction, it will be brought to rest very rapidly. This is referred to as "plugging." If a motor is plugged more than 5 times per minute, derating of the controller is necessary, due to the heating of the contacts.

Plugging can only be used if the driven machine and its load will not be damaged by the reversal of the motor torque.

SEQUENCE (INTERLOCKED) CONTROL

Many processes require a number of separate motors which must be started and stopped in a definite sequence, as in a system of conveyors. When starting up, the delivery conveyor must start first with the other conveyors starting in sequence, to avoid a pile up of material. When shutting down, the reverse sequence must be followed with time delays between the shutdowns (except for emergency stops) so that no material is left on the conveyors. This is an example of a simple sequence control. Separate starters could be used, but it is common to build a special controller which incorporates starters for each drive, with timers, control relays, etc.

NATIONAL CODES AND STANDARDS

Motor control equipment is designed to meet the provisions of the **National Electrical Code** (NEC).® Code sections applying to industrial control devices are Article 430 on motor controllers and Article 500 on hazardous locations.

The 1970 Occupational Safety and Health Act, as amended in 1972, requires that each employer furnish employment free from recognized hazards likely to cause serious harm. Provisions of the act are strictly enforced by inspection.

Standards established by the National Electrical Manufacturers Association (NEMA) assist users in the proper selection of control equipment. NEMA Standards provide practical information concerning construction, test, performance, and manufacture of motor control devices such as starters, relays, and contactors.

One of the organizations which actually tests for conformity to national codes and standards is Underwriters Laboratories (UL). Equipment tested and approved by UL is listed in an annual publication, which is kept current by means of bimonthly supplements which reflect the latest additions and deletions.

Enclosures: We refer you to Section F of Chapter 11 for a list of the various NEMA enclosure types.

C. Magnetic Starters

MAGNETIC CONTROL

A high percentage of applications require the controller to be capable of operation from remote locations, or to provide automatic operation in response to signals from pilot devices such as thermostats, pressure or float switches, limit switches, etc. Low voltage release or protection might also be desired. Manual starters cannot provide this type of control, and therefore magnetic starters are used.

The operating principle which distinguishes a magnetic from a manual starter is the use of an electromagnet. The electromagnet consists of a coil of wire placed on an iron core. When current flows through the coil, the iron of the magnet becomes magnetized, attracting the iron bar, called the armature. Interrupting the current flow through the coil of wire causes the armature to drop out due to the presence of an air gap in the magnetic circuit.

With manual control, the starter must be mounted so that it is easily accessible to the operator. With magnetic control, the push-button stations or other pilot devices can be mounted anywhere and connected by control wiring into the coil circuits of the remotely mounted starter.

MAGNETIC STARTERS—POWER CIRCUIT

The power circuit of a starter includes the stationary and movable contacts, and the thermal unit or heater portion of the overload relay assembly. The number of contacts (or "poles") is determined by the electrical service. In a three-phase three-wire system, for example, a three-pole starter is required (see Fig. 17-1).

MAGNETIC STARTERS—NEMA SIZES AND RATINGS

Power circuit contacts handle the MOTOR LOAD. The ability of the contacts to carry the full-load current without exceeding a rated temperature rise, and their isolation from adjacent contacts, corresponds to NEMA Standards established to categorize the NEMA size of the starter. The starter must also be capable of interrupting the motor circuit under prolonged locked rotor current conditions. To be suitable for a given motor application, the magnetic starter selected should equal or exceed the motor horsepower and full-load current ratings.

CHOOSING A STARTER

So far, we have progressed through the stage of choosing the circuit breaker for the problem given in Chapter 3, so it is now time to make the choice of the proper starting equipment for our 50 hp motor.

We have determined that the normal running current for the motor is about 63 amps at 460 volts. Referring now to Table 17-1, we can make our selection. First, we shall look at the column heading for standard service, which in this case bears the heading "Maximum Horsepower Rating-Nonplugging and Nonjogging Duty."

For polyphase motors drawing 63 amps, we find that a size 3 starter will suffice. However,

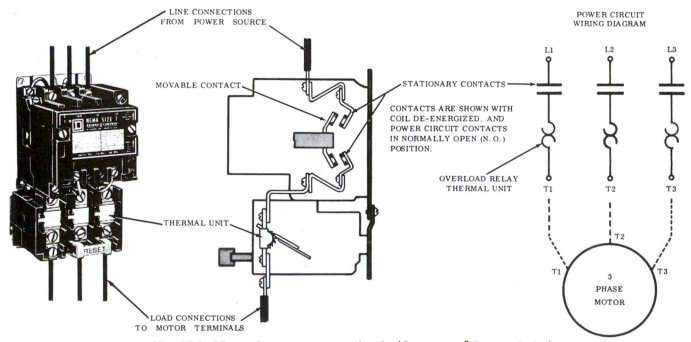

Fig. 17-1. Magnetic starter power circuit. (Courtesy of Square D Co.)

Table 17-1. Electrical Ratings for AC Magnetic Contractors and Starters

NEMA Size	Volts	Max HP Rating — Nonplugging and Nonjogging Duty, Single Phase	Poly-Phase	Max HP Rating — Plugging and Jogging Duty †, Single Phase	Poly-Phase	Continuous Current Rating, Amperes — 600 Volt Max.	Service-Limit Current Rating, Amperes *	Tungsten and Infrared Lamp Load, Amperes — 250 Volts Max. ★	Resistance Heating Loads, KW — other than Infrared Lamp Loads ‡, Single Phase	Poly-Phase	KVA Rating for Switching Transformer Primaries at 50 or 60 Cycles ▲, Single Phase	Poly-Phase	3 Phase Rating for Switching Capacitors ◐, Kvar
00	115	1/3				9	11	5					
	200		1 1/2			9	11	5					
	230	1	1 1/2			9	11	5					
	380		1 1/2			9	11						
	460		2			9	11						
	575		2			9	11						
0	115	1		1/2		18	21	10			0.9	1.2	
	200		3		1 1/2	18	21	10				1.4	
	230	2	3	1	1 1/2	18	21	10			1.4	1.7	
	380		5		1 1/2	18	21					2.0	
	460		5		2	18	21				1.9	2.5	
	575		5		2	18	21				1.9	2.5	
1	115	2		1		27	32	15	3	5	1.4	1.7	
	200		7 1/2		3	27	32	15		9.1		3.5	
	230	3	7 1/2	2	3	27	32	15	6	10	1.9	4.1	
	380		10		5	27	32			16.5		4.3	
	460		10		5	27	32		12	20	3	5.3	
	575		10		5	27	32		15	25	3	5.3	
1P	115	3		1 1/2		36	42	24					
	230	5		3		36	42	24					
2	115	3		2		45	52	30	5	8.5	1.9	4.1	
	200		10		7 1/2	45	52	30		15.4		6.6	11.3
	230	7 1/2	15	5	10	45	52	30	10	17	4.6	7.6	13
	380		25		15	45	52			28		9.9	21
	460		25		15	45	52		20	34	5.7	12	26
	575		25		15	45	52		25	43	5.7	12	33
3	115	7 1/2	25		15	90	104	60	10	17	4.6	7.6	
	200		30		20	90	104	60		31		13	23.4
	230	15	30		20	90	104	60	20	34	8.6	15	27
	380		50		30	90	104			56		19	43.7
	460		50		30	90	104		40	68	14	23	53
	575		50		30	90	104		50	86	14	23	67
4	200		40		25	135	156	120		45		20	34
	230		50		30	135	156	120	30	52	11	23	40
	380		75		50	135	156			86.7		38	66
	460		100		60	135	156		60	105	22	46	80
	575		100		60	135	156		75	130	22	46	100
5	200		75		60	270	311	240		91		40	69
	230		100		75	270	311	240	60	105	28	46	80
	380		150		125	270	311			173		75	132
	460		200		150	270	311		120	210	40	91	160
	575		200		150	270	311		150	260	40	91	200
6	200		150		125	540	621	480		182		79	139
	230		200		150	540	621	480	120	210	57	91	160
	380		300		250	540	621			342		148	264
	460		400		300	540	621		240	415	86	180	320
	575		400		300	540	621		300	515	86	180	400
7	230		300			810	932	720	180	315			240
	460		600			810	932		360	625			480
	575		600			810	932		450	775			600
8	230		450			1215	1400	1080					360
	460		900			1215	1400						720
	575		900			1215	1400						900

Tables and footnotes are taken from NEMA Standards Publication No. IC 1-1965 Section 2, Part 11 for Magnetic Contactors and Section 3, Parts 21B, 21C, 21D and 21F for Magnetic Starters and includes 1971 revisions for 200 V. and 380 V. ratings.

†Ratings shown are for applications requiring repeated interruption of stalled motor current or repeated closing of high transient currents encountered in rapid motor reversal, involving more than five openings per minute such as plug-stop, plug-reverse or jogging duty. Ratings apply to single speed and multi-speed controllers.

*Per NEMA Standards paragraph IC 1-21A.20, the service-limit current represents the maximum rms current, in amperes, which the controller may be expected to carry for protracted periods in normal service. At service-limit current ratings, temperature rises may exceed those obtained by testing the controller at its continuous current rating. The ultimate trip current of overcurrent (overload) relays or other motor protective devices shall not exceed the service-limit current ratings of the controller.

★FLUORESCENT LAMP LOADS — 300 VOLTS AND LESS — The characteristics of fluorescent lamps are such that it is not necessary to derate Class 8502 contactors below their normal continuous current rating. Class 8903 contactors may also be used with fluorescent lamp loads. For controlling tungsten and infrared lamp loads, Class 8903 ac lighting contactors are recommended. These contactors are specifically designed for such loads and are applied at their full rating as listed in the Class 8903 Section. Do not use Class 8903 contactors with motor loads or resistance heating loads.

‡Ratings apply to contactors which are employed to switch the load at the utilization voltage of the heat producing element with a duty which requires continuous operation of not more than five openings per minute.

▲Applies to contactors used with transformers having an inrush of not more than 20 times their rated full load current, irrespective of the nature of the secondary load.

◐Kilovar ratings of contactors employed to switch power capacitor loads. When capacitors are connected directly across the terminals of an alternating current motor for power factor correction, the motor manufacturer should be consulted as to the maximum size of the capacitor and the proper rating of the motor overcurrent protective device.

"CAUTION: For three phase motors having locked-rotor KVA per horsepower in excess of that for the motor code letters in the right table, do not apply the controller at its maximum rating without consulting the factory. In most cases, the next higher horsepower rated controller should be used."

Controller HP Rating	Maximum Allowable Motor Code Letter
1 1/2-2	L
3-5	K
7 1/2 & above	H

if the motor will have to be started and stopped very frequently for load or process adjustments, a duty known as "jogging," then we shall have to look under the appropriate heading for Plugging and Jogging Duty. Doing this, we find that a size 4 is required.

There still remains the details of enclosure, accessories, ambient conditions, etc., to be settled, which should not be too difficult after reading this chapter through from beginning to end.

If it is anticipated that considerable servicing may be required on the motor and the starting equipment, due to the load type and the ambient conditions, it is an excellent idea to include a fused disconnect switch in the same enclosure with the magnetic starter. This provides a convenient local switch for servicing, provided the combination fused disconnect switch and starter are within sight of the motor, which should be the case for proper installation. The fuses give additional protection, in addition to the circuit breakers in the motor control center.

The wiring diagrams for a typical combination of this type are shown in Fig. 17-9.

MAGNETIC STARTER—CONTROL CIRCUIT

The circuit to the magnet coil which causes a magnetic starter to pick-up and drop-out is distinct from the power circuit. Although the power circuit can be single phase or polyphase, the coil circuit is always a single-phase circuit. Elements of a coil circuit include the following:

1. the magnet coil;

2. the contact(s) of the overload relay assembly;

3. a momentary or maintained contact pilot device, such as a push-button station, pressure, temperature, liquid level or limit switch, etc.;

4. in lieu of a pilot device, the contact(s) of a relay or timer; and

5. an auxiliary contact on the starter, designated as a holding circuit interlock, which is required in certain control schemes.

The coil circuit is generally identified as the CONTROL CIRCUIT, and contacts in the CONTROL CIRCUIT handle the CONTROL LOAD.

In this section we shall describe the various methods of actuating a magnetic starter using a voltage supply taken from the common and one outside leg of a three-wire three-phase power supply to the starter. This is not the only way, as the control power may easily come from any outside source at any voltage, and this is quite often done, as stated in Section H.

ELEMENTARY DIAGRAM

The elementary diagram, Fig. 17-2, gives a fast, easily understood picture of the circuit. The devices and components are not shown in their actual positions. All the control circuit components are shown as directly as possible, between a pair of vertical lines, representing the control power supply. The arrangement of the components is designed to show the sequence of operation of the devices, and helps in understanding how the circuit operates. The effect of operating various interlocks, control devices, etc., can be readily seen—this helps in troubleshooting, particularly with the more complex controllers. This form of electrical diagram is sometimes referred to as a "schematic" or "line" diagram.

To help in analyzing the circuit in Fig. 17-2, remember that the magnetic coil, shown in the control circuit with an "M" over it, is the coil

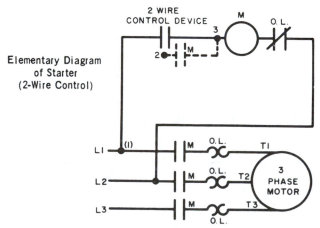

Fig. 17-2. Elementary diagram of starter (two-wire control). (Courtesy of Square D Co.)

which closes the three motor lead contacts marked "M." Thus, closing the 2-wire control device (manually or automatically, remote or local) powers the magnetic coil "M" from L1 to L2. The coil "M" then closes all three contacts "M" in leads L1, L2, and L3, applying full voltage to the motor.

The power to the control circuit does not have to come from the motor leads L1 and L2 as shown in Figs. 17-2 and 17-3. The power for the control circuit may come from any other reliable available source and at any voltage. Often, the control circuit is tied into an outside low voltage system, as explained in Section H. Of course, all elements in the control circuit must then be chosen to match the control voltage and current, including the starter coil, interlocks, switches, overloads, etc.

Figure 17-3 is a WIRING DIAGRAM, and it shows, as closely as possible, the actual location of all of the component parts of the starter. The dotted lines represent Power Circuit connections made to the starter by the user.

Since wiring connections and terminal markings are shown, this type of diagram is helpful when wiring the starter, or tracing wires when troubleshooting. Note that bold lines denote the Power Circuit, and thin lines are used to show the Control Circuit. Conventionally, in AC magnetic equipment, black wires are used in Power Circuits and red wiring is used for Control Circuits.

A wiring diagram is limited in its ability to convey a clear picture of the sequence of operation of a controller. Where an illustraton of the circuit in its simplest form is desired, the Elementary Diagram is used.

FULL-VOLTAGE (ACROSS-THE-LINE) STARTER

As the name implies, a full-voltage or across-the-line starter directly connects the motor to the lines. The starter can be either manual or magnetic.

A motor connected in this fashion draws full inrush current and develops maximum starting torque so that it accelerates the load to full speed in the shortest possible time. Across-the-line starting can be used wherever this high inrush current and starting torque are not objectionable.

With some loads, the high starting torque will damage belts, gears, and couplings and material being processed. High inrush current can produce line voltage dips which cause lamp flicker and disturbances to other loads. Lower starting currents and torques are therefore often required, and are achieved by reduced-voltage starting, covered in Section F.

TWO-WIRE CONTROL

In the wiring and elementary diagrams shown, Figs. 17-2 and 17-3, two wires connect the control device (which could be a thermostat, float switch, limit, or other maintained contact device) to the magnetic starter. When the contacts of the control device close, they complete the coil circuit of the starter, causing it to pick up and connect the motor to the lines. When the control device contacts open, the starter is de-energized, stopping the motor.

Two-wire control provides low voltage release, but not low voltage protection. Wired as illustrated, the starter will function automati-

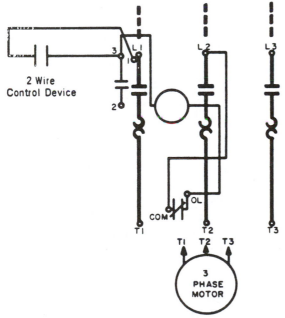

Fig. 17-3. Wiring diagram of starter (two-wire control). (Courtesy of Square D Co.)

cally in response to the direction of the control device, without the attention of an operator.

The dotted portion shown in the elementary diagram represents the holding circuit interlock furnished on the starter, but not used in two-wire control. For greater simplicity, this portion is omitted from the conventional two-wire elementary diagram.

If the two-wire control device in the diagrams in Figs. 17-2 and 17-3 is closed, a power failure or drop in voltage below the seal-in value will cause the starter to drop out, but as soon as power is restored, or the voltage returns to a level high enough to pick up and seal, the starter contacts will re-close and the motor will again run. This is an advantage in applications involving unattended pumps, refrigeration processes, ventilating fans, etc.

In many applications, however, the unexpected restarting of a motor after power failure is undesirable, as in a process where a number of motors must be restarted, or operations performed, in a prescribed sequence. In some applications, the automatic restart presents the possibility of danger to personnel, or damage to machinery or to work in progress.

If protection from the effects of a low voltage condition is required, the two-wire control scheme is not suitable, and three-wire control, which provides the desired protection, should be used.

THREE-WIRE CONTROL

A three-wire control circuit uses momentary contact Start-Stop buttons and a holding circuit interlock wired in parallel with the Start button, to maintain the circuit, see Figs. 17-4 and 17-5.

Pressing the N.O. Start button completes the circuit to the coil. The power circuit contacts in Lines 1, 2, and 3 close, completing the circuit to the motor, and the holding circuit contact (mechanically linked with the power contacts) also closes. Once the starter has picked up, the Start button can be released, as the now closed interlock contact provides an alternate current path around the re-opened Start contact.

Pressing the N.C. Stop button will open the

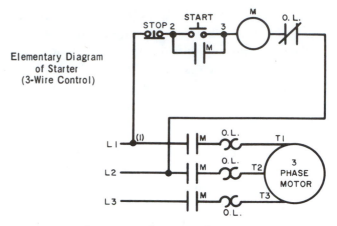

Elementary Diagram of Starter (3-Wire Control)

Fig. 17-4. Elementary diagram of starter (three-wire control). (Courtesy of Square D Co.)

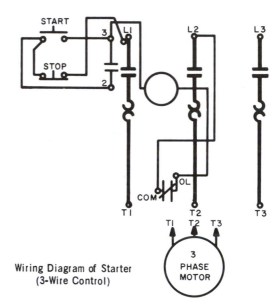

Wiring Diagram of Starter (3-Wire Control)

Fig. 17-5. Wiring diagram of starter (three-wire control). (Courtesy of Square D Co.)

circuit to the coil, causing the starter to drop out. An overload condition, which causes the overload contact to open, a power failure, or a drop in voltage to less than the seal-in value, would also de-energize the starter. When the starter drops out, the interlock contact re-opens, and both current paths to the coil, through the start button and the interlock, are now open.

When power is restored, however, the starter connected for three-wire control will not pick up, as the re-opened holding circuit contact and the N.O. Start button contact prevent current flow to the coil. To restart the motor after a power failure, the low voltage protection offered by three-wire control requires that the Start button be depressed. A deliberate action must be performed, ensuring greater safety than that provided by two-wire control.

Manual starters with low voltage protection offer this same type of protection.

Since three wires from the push-button station are connected into the starter—at points L1, L2, and L3—this wiring scheme is commonly referred to as three-wire control.

HOLDING CIRCUIT INTERLOCK

The holding circuit interlock is a normally open (N.O.) auxiliary contact provided on standard magnetic starters and contactors. It closes when the coil is energized to form a holding circuit for the starter after the "start" button has been released (see paragraph on three-wire control).

ELECTRICAL INTERLOCKS

In addition to the main or power contacts, which carry the motor current, and the holding circuit interlock, a starter can be provided with externally attached auxiliary contacts, commonly called electrical interlocks. Interlocks are rated to carry only control circuit currents, not motor currents. N.O. and N.C. versions are available.

Among a wide variety of applications, interlocks can be used to control other magnetic devices where sequence operation is desired; to electrically prevent another controller from being energized at the same time (see Reversing Starters) and to make and break circuits to indicating or alarm devices such as pilot lights, bells, or other signals.

A CONTROL DEVICE (PILOT DEVICE)

A device which is operated by some non-electrical means (such as the movement of a lever), and which has contacts in the control circuit of a starter, is called a "control device." Operation of the control device will control the starter and hence the motor. Typical control devices are control station limit switches, pressure switches, and float switches. The control device may be of the maintained contact or momentary contact type.

MAINTAINED CONTACT

A maintained contact control device is one which when operated will cause a set of contacts to open (or close) and stay open (or closed) until a deliberate reverse operation occurs. A conventional thermostat is a typical maintained contact device. Maintained contact control devices are used with two-wire control.

MOMENTARY CONTACT

A standard pushbutton is a typical momentary contact control device. Pushing the button will cause N.O. contacts to close and N.C. contacts to open. When the button is released, the contacts revert to their original states. Momentary contact devices are used with three-wire control or jogging service.

D. Overload Protection

This subject was discussed to a certain extent in Chapter 11, but we left portions of it to be covered in this chapter. As we are here discussing magnetic starters, the subject of overload relays is highly appropriate at this time.

An overload relay, as used here, is a device to automatically open the electrical circuit to a load upon an exceedingly high inrush of current, as would be caused by a short in the system. It may be one of several styles, but the main types only will be discussed here. They are the magnetic relay, the melting pot relay, and the bimetallic element relay.

The magnetic overload relay operates from the principle of a movable magnetic core inside a coil which carries the motor current. The flux induced in the coil pulls the core upward until it trips a set of contacts. The distance which it moves bears a direct relationship to the amount of the current being used by the motor, or load.

There are several variations of this principle, all designed to change the speed of operation, and other special characteristics desired for various applications.

MELTING ALLOY THERMAL OVERLOAD RELAY

In these overload relays (also referred to as "solder pot relays"), the motor current passes through a small heater winding. Under overload conditions, the heat causes a special solder to melt, allowing a ratchet wheel to spin free, opening the contacts. When this occurs, the relay is said to "trip." To obtain appropriate tripping current for motors of different sizes, of different full-load currents, a range of thermal units (heaters) is available. The heater coil and solder pot are combined in a one-piece, nontamperable unit. The heat transfer characteristic and the accuracy of the unit cannot be accidentally changed, as is possible when the heater is a separate component. Melting alloy thermal overload relays are "hand reset," thus, after they trip they must be reset by a deliberate hand operation. A reset button is usually mounted on the cover of enclosed starters. Thermal units are rated in amperes and are selected on the basis of motor full-load current, not horsepower, see Fig. 17-6.

BIMETALLIC THERMAL OVERLOAD RELAY

Bimetallic thermal overload relays, Fig. 17-7, employ a U-shape bimetal strip, associated with a current carrying heater element. When an overload occurs, the heat will cause the bimetal to deflect and open a contact. Different heaters give different trip points. In addition, most relays are adjustable over a range of 85% to 115% of the nominal heater rating.

These relays are field convertible from hand reset to automatic reset and vice versa. On automatic reset, the relay contacts, after tripping, will automatically reclose when the relay has cooled down. This is an advantage when the relays are inaccessible. However, automatic reset overload relays should not normally be used with two-wire control. With this arrangement, when the overload relay contacts reclose after an overload relay trip, the motor will restart, and, unless the cause of the overload has been removed, the overload relay will trip again. This cycle will repeat and eventually the motor will burn out due to the accumulated heat from the repeated inrush current. More important is the possibility of danger to personnel. The unexpected restarting of a machine may find the operator or maintenance man in a hazardous situation, as he attempts to find out why his machine has stopped.

When replacing the thermal heater elements in either the melting alloy or the bimetallic relay, it is only necessary to obtain the FLC of the motor, and refer to the catalog furnished by the

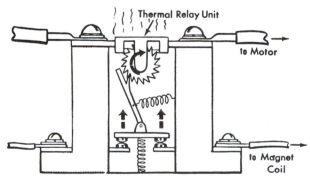

Fig. 17-6. Solder pot relay. (Courtesy of Square D Co.)

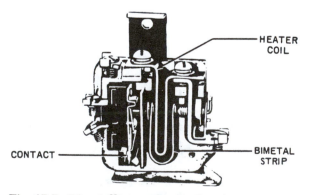

Fig. 17-7. Bimetallic overload relay. (Courtesy of Square D Co.)

manufacturer for the proper heater selection, as a starting point. Abnormal conditions may call for a higher thermal rating from the normal starting point of 125% of FLC listed in the selection tables.

AMBIENT COMPENSATION

Ambient-compensated bimetallic overload relays were designed for one particular situation, that is, when the motor is at a constant temperature and the controller is located separately in a varying temperature. In this case, if a standard thermal overload relay were used, it would not trip consistently at the same level of motor current if the controller temperature changed. This thermal overload relay is always affected by the surrounding temperature. To compensate for the temperature variations the controller may see, an ambient-compensated overload relay is applied. Its trip point is not affected by temperature and it performs consistently at the same value of current.

E. Magnetic Starter Variations

REVERSING STARTER

Reversing the direction of motor shaft rotation is often required. Three-phase squirrel cage motors can be reversed by reconnecting any two of the three line connections to the motor. By interwiring two contactors, an electromagnetic method of making the reconnection can be obtained.

As seen in the power circuit (Fig. 17-8), the contacts (F) of the Forward contactor, when closed, connect Lines 1, 2, and 3 to motor terminals T1, T2, and T3, respectively. As long as the Forward contacts are closed, mechanical and electrical interlocks prevent the Reverse contactor from being energized.

When the Forward contactor is de-energized, the second contactor can be picked up, closing its contacts (R) which reconnect the lines to the motor. Note that by running through the Re-

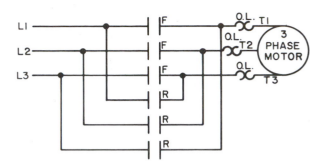

Fig. 17-8. Reversing starter. (Courtesy of Square D Co.)

verse contacts, Line 1 is connected to motor terminal T3, and Line 3 is connected to motor terminal T1. The motor will now run in the opposite direction.

Whether operating through either the Forward or Reverse contactor, the power connections are run through an overload relay assembly, which provides motor overload protection. A magnetic reversing starter, therefore, consists of a starter and contactor, suitably interwired, with electrical and mechanical interlocking to prevent the coil of both units from being energized at the same time.

Manual Reversing Starters (employing two manual starters) are also available. As in the magnetic version, the Forward and Reverse switching mechanisms are mechanically interlocked, but since coils are not used in the manually operated equipment, electrical interlocks are not furnished.

COMBINATION STARTER

A combination starter is so named since it combines a disconnect means, which might incorporate a short circuit protective device, and a magnetic starter in one enclosure. See Fig. 17-9.

Compared with a separately mounted disconnect and starter, the combination starter takes up less space, requires less time to install and wire, and provides greater safety.

Safety to personnel is assured because the door is mechanically interlocked, so that it cannot be opened without first opening the disconnect.

Combination starters can be furnished with

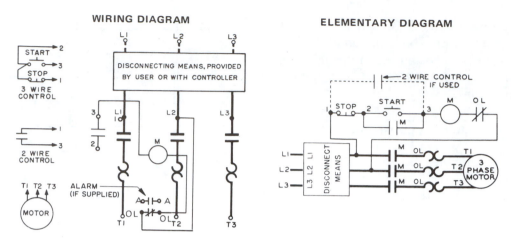

Fig. 17-9. Typical three-phase combination starters. (Courtesy of Square D Co.)

circuit breakers or fuses to provide overcurrent protection, and are available in non-reversing and reversing versions.

F. Reduced Voltage Motor Starting

When the switch is closed to start an induction motor, there is a surge of current into the motor to get it started, and this surge may peak to 600% or more of the normal starting current. Figure 17-10 indicates that this high current inflow rate will continue until the motor has reached about 70% of its normal speed, at which point the current draw reduces rapidly as the motor comes up to speed. As the graph shows, the motor is operating at virtually a locked rotor condition until the 40% speed point is reached.

This high current flow is very disruptive to the remaining circuits in the plant, especially on other equipment on the same load center. It is a surge like this that can cause voltage dips, current flows to increase, overloads to kick out, and stoppage of any high-tech electronic equipment not served by uninterruptable power sources, such as computers and sophisticated process control systems.

The usual induction motor being started by throwing it directly across the line under full

voltage is small enough to prevent the above upsets from occurring. Also, the motor usually comes up to speed rather quickly, so that the current surge lasts but a few seconds. Otherwise, the current surge would produce an abnormally high temperature in the motor, as while the motor is coming up to operating speed, the motor windings are heating up rapidly, from the following formula:

Heating effect in Watts = I^2 x R, from Chapter 2, Section E.

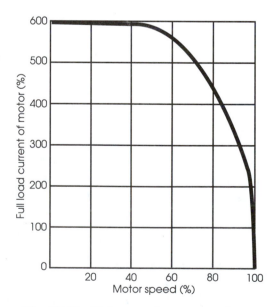

Fig. 17-10. Motor starting current draw.

After the motor is up to rated speed, and the cooling fan inside the motor dissipates the excess heat, the motor then reaches its normal operating speed, temperature, and current draw for the load and ambient conditions. This explains what happens when a motor is started and stopped too frequently, so frequently that it does not have enough time to cool off from the starting current, before the next starting surge hits it. To prevent this, there is a general rule that motors should not be started more often than about 5 or 6 times an hour, in sizes from $\frac{1}{4}$ hp up to about 30 hp. This depends upon the motor service factor, ambient conditions, and load characteristics. The motor manufacturer will provide assistance if it is not clear in his operating instructions how it will perform under your plant conditions.

One of the simplest ways to lessen the effects of the high starting surges from three-phase induction motors is to start them under reduced voltage, then increase the voltage in steps, until the motor is up to full speed and normal current draw. We shall cover the several methods in common use in the following discussion.

Figure 17-11 is of the reduced voltage primary resistance starting system, which pro-

vides one step of reduced voltage at about 70% of full line voltage during the initial starting phase. The timer is activated when the "start" button is pressed, and the starting current flows through the resistors in each of the three power legs to the motor, reducing the voltage to the motor. When the timer has run its set course, usually only one or two seconds, it closes the control relays CRI, thus bypassing the power around the resistors R1, R2, and R3, so that full voltage then flows to the motor. This description has been greatly simplified, and we urge the maintenance worker to contact the supplier of the control panel for his servicing literature.

For motors which must start very smoothly, the primary resistance method must be provided with more than one intermediate voltage step, which complicates the wiring diagram.

The discussion on reduced-voltage starting and the wiring diagrams include symbols not encountered before in this book. In Fig. 17-11 there are three small triangles above CR1, with zeros in the first two and an x in the third one, reading from left to right, which are the code symbols for the switch action inside CR1. Reading from left to right, the first triangle rep-

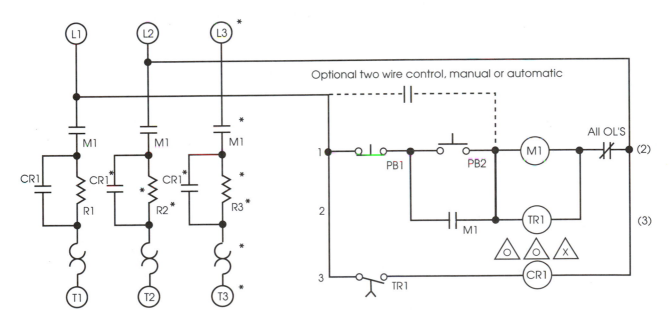

*Ommited for single phase motor

Fig. 17-11. Reduced-voltage primary resistance system.

resents the "Reset" portion of the cycle, the middle triangle represents the "Timing" portion of CR1, and the right-hand triangle represents the switch action after the timing cycle has ended. This is shown in detail in Table 17-2, which should help considerably in analyzing the circuits in this section. To further aid in your analysis, we give here the definitions for each portion of the timing cycle.

> RESET—The beginning of the timing cycle when no signals are being introduced into the circuit and all contacts are in their normal position, at rest.
>
> BEFORE TIMING—Period of time the action is allowed to continue after reset occurs, but before timing occurs.
>
> TIMING—The period of time from when the timer receives a signal until it makes a decision for action.
>
> AFTER TIMING—Period of time the action is allowed to continue before reset occurs.

When confronting manufacturer's diagrams for control systems having these elements, do not be surprised if the triangles giving the timer action are not included. You may have to insert them to assist you in your analysis and understanding of the controls. Servicing should preferably be done by the supplier's qualified serviceman.

We shall not go too deeply into the operations performed by the control systems in the following reduced-voltage methods, but will merely cover the characteristics of each, with some of the basic wiring diagrams.

The diagrams shown in this section have been drawn with manually operated "Start" and "Stop" buttons. These may also be operated by automatic means from sequential or programming panels.

Figure 17-12 is the diagram for an autotransformer reduced-voltage starting system which, as the name implies, utilizes an autotransformer inserted in each leg of the power supply to the motor. This is accomplished by timer TR1 operting C1, C2, and C3 to increase the voltage supplied to the motor in timed steps,

through the autotransformers in the power legs.

The system shown here is known as the closed circuit transition system, as the power to the motor is never stopped while the control is altering the voltage, so that there is a smooth power output, with less upset to the power system.

The other system, not shown here, known as the open circuit system, is cheaper and simpler, but it interrupts the power flow to the motor when the control changes the voltage steps.

Figure 17-13 contains the diagram for one of the simpler types of reduced-voltage starting: the part-winding method. It is not truly a reduced-voltage system, but is classifed as such because of its characteristics. This method requires that the motor have two sets of stator windings, as shown in the upper part of the diagram. The control is designed to cut out one set of windings during the first seconds of start-up, which reduces the starting current to about 65% of full-load current, followed by the timer action placing the second set of windings back into service, placing the motor on full-voltage operation. The result is a smooth flow of power during the starting cycle, as the power is never disconnected during the switching.

The motors for this method of starting must be designed for the service, or trouble will result. This is especially true when attempting to wire a dual voltage motor into this system without first consulting the motor manufacturer.

Figure 17-13 is for a Y-connected motor. For reduced-voltage starting of a delta-connected motor under the part-winding system, a special type of starter panel is required.

The Y-Delta starting system requires that the motor be wound as shown in the upper panel of Fig. 17-14. The reduced-voltage starting system merely starts the motor under the Y configuration, which reduces the motor winding voltage to 57% of the power voltage being supplied. Thus, 208 volts becomes 120 volts inside the motor, and other supply voltages are reduced in like manner. After the initial starting time, the control circuit then switches the motor connections into a Delta pattern, which supplies

Table 17-2. Symbols for Load Sequencing

Symbol	Description of Switch Action
o—⊣⊢—o	Instantaneous action, normally open (NO) when system is deactivated.
o—⊣/⊢—o	Instantaneous action, normally closed (NC) when system is deactivated.
o—o—⟍—o—o	On-Delay Timed Closed Contact (NOTC): timer contact is normally open (NO), and upon relay energizing the switch closes after a set time (TC). Switch opens immediately upon relay deenergizing.
o—o—⟍—o—o	On-Delay Timed Open Contact (NCTO); timer is normally closed (NC), and upon relay energizing the switch opens after a set time (TO). Switch closes immediately upon relay deenergizing.
o—o—⟍—o—o	Off-Delay Timed Closed Contact (NOTO); the timer contact is normally open (NO), and upon relay energizing the switch closes immediately. The switch opens after a set time upon relay deenergizing.
o—o—⟍—o—o	Off-Delay Timed Closed Contact (NCTC); the timer contact is normally closed (NC), but opens immediately upon relay energizing and upon relay deenergizing the switch closes after a set time (TC).

ON DELAY Load Sequencing Symbols[a]			
LOAD MARKING	**RESET**	**DURING TIMING**	**AFTER TIMING**
ENERGIZED	△x	△x	△x
DE-ENERGIZED	△o	△o	△o

OFF DELAY Load Sequencing Symbols[a]			
LOAD MARKING	**RESET**	**BEFORE TIMING**	**DURING TIMING**
ENERGIZED	▽x	▽x	▽x
DE-ENERGIZED	▽o	▽o	▽o

[a]If box is square instead of triangular, the relay contact is automatic reset style.

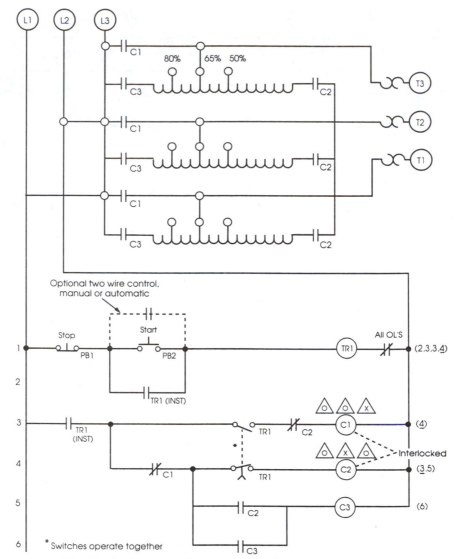

Fig. 17-12. Autotransformer reduced-voltage starting system.

full-line voltage to the motor windings. The diagram to provide this method of starting is given in Fig. 17-15. It appears to be more complex than any of those already presented. However, the principles embodied in the previous control diagrams are simply repeated here as normal progression, which should cause no consternation after mastering the control systems thus far. Of course, the control panel's supplier should be consulted for accuracy, as we have only shown the basic systems, and there are many variations to be found.

The latest method of reduced-voltage starting for motors is called Solid State Starting, and the diagram is shown in Fig. 17-16. The

heart of this system is the selenium controlled rectifier (SCR). We shall not go into the SCR's method of performing its function, other than to state than when used in this system, it controls motor voltage and current in a smooth flow of output during acceleration of the motor. It is a relatively trouble-free control, with exceedingly long life, but more expensive than conventional relays and timers used in the other systems given previously. They are used in starting circuits for motors requiring close, rapid, and smooth changes of power and speed to a motor, such as in overhead cranes, conveyor belts, etc. They are often used in variable speed motors.

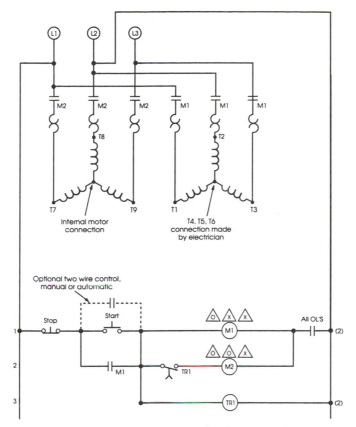

Fig. 17-13. Part-winding reduced-voltage starting system.

Maintenance of control systems utilizing the SCR is similar to any other control system containing plug-in or readily changed elements. The SCR does not lend itself to being repaired, and they should be replaced if they cause trouble, which should not happen under ordinary, proper use and care.

As you can see from Fig. 17-16, there are other elements involved in the solid state starting system, not shown on the diagram, but which must be included within the control panel.

As a brief description of the qualifications and characteristics of the various types of reduced-voltage starters which we have covered in this section, we give here Table 17-3.

G. Variable Speed Motor Control

In Chapter 13 we covered the controls for DC motors, and it was pointed out that DC motors

lend themselves very well to variable speed control. For that reason, they are often used to drive equipment or systems which require very close, accurate, and sensitive speed regulation. There are many industrial process operations which come under this classification. Consequently, DC motors will be found in uses which are controlled by highly sophisticated systems which are beyond the scope of this book.

In Chapter 15 we explained that the speed of an AC induction motor is fixed by the number of stator poles and the frequency of the AC supply. The formula for this relationship is:

$$\text{Synchronous speed, RPM} = \frac{120 \times F}{P},$$

in which;

F = Supply frequency, cycles/second
P = Number of poles in the stator winding.

This relationship limits the possible basic speed of an AC motor to one of the following

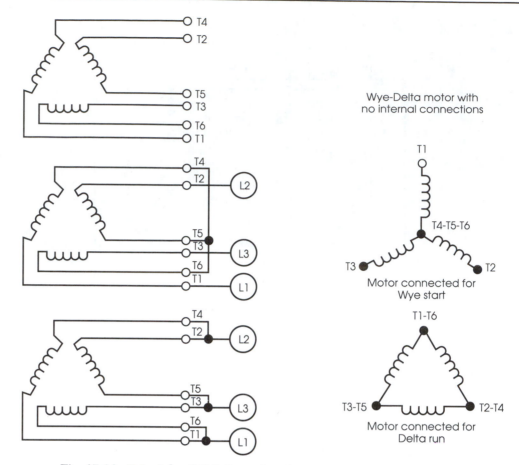

Fig. 17-14. Principle of Y-Delta reduced-voltage starting system.

speeds, assuming that the supply frequency is 60 Hz:

Poles	2	4	6	8	10	12	14	16
Speed, RPM	3600	1800	1200	900	720	600	514	450.

We also explained that the actual speed of the AC motor is slightly less than the above synchronous speeds, as in order to provide the rotating force, the motors must slip behind by about 2% to 10% of the synchronous speed.

Referring again to the above formula, we can see that we can only change the speed of an AC induction motor by changing either the number of poles or by changing the frequency. As might be expected, it is easier to change the number of poles in a motor when it is being designed, than it is to change the frequency of the applied voltage. The voltage is supplied by the electric utility company, and is usually at 60 Hz. Therefore, the discussion here will concen-

trate on the multiple pole speed regulation methods.

As we stated above, the AC induction motor being used for a variable speed drive will have to be designed to permit changing the number of poles in the stator, one pair of poles for each desired speed. Also, leads must be supplied to accommodate those poles, which results in some complicated terminal combinations at times.

Figure 17-17 is the basic diagram for wiring a two-speed motor, in both 240-volt and 120-volt systems. The starter to perform this function must start the motor at low speed by closing contacts "L," which would supply full voltage to all poles in the stator. If the motor was wired to give a speed of 3600 rpm top speed and 1800 rpm low speed, then the "L" contacts would activate all four poles, starting the motor at 1800 rpm. Pushing the "H" button will close contacts "H," starting the motor up

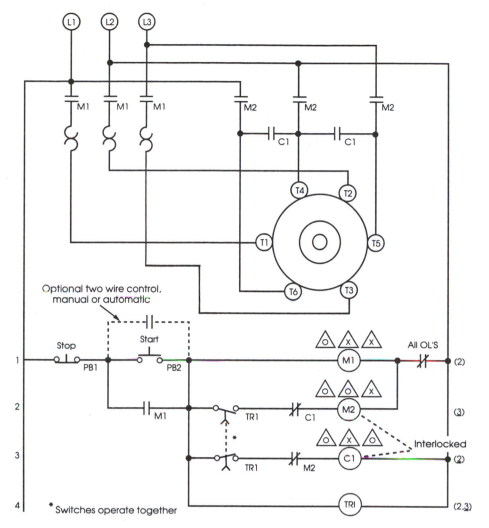

Fig. 17-15. Wye-Delta reduced-voltage starting system.

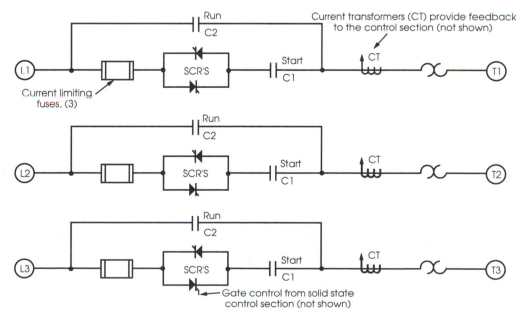

Fig. 17-16. Solid state reduced-voltage starting.

Table 17-3. Selection Table for Reduced Voltage Starters

Characteristic Wanted	Type of Starter to Use (Listed in order of desirability)	Comments
Smooth Acceleration	1. Solid State (Class 8660) 2. Primary Resistor (Class 8647) 3. Wye-Delta (Class 8630) 4. Autotransformer (Class 8606) 5. Part Winding (Class 8640)	•Little choice between 3 and 4.
Minimum Line Current	1. Autotransformer (Class 8606) 2. Solid State (Class 8660) 3. Wye-Delta (Class 8630) 4. Part Winding (Class 8640) 5. Primary Resistor (Class 8647)	
High Starting Torque	1. Autotransformer (Class 8606) 2. Solid State (Class 8660) 3. Primary Resistor (Class 8647) 4. Part Winding (Class 8640) 5. Wye-Delta (Class 8630)	
High Torque Efficiency (Torque vs. Line Current)	1. Autotransformer (Class 8606) 2. Wye-Delta (Class 8630) 3. Part Winding (Class 8640) 4. Solid State (Class 8660) 5. Primary Resistor (Class 8647)	•Little choice between 3, 4, and 5.
Suitability For Long Acceleration	1. Wye-Delta (Class 8630) 2. Autotransformer (Class 8606) 3. Solid State (Class 8660) 4. Primary Resistor (Class 8647)	•For acceleration time greater than 5 seconds, primary resistor requires non-standard resistors. •Part winding controllers are unsuitable for acceleration time greater than 2 seconds.
Suitability For Frequent Starting	1. Wye-Delta (Class 8630) 2. Solid State (Class 8660) 3. Primary Resistor (Class 8647) 4. Autotransformer (Class 8606)	•Part winding is unsuitable for frequent starts.
Flexibility in Selecting Starting Characteristics	1. Solid State (Class 8660) 2. Autotransformer (Class 8606) 3. Primary Resistor (Class 8647)	•For primary resistor, resistor change required to change starting characteristics. •Starting characteristics cannot be changed for Wye-Delta or part-winding controllers.

(Courtesy of Square D Co.)

at 3600 rpm by powering only two poles of the stator. The two-speed combination may be any pair of poles and speeds from the above table of poles versus rpm, depending upon the intended use of the motor.

The above description is brief, and leaves several possibilities for control, which we shall now explore. We shall not go into the analysis of the control diagrams, as by now we feel you should be capable of arriving at the sequencing logic built into the circuit.

We show manually operated Start and Stop, High and Low buttons, but these may be automatically actuated, either local or remote, or from a sequencing program panel.

Figure 17-18 is the first and probably the

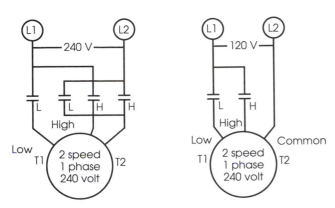

Fig. 17-17. Basic wiring diagram for a two-speed motor.

simplest type of control circuit for the two-speed motor. The motor may be started at either the low or the high speed, as the circuit shows two parallel circuits, either of which would activate the correct starting coil in the magnetic starter for that chosen speed. In all cases of step speed control given here, the control must contain a separate magnetic starter for each speed, as to change speed, a separate set of poles must be activated or deactivated. Also, this requires interlocks between the starters to prevent both circuits being activated simultaneously. This could damage the motor, and could also upset the process being driven.

In the starting method just described for Fig. 17-18, the motor must be brought to a stop before the speed change can take place. This may or not be a good thing, depending upon the driven equipment. For instance, the low speed may be for the purpose of jogging or positioning a piece of machinery which, once positioned, may then be started up at high speed.

If it is desired that the increase in speed be made without stopping the motor, then the control diagram will look like Fig. 17-19. The motor may still be started from rest to either low or high speed, and the motor may be stopped from either speed. However, in dropping speed from high to low, the Stop button must first be pushed, which is done to protect the motor from excessive current and shock to the driven equipment.

If it is required that no one be able to start a two-speed motor from stop to high speed, without going through low speed first, then the circuit of Fig. 17-20 is employed. Here again, it is necessary to push the stop button before dropping from high speed to low speed.

In Fig. 17-21, a timer has been added to the circuit. This serves to automatically accelerate the motor from low speed to high speed, only if the High Speed button has been pushed. If the Low Speed button is pushed, the motor will start at low speed, and will run there until the High Speed button is pushed. The timer controls the rate at which the motor is brought up to full speed, protecting the motor and the driven equipment. To change from high speed to low speed, the Stop button must be pushed first, then the Low Speed button pushed.

There are processes or equipment in use which require that the driving motor be brought from high speed to low speed partially by slowing down due to built-in inertia or braking to the low speed setting. Figure 17-22 is such a circuit, in which the timer TR1 controls the time alloted for slowing down from high speed to low speed. Pushing the Low Speed button causes the motor to start at low speed,

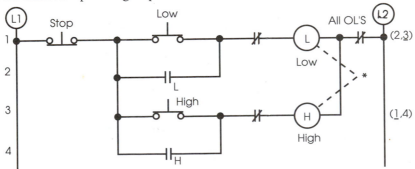

*Elements are interlocked

Fig. 17-18. Basic two-speed motor control circuit.

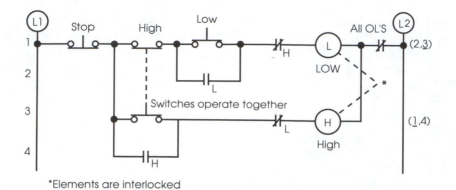

*Elements are interlocked

Fig. 17-19. Changing from low speed to high speed without stopping the motor.

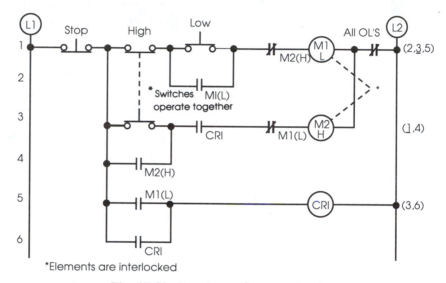

*Elements are interlocked

Fig. 17-20. Starting at low speed only.

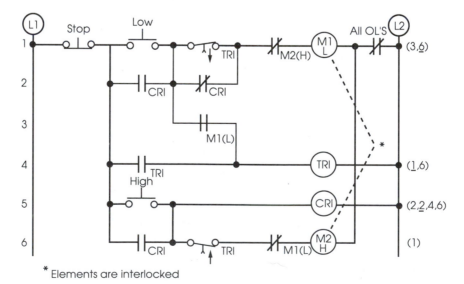

* Elements are interlocked

Fig. 17-21. Automatically accelerated system from low to high speed.

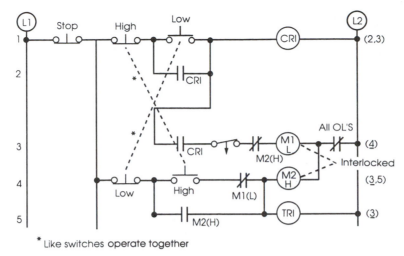

Fig. 17-22. Automatically decelerating system from high to low speed.

until the High Speed or the Stop button are pushed. The motor may be started at high speed without going through the low-speed step. The action of the timer is to disconnect all power to the motor when the Low Speed button is pushed while the motor is running at high speed. The timer then activates the low-speed circuit after its timed cycle has run its course, and the motor continues to run at low speed until the Stop button is pushed.

Maintenance of the starting controls described so far consists of normal servicing of components (already covered elsewhere in this book), as the circuits contain only standard items, such as control relays, contactors, solenoid coils, overloads, switches, and timers. The voltages and currents to be expected in each leg of the circuits are found in the usual manner, and the circuits are analyzed with the instruments described in Chapter 10.

There are several types of more sophisticated speed controls for AC motors, which are beyond the scope of this book. They consist of electronic packages which change the frequency of the voltage supplied to the motor. These controls should be serviced by experienced personnel trained specifically for each system, and this training will often be provided by the supplier's factory.

The AC motor speed may also be controlled by changing the voltage, under very carefully monitored conditions. There are some of these

in use, and we do not suggest that the beginning electrical maintenance worker attempt to service them until a high degree of skill has been attained on the more conventional plant electrical equipment.

H. Local Switching Devices

There are a large number of styles and arrangements of control switches, manually operated, for local control of motors and other electrical equipment. These switches are for the convenience of the plant personnel, either in regular operation of the equipment, or else for testing or maintenance work on the electrical apparatus being controlled.

Many of these local switches control the power to the equipment directly, providing the switch contacts are built strong enough to take the current demand. However, most of the local switches control the load by serving as a pilot control through the magnetic starters previously described.

In this section we shall cover a number of the more common methods of local control, enough to provide a general overview of the methods being used. This will serve to guide you when a type of local control station not described here is encountered, and will help you to understand the functions involved.

MANUAL STARTERS

A manual starter is a motor controller whose contact mechanism is operated by a mechanical linkage from a toggle handle or push button which is in turn operated by hand. A thermal unit and direct acting overload mechanism provides motor running overload protection. Basically, a manual starter is an "ON-OFF" switch with overload relays.

Manual starters are generally used on small machine tools, fans and blowers, pumps, compressors, and conveyors. They are the lowest cost of all motor starters, have a simple mechanism, and provide quiet operation with no AC magnet hum. Moving a handle, or pushing the START button, closes the contacts which remain closed until the handle is moved to "OFF," or the STOP button is pushed, or the overload relay thermal units trips.

FRACTIONAL HORSEPOWER (FHP) MANUAL STARTERS

FHP manual starters are designed to control and provide overload protection for motors of 1 hp or less on 115 or 230 volts single phase. They are available in single- and two-pole versions and are operated by a toggle handle on the front. When a serious overload occurs, the thermal unit "trips" to open the starter contacts, disconnecting the motor from the line. The contacts cannot be reclosed until the overload relay has been reset by moving the handle to the full OFF position, after allowing about two minutes for the thermal unit to cool (see Fig. 17-23).

MANUAL MOTOR STARTING SWITCHES

Manual motor starting switches provide on-off control of single-phase or three-phase AC motors where overload protection is not required, or is separately provided. Two- or three-pole switches are available with ratings up to 10 hp, 600 volts, 3 phase. The continuous current rating is 30 amperes at 250 volts maximum and 20 amperes at 600 volts maximum.

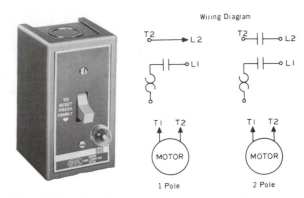

Fig. 17-23. Single-phase motor starter. (Courtesy of Square D Co.)

The toggle operation of the manual switch is similar to the FHP starter, and typical applications of the switch include small machine tools, pumps, fans, conveyors, and other electrical machinery which have separate motor protection. They are particularly suited to switch non-motor loads, such as resistance heaters.

INTEGRAL HORSEPOWER MANUAL STARTER

The integral horsepower manual starter is available in two- and three-pole versions, to control single-phase motors up to 5 hp and polyphase motors up to 10 hp, respectively. See Figs. 17-23 and 17-24.

The two-pole starters have one overload relay, and three-pole starters usually have three overload relays. When an overload relay trips, the starter mechanism unlatches, opening the contacts to stop the motor. The contacts cannot

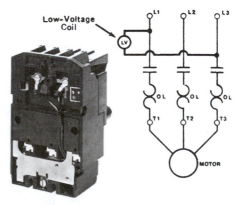

Fig. 17-24. Three-phase motor starter. (Courtesy of Square D Co.)

be reclosed until the starter mechanism has been reset by pressing the STOP button or moving the handle to the RESET position, after allowing time for the thermal unit to cool.

MANUAL STARTER WITH LOW-VOLTAGE PROTECTION

Integral horsepower manual starters with Low Voltage Protection (LVP) prevent automatic start-up of motors after a power loss. This is accomplished with a continuous-duty solenoid which is energized whenever the line-side voltage is present. If the line voltage is lost or disconnected, the solenoid de-energizes, opening the starter contacts. The contacts will not automatically close when the line voltage is restored. To close the contact, the device must be manually reset. This manual starter will not function unless the line terminals are energized.

Typical applications include conveyors, grinders, metal-working machinery, mixers, woodworking machinery, and wherever standards require low-voltage protection.

DISCONNECT SWITCHES

We have already mentioned previously the existence of a disconnect switch in the motor control circuits. This switch is also known as a Safety Switch, as that is its primary function, to provide the manual, positive shut-off of current to the load. Figure 17-9 shows it as being mounted ahead of the magnetic starter, but it may also be installed between the magnetic starter and the load. The first arrangement is usually found when the disconnect switch is incorporated with fuses, giving the device the name, Fused Disconnect Switch. This provides double protection, as the fuses are selected to permit the initial starting current inrush to get the motor up to speed, but to "blow" when the current inrush persists, or if there is a short or a grounded circuit on the load side of the switch. By placing the fused disconnect switch ahead of the magnetic starter and the load, the fuses also protect the magnetic starter from the extended current rush.

What is being accomplished here is the over-current protection of the motor branch circuit conductors, control apparatus, and the motor, caused by short circuits or grounds. Notice in Fig. 17-9 that the disconnect switch does not contain fuses or any overcurrent protective device, as there are thermal overloads in the magnetic starter. The **NEC**® requires, with a few exceptions, a means to disconnect the motor and controller from the line, in addition to an overcurrent protective device to clear short circuit faults. The short circuit device shall be capable of carrying the starting current of the motor, but its setting shall not exceed the following values:

Time Delay Fuses	175% FLA
Non-Time Delay Fuses	300% FLA
Inverse Time Circuit Breakers	250% FLA
Instantaneous Trip Circuit Breakers	700% FLA

The disconnect switch, whether fused or not, shall be placed within sight of the motor or device being served, so that the maintenance workers may see the position of the switch handle. Consequently, the modern, well-designed disconnect or safety switch is designed with that idea in mind, with a plainly visible switch handle, operating directly through the wall of the cabinet to the switch mechanism. Additional safety features are interlocks which prevent switch closing while the cover is open, switch blades and fuses in plain view when the cover is open, and locking arrangements on the switch handle. They come in any of the more common NEMA enclosures, and in NEMA sizes to handle the expected currents within each size.

The input to the disconnect switch is always down through the top of the cabinet or switch mechanism, and the connection points to the load are always at the bottom, on the output side of any overcurrent device within the cabinet. See Fig. 17-25. This fact should be kept in mind when servicing a strange disconnect switch for the first time. Before touching anything inside the cabinet, always check to be sure that when the switch is open, there is no current being supplied from the bottom into the overcurrent device. If there is, the switch is improperly connected, as opening the switch should stop all power to the load. So beware, and check those bottom terminals with a voltmeter or ammeter before doing anything else.

CONTACTORS

The general classification of "CONTACTOR" covers a type of electromagnetic apparatus designed to handle relatively high currents. A special form of contactor exists for lighting load applications, and will be covered separately.

The conventional contactor is identical in appearance, construction, and current carrying ability to the equivalent NEMA size magnetic starter.

The magnet assembly and coil, contacts, holding circuit interlock, and other structural features are the same.

The significant difference is that the contactor does not provide overload protection. Contactors, therefore, are used to switch high current, non-motor loads, or are used in motor circuits if overload protection is separately provided. A typical application of the latter is in a Reversing Starter, Fig. 17-8.

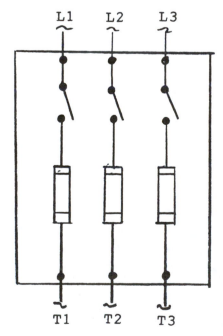

Fig. 17-25. Wiring diagram of fused disconnect switch.

LIGHTING CONTACTORS

Filament type lamps (tungsten, infra-red, quartz) have inrush currents of approximately 15–17 times the normal operating currents. Standard motor control contactors must be derated if used to control this type of load, to prevent welding of the contacts on the high initial current. The standard contactor, however, need not be derated for resistance heating or fluorescent lamp loads, which do not impose as high an inrush current.

Lighting contactors differ from standard contactors in that the contact tip material is a silver tungsten carbide which resists welding on high initial currents. A holding circuit interlock is not normally provided, since this type of contactor is frequently controlled by a two-wire pilot device such as a time clock or photoelectric relay. It should be noted that lighting contactors are specialized in their application, and should not be used on motor loads.

MECHANICALLY HELD CONTACTS

In a conventional contactor, current through the coil creates a magnetic pull to seal in the armature and maintain the contacts in a switched position. (N.O. contacts will be held closed, N.C. will be held open.) Because the contactor action is dependent on the current flow through the coil, the contactor is described as ELECTRICALLY HELD. As soon as the coil is de-energized, the contacts will revert to their initial position.

Mechanically held versions of contactors (and relays) are also available. The action is accomplished through use of two coils and a latching mechanism. Energizing one coil (latch coil) through a momentary signal causes the contacts to switch, and a mechanical latch holds the contacts in this position, even though the initiating signal is removed, and the coil is de-energized. To restore the contacts to their initial position, a second coil (unlatch coil) is momentarily energized.

MECHANICALLY HELD contactors and relays are used where the slight hum of an electrically held device would be objectionable, as in auditoriums, hospitals, churches, etc.

RELAY/CONTACTOR COMPARISON

A control relay is an electromagnetic device, similar in operating characteristics to a contactor. The contactor, however, is generally employed to switch power circuits or relatively high current loads.

Relays, with few exceptions, are used in control circuits, and consequently their lower ratings (15 amperes maximum at 600 volts) reflect the reduced levels at which they operate.

Contactors generally have from one to five poles. Although normally open and normally closed contacts can be provided, the great majority of applications use the normally open contact configuration, and there is little, if any, conversion of contact operation in the field.

I. Control Relays

A relay is an electromagnetic device whose contacts are used in control circuits of magnetic starters, contactors, solenoids, timers, and other relays. Relays are generally used to amplify the contact or multiply the switching functions of a pilot device.

Diagrams A and B of Fig. 17-26 demonstrate how a relay amplifies contact capacity. Diagram A represents a CURRENT AMPLIFICATION.

Relay and starter coil voltages are the same (220 volts), but the ampere rating of the temperature switch is too low to handle the current drawn by the starter coil (M). A relay is interposed between the temperature switch and starter coil. The current drawn by the relay coil (CR) is within the rating of the temperature switch, and the relay contact (CR) has a rating adequate for the current drawn by the starter coil.

Diagram B represents a VOLTAGE AMPLIFICATION. A condition may exist in which the voltage rating of the temperature switch is too low to permit its direct use in a starter control circuit operating at some higher voltage. In this application, the coil of the interposing relay and the pilot device are wired to a low voltage source of power compatible with the rating of the pilot device. The relay contact, with its higher voltage rating, is then used to control the operation of the starter.

Figure 17-27 represents another use of relays, which is to multiply the switching functions of a pilot device with a single or limited number of contacts.

In the circuit shown, a single-pole push-button contact can, through the use of an interposing six-pole relay, control the operation of a number of different loads such as a pilot light, starter, contactor, solenoid, and timing delay.

Depressing the "ON" button in this control

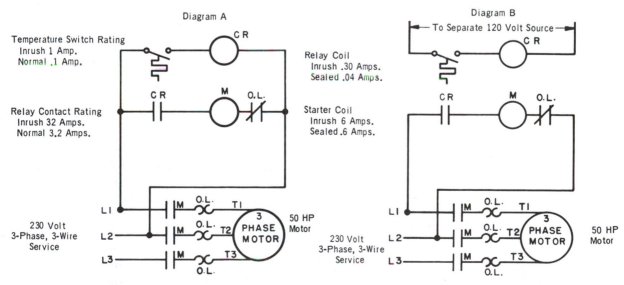

Fig. 17-26. Control relays in motor starting circuits. (Courtesy of Square D Co.)

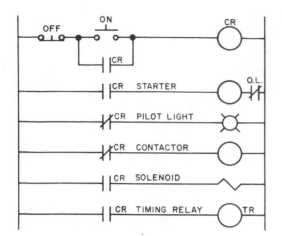

Fig. 17-27. Typical multiple relay circuit. (Courtesy of Square D Co.)

circuit energizes the relay coil (CR). Its normally open contacts close to complete the control circuits to the starter, solenoid, and timing relay, and one contact forms a holding circuit around the "ON" button. The normally closed contacts open to de-energize the contactor and turn off the pilot light.

Relays are commonly used in complex controllers to provide the logic or "brains" to set up and initiate the proper sequencing and control of a number of interrelated operations.

TIMERS AND TIMING RELAYS

A pneumatic timer or timing relay is similar to a control relay, except that certain of its contacts are designed to operate at a pre-set time interval after the coil is energized or de-energized. A delay on energization is also referred to as "On Delay." A time delay on the de-energization is also called "Off Delay."

A timed function is useful in applications such as the lubricating system of a large machine, in which a small oil pump must deliver lubricant to the bearings of the main motor for a set period of time before the main motor starts.

DRUM SWITCH

A drum switch is a manually operated three-position three-pole switch which carries a horsepower rating and is used for manual re-

versing of single- or three-phase motors. Drum switches are available in several sizes, and can be spring-return-to-off (momentary contact) or maintained contact. A typical example is the "Dead Man Throttle" used on electric trains and streetcars. Separate overload protection, by manual or magnetic starters, must usually be provided, as drum switches do not include this feature. See Fig. 17-28 for a typical arrangement.

CONTROL STATION (PUSH-BUTTON STATION)

A control station may contain push buttons, selector switches, and pilot lights. Push buttons may be momentary or maintained contact. Selector switches are usually maintained contact, or can be spring return to give momentary contact operation for jogging or positioning service.

LIMIT SWITCH

A limit switch is a control device which converts mechanical motion into an electrical control signal. Its main function is to limit movement, usually by opening a control circuit when the limit of travel is reached. Limit switches may be momentary contact (spring return) or maintained contact types. Among other applications, limit switches can be used to start, reverse, slow down, speed up, or recycle machine operations.

SNAP SWITCH

Snap switches for motor control purposes are enclosed, precision switches which require low operating forces and have a high repeat accu-

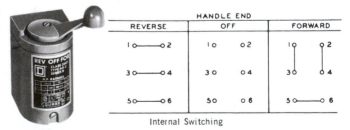

Fig. 17-28. Drum switch and connections. (Courtesy of Square D Co.)

racy. They are used as interlocks, and as the switch mechanism for control devices such as precision limit switches and pressure switches. They are available also with integral operators for use as compact limit switches, door operated interlocks, etc. Single-pole double throw and two-pole double throw versions are available.

The above covers the majority of the special purpose switches used either for direct local control of electrical loads or control through auxiliary starting devices. There are a multitude of others on the market, but many of them have one thing in common, and that is they are two-pole style when used for voltages higher than 100 volts AC.

A warning note on this last statement is in order.

Most electrical codes now require that when switching devices are used on AC voltages over about 100 volts, both lines of a two-wire system must be switched at the same time and by the same action, so that both lines are controlled. This does not apply to lighting circuits, but does apply to any circuit in which there is a possibility of a short to ground bypassing the switch, and resulting in a continual flow of current after the switch has been opened.

J. Low Voltage Control Systems

So far in this chapter we have assumed that the control systems for the magnetic starters are either 120 volt or 240 volt, and all wiring diagrams have indicated this. This does not mean that lower voltages cannot be used, and in fact, there are many advantages to using voltages as low as 24 volts for the control circuits in magnetic starters, as well as in other areas in the plant.

One of the main advantages for 24 volts and lower is safety, as there is less danger from sparking and arcing in the control systems, which is definitely to be considered when working in hazardous areas. The low voltages also require simpler transmission facilities for the

wiring, so that in long control wiring runs, there is a savings in wiring, conduit, supports, and fittings.

Another very important advantage is the flexibility in auxiliary control apparatus, as many more sophisticated control elements are available in low voltages. This includes timers, delays, programming and sequencing modules. For these reasons, low voltage control systems are often used in the self-contained, modular, control panels designed for specific functions in today's more modern plants.

We give here two wiring diagrams to illustrate the basics only of 24 volts AC or DC circuits, utilizing manual switches as the activating elements. However, they may be replaced with any compatible automatic switch mechanism, or combinations. Notice that the only difference in the wiring diagrams between DC and AC is the use of a rectifier for DC, and this rectifier may be one of the solid state selenium rectifiers (SCR) which have proven to be so desirable.

The two wiring diagrams in Figs. 17-29 and 17-30 include relays for operating loads at full voltage, in this case, electric lights.

If these systems were to be used for actuating the starter coils in a motor starter, the 120/24-volt relay and rectifier may be omitted, and the coil loads would then be supplied through the low voltage system directly to the low voltage starter coils, and not from the 120-volt lines as shown.

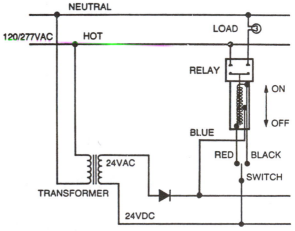

Fig. 17-29. Basic low voltage control circuit. (Courtesy of Pass and Seymour Co.)

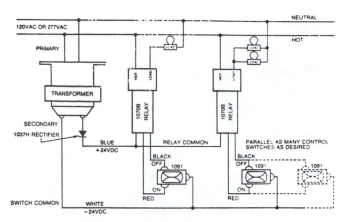

Fig. 17-30. Multiple low voltage control switches in parallel. (Courtesy of Pass and Seymour Co.)

Maintenance of low voltage systems follows the same principles as for the normal voltages, with the exception that often servicing is simplified by the use of plug-in modules, and working around the low voltage system is much safer than when working with higher voltages.

K. Repair of Controlling Equipment

As the electrical control equipment probably represents the one single most troublesome class of equipment in the electrical system, we shall go fairly deeply into the repair.

Generators and motors are useless unless their controls function properly. For proper functioning, the controller must first of all be suited to the application and the conditions of operation. After this, satisfactory functioning of control equipment requires routine inspection and maintenance. The material in this section is arranged to give the repairman a ready reference to information on the inspection, servicing, cleaning, and replacement of parts of the various types of controllers.

CAUSES OF NONOPERATION

Inspection and servicing of controllers are important functions of the repair shop since many motor and generator troubles can be traced to the improper functioning of control equipment resulting from:

1) Loose connections.

2) Broken connections.

3) Opens or shorts in magnetic coils.

4) Loose contacts.

5) Worn contacts.

6) Dirty contacts.

7) Dirty pole faces.

8) Improper tensions (poor adjustments).

9) Improper insulation.

10) Operation in high ambient temperatures.

11) Improper placement of arc chutes.

12) Moisture and dirt.

CLEANING AND DRYING

A controller, to function properly, must be clean in all its parts. The exterior and interior must be absolutely dry, except when the controller is of the oil-immersed type, where the entire apparatus is immersed in an oil tank.

Cleaning the Controller. Dry compressed air under 30 to 50 pounds pressure is used to free controllers of accumulations of dirt and dust. If this blown-air method is ineffective, the controller is wiped or washed with a safety-type cleaning fluid. Compressed air must be used with caution, since excessive air pressures may drive metallic dust and dirt into insulations or lodge particles between moving parts of contactors and relays.

To dry electrical apparatus, care should be taken to see that the heat is not too extreme, such as would damage paint, solder, plastics, or similar material. Generally, keeping the heat below 190° F will be satisfactory for all metal components, but much lower for plastics and paints. Warm air being circulated by a fan at about 140° F should do in that case.

If an oven is not available, any convenient enclosure of canvas, sheet metal, etc., will do very well.

Cleaning Contacts. The method of cleaning contacts is important. Cleaning is usually done with sandpaper or a buffing wheel. However, a fine file is permissible if the contact shape is maintained. Silver contacts seldom require cleaning although they look black and dirty because of the silver oxide. Since silver oxide is a conductor, cleaning is not necessary. When contacts are replaced, the surface against which they are bolted is thoroughly cleaned. This surface is usually a current-carrying joint, and a clean contact bolted to a dirty surface will cause future trouble. Traces of copper oxide should be removed.

Harmful Cleaning Methods. Emery cloth or coarse files should not be used. Emery particles may stick to moving parts and contact surfaces, causing unnecessary wear. Coarse, crude filing causes waste of contact material and makes it difficult to maintain original contact shape. The use of sandpaper (dry) is recommended.

Controller Locations. It is necessary to check the temperatures of the locations in which thermal overload relays are to be used. It is also necessary to check the location of the thermal overload relays so they do not affect, and are not affected by, other thermal control apparatus within their immediate vicinity. With ambient temperatures known, it is possible, for example, to compensate for higher temperatures by selecting a thermal heater which will operate properly in above-normal temperatures. When a motor is reconnected for use, the heater ratings should be checked against the motor nameplate data to be certain that the thermal overload heater rating is not too high or too low.

The testing and inspection of controllers is an important phase of troubleshooting. When inspecting, determine whether the controller is mounted in a dry area away from heat sources such as steampipes or operating machinery which gives off heat. Check the voltage at the power supply and at the motor lead terminals. If it is a polyphase power source, check the voltage of each phase (AB, BC, and AC). It is also important to check the setting or value of overload heaters and fuses for proper ratings. Follow the procedures below for these inspections

and checks, and refer to the troubleshooting at the end of this section.

AREA IN WHICH A CONTROLLER IS MOUNTED

If it appears to be unusually warm in an area in which a controller is mounted, check the temperature with a standard thermometer placed near the controller.

If it is too warm, the controller should be relocated in a lower temperature area, if possible. Otherwise, the rise in temperature can be compensated for by selecting an overload relay heater rated for operation at the higher temperature.

If the controller is mounted in an area which vibrates because of running machinery, and the controller continually trips, it will be necessary to shock-mount the controller. Contact the equipment supplier for his recommendations.

VOLTAGE CHECK

Voltages should be checked at the source first and then at the motor terminals. If it is a polyphase feeder source, check voltage of each phase. If voltage is not up to normal operating level, the causes for the drop in voltage must be corrected.

Check voltage at the line connections in the controller, and if the voltage is below normal, follow this procedure.

a) Shut off the power.

b) Tighten the line connections.

c) Check all other connections for tightness. Do not judge tightness of connections by appearance. Use a tool for tightening to make certain connections are tight.

d) Remove all tools and check equipment from the controller.

e) Energize the controller.

f) Check voltages. If the voltages are still below normal, perform the operations stated in 3) and 4) below.

g) De-energize the controller.

h) Check the line conductors to determine whether they are of proper size for the application. If they are not, they must be replaced.

RELAY HEATERS

Repeated starting after relays trip out may cause a motor to burn out. The following checks may reveal the reason for repeated tripping out of relays.

1) Compare heater's rating with that recommended on heater data sheet or with the current rating on the motor nameplate.

2) If heater rating is correct, measure the current taken by the motor. If the current is higher than that shown on the motor nameplate (plus the overload factor), the relay is operating correctly and mechanical load is to be checked for overload or possible motor trouble.

3) Low line voltage causes high current and trips the relay. Check line voltage. When low voltage is discovered, check for inadequate wire size of power leads to the controller from the power supply or overloading of the circuit.

4) Check ambient temperatures surrounding the enclosure of thermal relays for a unit or units which are causing a high temperature rise in the controller. Any rise in surrounding temperature will cause the thermal overload to trip sooner than intended. A thermal relay controller should be in the same area as the motor, unless the overload-relay heater has been selected to operate under unusual temperature conditions. Fuses also fail under high temperature conditions, as outlined.

DEFECTIVE WIRING

Wiring between the motor and controller may be open, grounded, or short circuited. These faults may be checked with an ohmmeter or megger between conductors and ground. Accidental grounds occurring in control circuits produce false circuits that may cause unexpected starts, prevent normal stopping, eliminate overload protection, and cause erratic operation. The remedy for this is a new wiring installation for remote control wiring, and repair and replacement of defective internal wiring. Excessive vibration is often a cause of these troubles; in this case, the controller is either relocated or shock-mounted.

DEFECTIVE CONTROL EQUIPMENT

Mechanical troubles which may develop after a period of operation of controller equipment can be a result of surrounding conditions. Tight or worn bearings may be due to an atmosphere containing oily vapors, which cause gummy deposits, or an atmosphere containing abrasive dust particles, which cause premature wear. The remedy is to clean and replace defective parts and correct the surrounding conditions. Vibration may cause loose or broken connections, misalignment of mating parts, and also excessive wear.

Mechanical troubles may also result in electrical troubles, such as defective overload devices, open coils, short-circuited coils, abraded conductors causing grounds, and faulty contact surfaces due to misalignment. Push-button and pilot circuits are to be suspected when the motor cannot be de-energized or the motor makes false starts. When excessive noise and vibration are evident, the unit must be shut down and the trouble remedied. Refer to the section on "AC HUM" later in this chapter.

Both overvoltage and undervoltage can damage electric controls. Overvoltage can shorten coil life and cause excessive wear in contactors and relays. Overvoltage also operates a contactor or relay with more mechanical force than needed. This tends to shorten the mechanical life of the control if allowed to continue. Low voltage causes sluggish action because tips may touch without being forced completely closed against the contact spring pressure. Under such conditions, contact tips overheat and will probably weld together. Contacts must always close completely.

AC controller coils are designed to function at 15 percent undervoltage and to withstand 10 percent overvoltage.

Three-phase controllers are designed with overload protection on two phases. If a phase which is not electrically connected to the control circuit happens to go out, the motor will continue to operate at a lower efficiency, but only long enough for the overload devices or the other devices to become overheated and trip the circuit. This is caused by a higher current drain on the remaining two phases. Power is restored to the open phase to remedy this condition.

In many instances, low voltage conditions are corrected by redistribution of feeder circuits. A study and analysis of distribution is made before redistribution.

High voltage conditions are to be corrected at the generating point, or at the plant transformer station. See Chapter 9, Section D.

REPAIR PROCEDURES

The majority of repairs to control equipment consist of the replacement of defective parts. It is important to obtain complete information about the controller under repair before ordering replacement parts. The following paragraphs will cover starting and speed-regulating rheostats for AC and DC motors, and starters for squirrel-cage induction motors. A few representative types are selected, and repair procedures peculiar to these controls are given.

It is possible on some tyes of controllers to reverse the contact tips. This provides a new surface for use. When doing this, it is necessary to:

1) de-energize the circuit;

2) make certain the contact tips have not been reversed before;

3) make certain that a new contact-tip surface mates with another; and

4) after reversal, make certain that the contact tips are mechanically tight.

When checking a controller before energizing it for operation, the following must be performed to determine if it will operate properly.

1) Operate all movable parts by hand and note that they return to normal nonoperating positions without binding or strain.

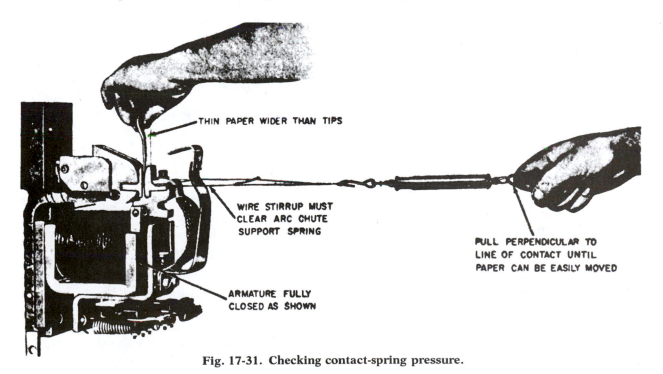

THIN PAPER WIDER THAN TIPS

WIRE STIRRUP MUST CLEAR ARC CHUTE SUPPORT SPRING

ARMATURE FULLY CLOSED AS SHOWN

PULL PERPENDICULAR TO LINE OF CONTACT UNTIL PAPER CAN BE EASILY MOVED

Fig. 17-31. Checking contact-spring pressure.

2) Check contact pressures by the following procedure (Fig. 17-31).

   a) Place a piece of thin paper 0.003–0.05 inch thick (newspaper will do), slightly wider than the contact tips, between the contact tip faces.

   b) Manually close the contactor until the electromagnetic pole faces come together.

   c) Try to remove the paper; it should be difficult to remove the paper with the contactor in this position. In fact, the paper should tear if force is used to remove it.

   d) If the paper is easily removed, adjust the contacts to give the condition of pressure in c) above.

   e) Check every set of contacts in the same manner.

It is often necessary to replace damaged or burned out resistors. The procedures used should follow strictly the instructions for removal and replacement of electrical circuit parts, repeated here.

1) Use caution in removal so that other parts will not be damaged when removing a resistor.

2) Record circuits or tag disconnected wires.

3) Be certain that all contact and connection parts are clean and bright.

4) Make electrically tight all permanent connections on the replacement resistor.

The methods of removal and replacement of magnet coils vary with the type and manufacturer of the controller device. Generally, the procedure outlined below is to be followed for all magnet coils.

1) Loosen the coil from the core (after determining method of fastening).

2) Remove from the core.

3) Replace the proper coil for the controller according to code designation.

4) Be careful when slipping the coil into place not to abrade the insulation or bend the connecting lugs.

5) Fasten the coil securely, because a mechanical force is developed when the coils are energized. This force may cause them to slip back and forth on the frame, resulting in excessive wear on the coil insulation, and eventual trouble.

6) By means of a manual closure, check for complete contact of pole faces; if necessary, readjust the coil until the complete closure is obtained. Coils become overheated if mechanical interference prevents complete closure of the magnetic air gap. Overheating weakens insulation.

Manual and magnetic across-the-line starters are used mostly for fractional-horsepower motors. The current drawn at starting will range from four to six times the full-load current. If an induction motor is very large, however, the high starting current may cause too much voltage fluctuation in the power line or place too much stress on the driven machinery. Under these conditions, the starting voltage is reduced during the starting period by the use of resistors or autotransformers. Resistance starters are also used with AC wound-rotor induction motors for starting and speed control. The points to be checked, and precautions to be observed, are listed below.

1) **Manual Starters.**

   a) Check for freedom of motion of moving parts.

   b) Check contact surfaces for wear, pitting, and proper pressure.

   c) Check overload devices for proper match to the motor unit. This information is listed on the inside surface of the cover.

2) **Magnetic Starters.** Refer to earlier paragraphs of this section for procedures on troubleshooting this equipment.

3) **Reduced-Voltage Starters.** After having checked all points as outlined above, make continuity checks of resistors or transformer windings to check for grounds, opens, and shorts.

4) **Resistance Type Starters.** Make checks as stated above.

AC HUM

All devices which incorporate a magnetic effect produce a characteristic hum. This hum or noise is due mainly to the changing magnetic pull (as the flux changes) inducing mechanical vibrations. Contactors, starters, and relays could become excessively noisy as a result of some of the following operating conditions.

Broken shading coil.

Operating voltage too low.

Wrong coil.

Misalignment between the armature and magnet assembly—the armature is then unable to seat properly.

Dirt, rust, filings, etc., on the magnet faces—the armature is unable to seal in completely.

Jamming or binding of moving parts (contacts, springs, guides, yoke bars) so that full travel of the armature is prevented.

Incorrect mounting of the controller, as on a thin piece of plywood fastened to a wall, for example, so that a "sounding board" effect is produced.

C H A P T E R 18
RECORDS, REPORTS, AND DOCUMENTATION

A. Extent and Purpose

No matter how small your plant may be, it is very important that you keep complete records of the activities and the equipment which make up your livelihood. No one's memory is good enough to be absolutely sure of remembering when definite maintenance procedures and repairs were performed. Also, in today's mobile work force, it is only a matter of forethought and common courtesy that you should keep sufficient written records to permit any successor to continue your work in the plant, with the least amount of upset to the remainder of the organization. How would you like to take over a going concern, with everything in place, and not know the age and history of the equipment, and the last time it had been checked and cleaned and lubricated?

Of course, if your's is a branch of a larger plant complex, then you may be under strict company rules to produce complete reports and documentation on the activities of the electrical plant operation. The firm may have already supplied you with their standard forms for that purpose. If not, it is a simple matter for you to make up your own forms to be filled out regularly and kept in a convenient location for your retrieval, as well as being available for any one else who is authorized to look at them.

Even if only for your own convenience, it is highly recommended that you keep the manufacturer's data, maintenance records, and other such data as we shall list in this chapter.

There is a more practical reason for keeping complete records of your daily activities and experiences in maintaining and operating the electrical plant. In case of trouble, such as an accident or a fire, for your own protection it is well to have complete records of the plant operation for the insurance firm which will probably be investigating the cause of the trouble. If there are no records for them to refer to, the first impression will be that the plant was improperly operated. You can be sure that the manufacturer has covered himself with all kinds of data and testing records on the electrical equipment, so it is wise for you to do the same.

Factory certificates are a part of the package of data which the manufacturer normally sends the customer either before shipping the electrical equipment, or included in the shipment. This package of data should be reviewed by someone responsible for the integrity of the plant, and then it should be kept in the company files in a safe and accessible location. The engineering department or the plant facilities department are the usual repository for this material.

Other data and literature available during the design, specification, and purchase of the plant equipment, and which should be retained in the same file, are the following.

Manufacturer's Data Report Forms, a signed certificate attesting to the in-shop inspection of the quality and integrity of all motors, transformers, protective equipment, and other important pieces of electrical equipment.

General arrangement drawings, with all pertinent dimensions of the complete package.

Bill of materials, including all accessories furnished by other than the basic manufacturer. Contains complete model numbers and specifications of the accessories.

Installation instructions for the complete electrical package.

Start-up instructions, including cleaning, testing in place (when used), and lubrication prior to start-up.

Operating instructions for the complete package.

Maintenance instructions for electrical equipment and its accessories. Includes literature from the makers of the accessories.

Spare parts list, with prices, of the major items in the package.

Copies of all factory test data.

Descriptive literature of the equipment and its accessories.

Nameplate data for all electrical equipment.

One of the most important items listed above is the recording of all nameplate data. In the case of electric motors, this will usually consist of all, or nearly all, of the following items.

1. The manufacturer's name.

2. The style of the motor, either by name or by code.

3. The horsepower at rated speed.

4. Motor NEMA frame size.

5. Speed, in revolutions per minute, at the rated power output.

6. Rated voltage, either in single, dual, or special voltage.

7. Motor phase, either single, two, or three phase.

8. Design Hertz (Hz), which in most cases will be 60 Hz.

9. Full-Load Current rating at standard conditions.

10. Degrees Celsius above ambient temperature the motor is designed to operate. A 40°C motor is designed to operate at 72°F above ambient temperature. A 50^C motor is designed to operate at 90°F above ambient temperature.

11. Duty, either continuous or intermittent, at rated conditions. Some motors will have other duty ratings.

12. Serial number of the motor, not always a unique identifying number. It may only designate a certain series of design or date of manufacture.

13. Locked rotor kVA draw at rated conditions.

14. Service factor, which is a number indicating at what overload the motor may operate continuously, but with increased temperature over its rating. A 1.00 service factor indicates no overload capability. A 1.15 service factor indicates 15% overload capability.

15. Type code, indicating split phase, capacitor, polyphase power, etc.

16. Insulation type, by NEMA code, indicating resistance to special operating atmospheres or conditions.

17. Motor enclosure, such as open drip-proof, explosion-proof, totally enclosed fan cooled (TEFC), etc.

Also, of course, similar data for all of the other miscellaneous plant electrical equipment purchased for operating the plant should be included.

B. Contract Data

If the heavy electrical equipment was furnished in the usual manner, it was probably bid by several suppliers to an inquiry by your firm's purchasing department, based on specifications written by the plant engineering department or an outside engineering firm. The successful bidder then was issued a purchase order by your purchasing department. The installation was probably handled in a similar manner, by an outside contractor experienced in this class of equipment installation.

It is an excellent idea to retain files of all of the major milestones in the above procedure, in case of questions or problems appearing later, such as method of installation, testing and cleaning procedures, etc. The minimum material to keep in your files for the project should consist of the following.

Specifications, with all addenda.

Successful bidder's proposal.

Purchase order to the contractor, with all supplements.

Copies of all attendant correspondence.

Final Letter of Acceptance.

C. Maintenance Records

Of equal importance to the above are the plant equipment maintenance records. On this form should be recorded all of the time and money spent to maintain each piece of equipment.

There are several good reasons why this record should be kept, and we give here some of the most obvious ones.

1. It will help pinpoint any constant source of difficulty which should be corrected.

2. In the case of recurring problems, it will serve as a reference source for any new personnel working on the equipment.

3. If the time spent and the parts used, with their cost, are all recorded, it will enable the management to determine the cost-effectiveness of the equipment, and help in any decisions concerning its retention or replacement.

4. It may at times help straighten out any discrepancies in spare parts inventories.

5. Properly organized and kept, the maintenance record forms will help in alerting you to the approximate times when further maintenance will be required.

It is not necessary that you invest in a lot of money to set up a system of record keeping, as long as the system you adopt is complete enough for your purposes, and is properly maintained. Even a simple notebook with pages devoted to the equipment will be satisfactory, as long as the relevant data and records are properly entered. Also, of course, the location for the records must be such as to ensure safekeeping and ready accessibility.

We will list here a few of the more obvious items of maintenance which should be included in the entries in the maintenance logs.

1. Name and serial number of the equipment.

2. Date installed.

3. List of accessories, and their make and model numbers.

4. Electrical characteristics.

5. Inspection dates, and name of inspector.

6. Dates of preventive maintenance performed, and name of personnel involved.

7. Spare parts used.

8. Time spent in performing the work.

9. Time that equipment went out of service and time placed back in service.

D. Spare Parts Inventories

Closely allied to the various records and documentation we have covered so far in this chapter, is the spare parts inventory. Any plant using mechanical equipment, such as motors, or other operating equipment which tends to wear out and require servicing, will have to keep on hand spare parts to service that equipment. As the spare parts are part of the plant operating supplies, records should be kept of their cost, the source of supply, and their consumption, as well as the complete description for purposes of reordering.

There are many inventory systems for spare parts available, and some even are computerized. That is very well for the larger plants, or those which are a branch of a national network, but for the small- to medium-sized plant, such as we are covering here, usually a simple form, kept in a file drawer, or a card file, is all that is normally required. It does not matter how complex or simple the system is, as long as it is adequate for your purposes, and is compatible with the records and inventory system used in the remainder of the plant.

Some of the items of information which should be entered on the form are:

1. Make and model number of the equipment the part is for.

2. Part identification by number and name.

3. Source of supply.

4. Cost.

5. Approximate delivery time required.

6. Number of parts required for each item of equipment.

7. Number to keep on hand for emergencies.

8. Where the part is stored.

9. Ordering dates, i.e., history and order numbers, recorded chronologically.

10. Dates parts are withdrawn and placed into storage.

That completes the chapter on record keeping in the plant. We hope this will give you sufficient information to enable you to organize your plant operations in an efficient manner and thus help you out of undue hardships and chaotic situations. Remember, these are only guides. You are free to improvise as necessary.

C H A P T E R 19
ELECTRICAL WIRING TECHNIQUES

A. Splices and Connections

Conductor splices and connections are an essential part of any electric circuit. When conductors join each other, or connect to a load, splices or terminals must be used. It is important that they be properly made, since any electric connection or joint must be both mechanically and electrically as strong as the conductor or device with which it is used. High-quality workmanship and materials must be employed to ensure lasting electrical contact, physical strength, and insulation (if required). The most common methods of making splices and connections in electric cables will now be discussed.

The first step in making a splice is preparing the wires or conductor. Insulation must be removed from the end of the conductor and the exposed metal cleaned. In removing the insulation from the wire, a sharp knife is used in much the same manner as in sharpening a pencil. That is, the knife blade is moved at a small angle with the wire to avoid "nicking" the wire. This produces a taper on the cut insulation, as shown in Fig. 19-1. The insulation may also be removed by a hand-operated wire stripper. After the insulation is removed, the bare wire ends should then be scraped bright with the back of a knife blade or rubbed clean with fine sandpaper.

WESTERN UNION SPLICE

Small, solid conductors may be joined together by a simple connection known as the WESTERN UNION SPLICE. In most instances, the wires may be twisted together with the fingers and the ends clamped into position with a pair of pliers.

Figure 19-2 shows the steps in making a Western Union splice. First, the wires are prepared for splicing by removing sufficient insulation and cleaning the conductor. Next, the wires are brought to a crossed position and a long twist or bend is made in each wire. Then one of the wire ends is wrapped four or five times around the straight portion of the wire. The other end of the wire is wrapped in a similar manner. Finally, the ends of the wires should be pressed down as close as possible to the straight portion of the wire. This prevents the sharp ends from puncturing the tape covering that is wrapped over the splice. See Fig. 19-2.

STAGGERED SPLICE

Joining small, multiconductor cables presents somewhat of a problem. Each conductor must be spliced and taped; and if the splices are directly opposite each other, the overall size of the joint becomes large and bulky. A

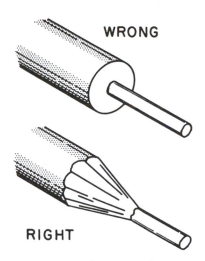

Fig. 19-1. End preparation.

smoother and less bulky joint may be made by staggered the splices.

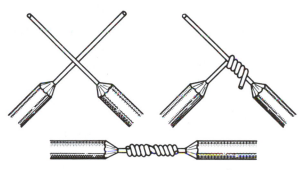

Fig. 19-2. Western Union Splice.

Figure 19-3 shows how a two-conductor cable is joined to a similar cable by means of the staggered splice. Care should be exercised to ensure that a short wire is connected to a long wire, and that the sharp ends are clamped firmly down on the conductor.

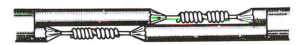

Fig. 19-3. Staggered splice.

RATTAIL JOINT

Wiring that is installed in buildings is usually placed inside long lengths of steel conduit. Whenever branch circuits are required, junction or pull boxes are inserted in the conduit. One type of splice that is used for branch circuits is the rattail joint shown in Fig. 19-4.

The ends of the conductors to be joined are stripped of insulation. The wires are then twisted to form the rattail effect.

FIXTURE JOINT

A fixture joint is used to connect a light fixture to the branch circuit of an electrical system where the fixture wire is smaller in diameter than the branch wire. Like the rattail joint, it will not stand much mechanical strain.

The first step is to remove the insulation from the wires to be joined. Figure 19-5 shows the steps in making the fixture joint.

After the wires are prepared, the fixture wire is wrapped a few times around the branch wire, as shown in the figure. The wires are not twisted, as in the rattail joint. The end of the branch wire is then bent over the completed turns. The remainder of the bare fixture wire is then wrapped over the bent branch wire. Soldering and taping completes the job.

KNOTTED TAP JOINT

All of the splices considered up to this point are known as BUTTED splices. Each was made by joining the FREE ends of the conductors together. Sometimes, however, it is necessary to join a conductor to a CONTINUOUS wire, and such a junction is called a TAP joint.

The main wire, to which the branch wire is to be tapped, has about one inch of insulation removed. The branch wire is stripped of about three inches of insulation. The steps in making the tap are shown in Fig. 19-6.

The branch wire is crossed over the main

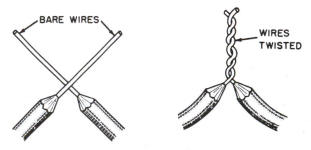

Fig. 19-4. Rattail joint.

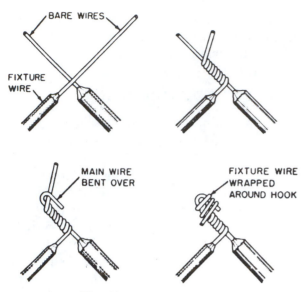

Fig. 19-5. Fixture joint.

wire, as shown in the figure, with about three-fourths of the bare portion of the branch wire extending above the main wire. The end of the branch wire is bent over the main wire, brought under the main wire, and then over the main wire to form a knot. It is then wrapped around the main conductor in short, tight turns and the end is trimmed off.

The knotted tap is used where the splice is subjected to strain or slip. When there is no mechanical strain, the knot may be eliminated.

B. Soldering Process

Cleanliness is a prime prerequisite for efficient, effective soldering. Solder will not adhere to dirty, greasy, or oxidized surfaces. Heated metals tend to oxidize rapidly, and the oxide must be removed prior to soldering. Oxides, scale, and dirt can be removed by mechanical means (such as scraping or cutting

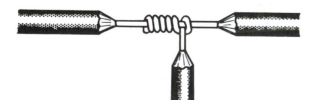

Fig. 19-6. Knotted tap joint.

with an abrasive) or by chemical means. Grease or oil films can be removed by a suitable solvent. Cleaning should be accomplished immediately prior to the actual soldering operation.

Items to be soldered should normally be tinned before making mechanical connection. When the surface has been properly cleaned, a thin, even coating of flux may be placed over the surface to be tinned to prevent oxidation while the part is being heated to soldering temperature. Rosin core solder is usually preferred in electrical work, but a separate rosin flux may be used instead. Separate rosin flux is frequently used when tinning wires in cable fabrication. Tinning is the coating of the material to be soldered with a light coat of solder.

The tinning on a wire should extend only far enough to take advantage of the depth of the terminal or receptacle. Tinning or solder on wires subject to flexing causes stiffness, and may result in breakage.

The tinned surfaces to be joined should be shaped and fitted, then mechanically joined to make good mechanical and electrical contact. They must be held still with no relative movement of the parts. Any motion between parts will likely result in a poor solder connection.

C. Soldering Tools

SOLDERING IRONS

All high-quality irons operate in the temperature range of 500° to 600° F. Even the little 25-watt midget irons produce this temperature. The important difference in iron sizes is not temperature, but thermal inertia (the capacity of the iron to generate and maintain a satisfactory soldering temperature while giving up heat to the joint to be soldered). Although it is not practical to try to solder a heavy metal box with the 25-watt iron, that iron is quite suitable for replacing a half-watt resistor in a printed circuit. An iron with a rating as large as 150 watts would be satisfactory for use on a printed circuit, provided that suitable soldering techniques are used. One advantage of using a small iron for small work is that it is light and easy to handle and has a small tip which is easily inserted into close places. Also, even though its temperature is high, it does not have the capacity to transfer large quantities of heat.

Some irons have built-in thermostats. Others are provided with thermostatically controlled stands. These devices control the temperature of the soldering iron, but are a source of trouble. A well-designed iron is self-regulating by virtue of the fact that the resistance of its element increases with rising temperature, thus limiting the flow of current. For critical work, it is convenient to have a variable transformer for fine adjustment of heat; but for general-purpose work, no temperature regulation is needed.

SOLDERING GUN

A transformer in the soldering gun supplies approximately 1 volt at high current to a loop of copper which acts as the tip. It heats to soldering temperature in 3 to 5 seconds, but may overheat to the point of incandescence if left on over 30 seconds. The gun is operated with a finger switch so that the gun heats only while the switch is depressed.

Since the gun normally operates only for short periods at a time, it is comparatively easy to keep clean and well tinned; thus, little oxidation is allowed to form. However, the tip is made of pure copper, and is susceptible to pitting which results from the dissolving action of the solder.

Tinning of the tip is always desirable unless it has already been done. The gun or iron should always be kept tinned in order to permit proper heat transfer to the work to be soldered. Tinning also provides adequate control of the heat to prevent thermal spillover to nearby materials. Tinning the tip of a gun may be somewhat more difficult than tinning the tip of an iron. Maintaining the proper tinning on either type, however, may be made easier by tinning with silver solder. The temperature at which the bond is formed between the copper tip and the silver solder is considerably higher than

with lead-tin solder. This tends to decrease the pitting action of the solder on the copper tip.

Pitting of the tip indicates the need for retinning, after first filing away a portion of the tip. Retinning too often results in using up the tip too fast.

Overheating can easily occur when using the gun to solder delicate wiring. With practice, however, the heat can be accurately controlled by pulsing the gun on and off with its trigger. For most jobs, even the LOW position of the trigger overheats the soldering gun after 10 seconds; the HIGH position is used only for fast heating and for soldering heavy connections.

Heating and cooling cycles tend to loosen the nuts or screws which retain the replaceable tips on soldering irons or guns. When the nut on a gun is loosened, the resistance of the tip connection increases, and the temperature of the connection is increased. Continued loosening may eventually cause an open circuit. Therefore, the nut should be tightened periodically.

RESISTANCE SOLDERING

A time-controlled resistance soldering set is now available. The set consists of a transformer that supplies 3 or 6 volts at high current to stainless steel or carbon tips. The transformer is turned ON by a foot switch and OFF by an electronic timer. The time can be adjusted for as long as 3 seconds soldering time. This set is especially useful for soldering cables to plugs and similar connectors—even the smallest types available.

In use, the double-tip probes of the soldering unit are adjusted to straddle the connector cup to be soldered. One pulse of current heats it for tinning and, after the wire is inserted, a second pulse of current completes the job. Since the soldering tips are hot only during the brief period of actual soldering, burning of wire insulation and melting of connector inserts are greatly minimized.

The greatest difficulty with this device is keeping the probe tips free of rosin and corrosion. A cleaning block is mounted on the transformer case for this purpose. Some technicians prefer fine sandpaper for cleaning the double tips. CAUTION: Do not use steel wool. It is dangerous when used around electrical equipment.

PENCIL IRON AND SPECIAL TIPS

An almost indispensable item is the pencil type soldering iron with an assortment of tips (Fig. 19-7). Miniature soldering irons, with wattage ratings of less than 40 watts, are easy to use and are recommended. In an emergency, larger irons can be converted and used on subminiature equipment as described later in this section.

One type of iron is equipped with several different tips that range from one-fourth to one-half inch in size (diameter) and are of various shapes. This feature makes it adaptable to a variety of jobs. Unlike most tips which are held in place by setscrews, these tips have threads and screw into the barrel. This feature provides excellent contact with the heating element, thus improving heat transfer efficiency. A pad of "antiseize" compound is supplied with each iron. This compound is applied to the threads each time a tip is installed in the iron, thereby enabling the tip to be easily removed when another is to be inserted.

A special feature of this iron is the soldering pot that screws in like a tip and holds about a thimbleful of solder. It is useful for tinning the ends of large numbers of wires.

The interchangeable tips are of various sizes and shapes for specific applications. Extra tips may be obtained and shaped to serve special purposes. The thread-in units are useful in soldering subminiature items. The desoldering units are specifically designed for performing special and individual functions.

Another advantage of the pencil soldering iron is its possible use as an improvised light source for inspections. Simply remove the soldering tip and insert a 120-volt, 6-watt, type 6S6, candelabra screwbase lamp bulb into the socket.

If leads, tabs, or small wires are bent against a board or terminal, slotted tips may be used to simultaneously melt the solder and straighten the leads.

A hollow tip, which fits over a pin terminal, may be used to desolder and resolder wiring at cables or feed-through terminal.

Many miniature components have multiple connections, all of which must be desoldered to permit removal of the component in one operation. These connections may be desoldered individually by heating each connection and brushing away the solder. With this method, particular care must be taken to ensure that loose solder does not stick to other parts or become lodged where it may cause a short circuit. A more efficient method is to use the specially shaped desoldering units. Select the proper size and shape tip that will contact all terminals to be desoldered—and nothing else. Do not permit the tip to remain in contact with the terminals too long at one time.

If no suitable tip is available for a particular operation, an improvised tip may be made. Wrap a length of copper wire around one of the regular tips and bend the wire into the proper shape for the purpose. This method also serves to reduce tip temperature when a larger iron must be used on miniature components (see Fig. 19-8).

In connection with the discussion of soldering tools and devices, the selection of solder

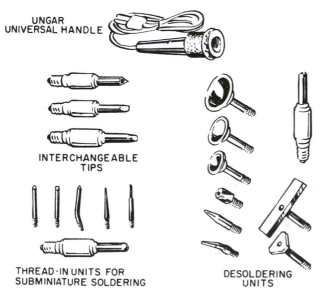

UNGAR
UNIVERSAL HANDLE

INTERCHANGEABLE
TIPS

THREAD-IN UNITS FOR
SUBMINIATURE SOLDERING

DESOLDERING
UNITS

Fig. 19-7. Pencil iron kit.

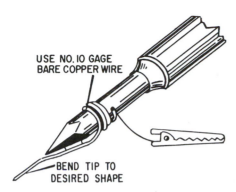

USE NO. IO GAGE
BARE COPPER WIRE

BEND TIP TO
DESIRED SHAPE

Fig. 19-8. Reducing tip temperature.

and flux is also critical. A small diameter rosin core solder with a high tin–lead ratio (60/40) is normally preferred in miniature circuits where heat is critical.

SOLDERING AIDS

Several devices other than the soldering iron and its tips are required in soldering miniature circuits. Several of these (brushes, probes, scrapers, knives, etc.) have been mentioned previously.

Some type of thermal shunt is essential in all soldering operations which involve heat-sensitive components. Pliers, tweezers, or hemostats may be used for some applications, but their effectiveness is limited. A superior heat shunt, as shown in Fig. 19-9, permits soldering the leads of component parts without overheating the part itself.

For maximum effectiveness, any protective coating should be removed before applying the heat shunt. The shunt should be attached carefully to prevent damage to the leads, terminals, or component parts. The shunt should be clipped to the lead, between the joint and the part being protected. As the joint is heated, the shunt absorbs the excess heat before it can react with the part and cause damage.

A small piece of beeswax may be placed between the protected unit and the heat shunt. When the beeswax begins to melt, the temperature limit has been reached. The heat source should be removed immediately, but the shunt should be left in place.

Premature removal of the heat shunt permits the unrestricted flow of heat from the melted solder into the component. The shunt should be allowed to remain in place until it cools to room temperature. A clip-on type shunt is preferred because it requires positive action to remove the shunt, but does not require that the technician maintain pressure to hold it in place.

Another invaluable soldering aid is the "solder sucker" syringe. One type is shown in Fig. 19-10. Its purpose is to "suck up" excess solder (and incidentally the excess heat) from a joint. The only requirements of an efficient solder sucker are a controllable source of vacuum (squeeze bulb), a solder receiver, and a tip. The tip must be able to withstand the heat of molten solder. Teflon is ideal, but may be difficult to acquire. A silicon rubber-covered Fiberglas sleeving with an inner diameter of 0.162 inch and the bulb from a medicine dropper makes a suitable syringe. (The glass or plastic tip of the medicine dropper cannot withstand the heat.)

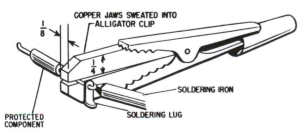

COPPER JAWS SWEATED INTO
ALLIGATOR CLIP

$\frac{1}{8}$

$\frac{1}{4}$

SOLDERING IRON

PROTECTED
COMPONENT

SOLDERING LUG

Fig. 19-9. Heat shunt.

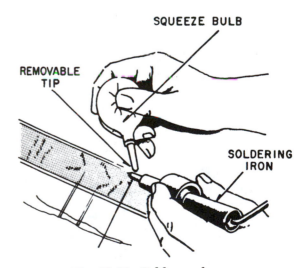

SQUEEZE BULB

REMOVABLE
TIP

SOLDERING
IRON

Fig. 19-10. Solder sucker.

D. Solder Connections

As a result of tests, the joints illustrated in Fig. 19-11 are recommended. Wrappings of three-eights to three-fourths turn are usually recommended so that the joint need not be held during the application and cooling of the solder.

The areas to be joined must be heated to, or slightly above, the flow temperature of the solder. The application of heat must be carefully controlled to prevent damage to components of the assembly, insulation, or nearby materials. Solder is then applied to the heated area. Only enough solder should be used to make a satisfactory joint. Heavy fillets or beads must be avoided.

Solder should not be melted with the soldering tip and allowed to flow onto the joint. The joint should be heated and the solder applied to the joint. When the joint is adequately heated, the solder will flow evenly. Excessive temperature tends to carbonize flux, thus hindering the soldering operation.

No liquid should be used to cool a solder joint. By using the proper tools and soldering technique, a joint should not become so hot that rapid cooling is needed.

If, for any reason, a satisfactory joint is not initially obtained, the joint must be taken apart, the surfaces cleaned, excess solder removed, and the entire soldering operation (except tinning) repeated.

After the joint has cooled, all flux residues should be removed. Any flux residue remaining on the surface of electrical contacts may collect dirt and promote arcing at a later time. This cleaning is necessary even when rosin-core solder is used.

Connections should never be soldered or desoldered while equipment power is on or while the circuit is under test. Always discharge any capacitors in the circuit prior to any soldering operation.

E. Solder Splicers

The solder-type splicer is essentially a short piece of metal tube. Its inside diameter is just large enough to allow the tip of a stranded conductor to be inserted in either end, after the conductor tip has been stripped of insulation. This type of splice is shown in Fig. 19-12.

The splice is first heated and filled with solder. While still molten, the solder is then poured out, leaving the inner surfaces tinned. When the conductor tips are stripped, the length of exposed strands should be long enough so that the insulation butts against the splicer when the conductors are tinned and fully inserted (see Fig. 19-12 (B)). When heat is applied to the connection and the solder melts, excess solder will be squeezed out through the vents. This must be cleaned away. After the splice has cooled, insulating material must be wrapped or tied over the joint.

F. Solder Terminal Lugs

In addition to being joined or spliced to one another, conductors are often connected to

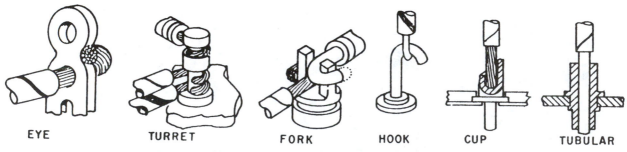

EYE   TURRET   FORK   HOOK   CUP   TUBULAR

Fig. 19-11. Wrapping of terminals for soldering.

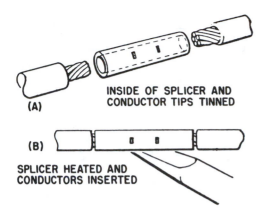

Fig. 19-12. Steps in using solder splicer.

other objects, such as motors and switches. Since this is where a length of conductor ends (terminates), such connections are referred to as terminal points. In some cases, it is allowable to bend the end of the conductor into a small "eye" and put it around a terminal binding post. Where a mounting screw is used, the screw is passed through the eye. The conductor tip which forms the eye should be bent as shown in Fig. 19-13. Note that when the screw or binding nut is tightened it also tends to tighten the conductor eye.

This method of connection is sometimes not desirable. When design requirements are more rigid, terminal connections are made by using special hardware devices called terminal lugs. There are terminal lugs of many different sizes and shapes, but all are essentially the same as the type shown in Fig. 19-14.

Each type of lug has a barrel (sleeve) which is wedged, crimped, or soldered to its conductor. There is also a tongue with a hole or slot in it

to receive the terminal post or screw. When mounting a solder-type terminal lug to a conductor, first tin the inside of the barrel. The conductor tip is stripped and also tinned, then inserted in the preheated lug. When mounted, the conductor insulation should butt against the lug barrel, so that there is no exposed conductor.

G. Solderless Connectors

Solderless connectors are made in a great variety of sizes and shapes, and for many different purposes. Only a few are discussed here.

SPLIT-SLEEVE SPLICER

A split-sleeve splicer is shown in Fig. 19-15. To connect this splicer to its conductor, the stripped conductor tip is first inserted between the split-sleeve jaws. Using a tool designed for that purpose, the slide ring is forced toward the end of the sleeve. The sleeve jaws are closed tightly on the conductor, and the slide ring holds them securely.

SPLIT-TAPERED SLEEVE SPLICER

A cross-sectional view of a split-tapered-sleeve splicer is shown in Fig. 19-16 (A). To mount this type of splicer, the conductor is stripped and inserted in the split-tapered sleeve. The threaded sleeve is turned or screwed into the tapered bore of the body. As the sleeve is turned, the split segments are squeezed tightly around the conductor by the

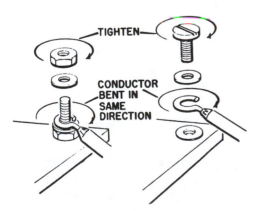

Fig. 19-13. Conductor terminal connection.

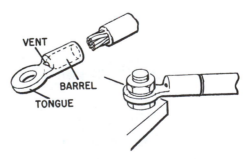

Fig. 19-14. Solder-type terminal lug.

narrowing bore. The finished splice (Fig. 19-16 (B)) must be covered with insulation.

CRIMP-ON-SPLICER

The crimp-on splicer (Fig. 19-17) is the simplest of the splicers discussed. The type shown is preinsulated, though uninsulated types are manufactured. These splicers are mounted with a special plier-like hand-crimping tool designed for that purpose. The stripped conductor tips are inserted in the splicer, which is then squeezed tightly closed. The insulating sleeve grips the outer insulated conductor, and the metallic internal splicer grips the bare conductor strands.

SPLIT-TAPERED-SLEEVE TERMINAL LUG (WEDGE)

This type of lug is shown in Fig. 19-18. It is commonly referred to as a "wedge on" because of the manner in which it is secured to a conductor. The stripped conductor is inserted through the hole in the split sleeve. When the sleeve is forced or "wedged" down into the barrel, its tapered segments are squeezed tightly around the conductor.

SPLIT-TAPERED-SLEEVE TERMINAL LUG (THREADED)

This lug (Fig. 19-19) is attached to a conductor in exactly the same manner as a split-sleeve

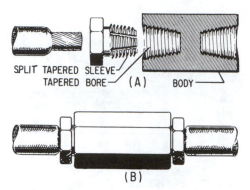

Fig. 19-16. Split-tapered-sleeve splice.

splicer. The segments of the threaded split sleeve squeezes tightly around the conductor as it is turned into the tapered bore of the barrel. For this reason, the lug is commonly referred to as a "screw-wedge."

CRIMP-ON TERMINAL LUG

The crimp-on lug is shown in Fig. 19-20. This lug is simply squeezed or "crimped" tightly onto a conductor. This is done by using the same tool used with the crimp-on splicer. The lug shown is preinsulated, but uninsulated types are manufactured. When mounted, both the conductor and its insulation are gripped by the lug.

H. Taping a Splice

The final step in completing a splice or joint is the placing of insulation over the bare wire. The insulation should be of the same basic sub-

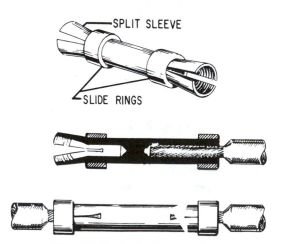

Fig. 19-15. Split-sleeve splicer.

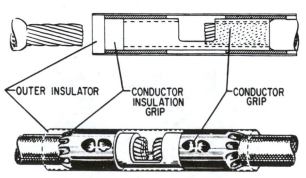

Fig. 19-17. Crimp-on splicer.

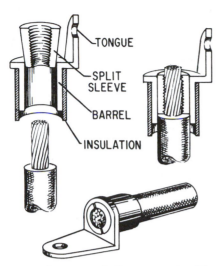

Fig. 19-18. Split-tapered-sleeve terminal lug, wedge type.

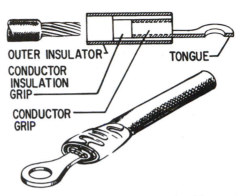

Fig. 19-20. Crimp-on terminal lug.

stance as the original insulation. Usually a rubber splicing compound is used.

Latex (rubber) tape is a splicing compound. It is used where the original insulation was rubber. The tape is applied to the splice with a light tension so that each layer presses tightly against the one underneath it. This pressure causes the rubber tape to blend into a solid mass. When the application is completed, an insulation similar to the original has been restored.

Between each layer of latex tape, when it is in roll form, there is a layer of paper or treated cloth. This layer prevents the latex from fusing while still on the roll. The paper or cloth is peeled off and discarded before the tape is applied to the splice.

Figure 19-21 shows the correct way to cover a splice with rubber insulation. The rubber

splicing tape should be applied smoothly and under tension so that there will be no air spaces between the layers. In putting on the first layer, start near the middle of the joint instead of the end. The diameter of the completed insulated joint should be somewhat greater than the overall diameter of the original cable, including the insulation.

Putting rubber tape over the splice means that the insulation has been restored to a great degree. It is also necessary to restore the protective covering. Friction tape is used for this purpose; it also affords a minor degree of electrical insulation.

Friction tape is a cotton cloth that has been treated with a sticky rubber compound. It comes in rolls similar to rubber tape except that no paper or cloth separator is used. Friction tape is applied like rubber tape; however, it does not stretch.

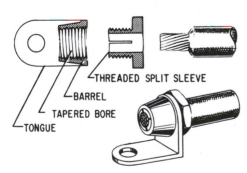

Fig. 19-19. Split-tapered-sleeve terminal lug, threaded.

Fig. 19-21. Applying rubber tape.

The friction tape should be started slightly back on the original braid covering. Wind the tape so that each turn overlaps the one before it; and extend the tape over onto the braid covering at the other end of the splice. From this point, a second layer is wound back along the splice until the original starting point is reached. Cutting the tape and firmly pressing down the end complete the job. When proper care is taken, the splice can take as much abuse as the rest of the wire.

Weatherproof wire has no rubber insulation, just a braid covering. In that case, no rubber tape is necessary, only friction tape need be used.

Plastic electrical tape has come into wide use in recent years. It has certain advantages over rubber and friction tape. For example, it will withstand higher voltages for a given thickness. Single thin layers of certain commercially available plastic tape will stand several thousand volts without breaking down. However, to provide an extra margin of safety, several layers are usually wound over the splice. Because the tape is very thin, the extra layers add only a very small amount of bulk; but at the same time, the added protection, normally furnished by friction tape, is provided by the additional layers of plastic tape. In the choice of plastic tape, the factor of expense must be balanced against the other factors involved.

Plastic electric tape normally has a certain amount of stretch so that it easily conforms to the contour of the splice without adding unnecessary bulk. The lack of bulkiness is especially important in some junction boxes where space is at a premium.

For high temperatures—for example, above 175° F—a special type of tape backed with glass cloth is used.

C H A P T E R 20
MAINTENANCE POLICIES

A. Types

By maintenance policies, we mean the general approach to the plant maintenance as set down by management.

But what types of policies are there?

The maintenance methods being followed by industrial plants and other organizations with large amounts of electrical and mechanical equipment, which must be kept operating, generally follow one of two systems; Preventive Maintenance or Emergency Maintenance.

Preventive maintenance merely means that a system of equipment checking and replacement of parts is followed on a periodic basis. The theory behind this method is that it is best to inspect the equipment and repair or replace parts before they fail in service, which is usually at a very inconvenient time, and thus very expensive in down time for the plant's production schedule. The regular vacation time for the production personnel is often devoted to maintenance of the plant equipment.

In practice, most plants usually follow the preventive maintenance system, the goal being to reduce, to as near zero as possible, the number of emergency shut-downs for repair. The number of emergency shut-downs is then taken as a measure of the effectiveness of the maintenance program.

The maintenance functions to be performed by the electrical staff cover the following items as a minimum.

Checking resistances of the critical circuits in equipment control systems.

Checking all mechanically operated switching mechanisms for damage and pitted contacts, and correcting all items close to failure.

Checking all ground systems and grounding connections.

Calibrating all meters and permanently installed instruments.

Inspecting all terminals and connections subject to abrasion and wear.

Actuating all safety devices, directly or by simulation.

Taking power readings on all operating equipment in conjunction with calibration of the instruments.

Cleaning dust, dirt, and debris from around the controls, panels, and operating equipment, such as motors, transformers, etc.

Checking lubrication of all motors.

Visually checking every foot of transmission cable, conduit, or leads for existing or potential damage.

Entering all results and activities in suitable record forms for action and filing.

When a plant sets up a PM program, it is usually formalized throughout the organization, with each department submitting its own plan, together with all proposed record forms to be maintained. This may appear to be an enormous amount of paper work for the electrical maintenance department to engage in, but experience has proven that it pays off in the long run, expecially in those plants on high speed production work where any shutdown is very expensive. See Section C for a procedure for setting up a PM program in your plant.

In any industrial plant, experience has disclosed which elements in the electrical controls and panels are most apt to fail. This is one of the advantages of having an accurate and current maintenance log of all equipment in the plant. Once these elements have been identified, and sufficient spares kept in stock, it is usually best to replace them on a regular basis, before they fail in service and at an inopportune time.

Emergency maintenance programs are generally based on the theory that if a piece of equipment is operating satisfactorily, leave it alone until it breaks down, then repair it. The implementers of this policy do not usually call it "emergency maintenance," but will attach some other more acceptable term to it.

Emergency maintenance methods require more spare parts to be kept on hand, more expertise in the maintenance personnel, and the willingness on the part of the maintenance personnel to pitch in when the emergency comes and stay with the job until the equipment is back on the line. This policy is usually found in high speed, high volume expensive production lines of a highly complex nature. It is not easy to predict where this class of equipment is going to break down, unless it has been in service for years and an accurate history of its breakdowns has been accumulated.

We are not attempting to recommend one policy over the other, but the mode of operation of most plants seems to fit the preventive maintenance pattern very well, and it is usually the one recommended by equipment manufacturers. Of course, in actual reality, even the preventive maintenance program is interrupted at times by serious break-downs.

The preventive maintenance policy lends itself very well also to the use of contract maintenance firms, as covered in the next section.

B. Contract Maintenance

So far, we have discussed only those normal maintenance problems which crop up frequently during the operating life of an ordinary industrial plant. These are items which the maintenance personnel are expected to be able to handle by themselves. They probably amount to at least 75% of the total maintenance cost of the electrical equipment over a period of years.

There are a number of major repair jobs which must be done on large electrical equipment from time to time, the frequency depending upon how well the maintenance staff have done their job, coupled with a fair amount of luck. Some plants go through their entire lifetime without having to have major repairs done on the electrical equipment, while other plants may have to have major repairs made every few months.

The type of repairs we are speaking of are best left up to one of the many contracting firms which make a specialty of this type of repair work. They do an excellent job, usually, as they have over the years in business made many repairs of the same order of magnitude as yours. To expect the average plant staff to be able to effect a quality repair on the major plant electrical equipment, and do it cheaply and quickly, is not practical. The average plant staff does not have the experience to do it, unless there is ample time for them to study the procedures and take their time learning the methods. In the long run, it is more economical to call in an outside contractor to perform the work, and get a guarantee of the quality in the bargain. It all reduces to the fact that the contracting firms are probably specialists, whereas the plant staffs are amateurs at the more complex repair jobs.

A typical example of an occasional maintenance job which should not be tackled by the small plant electrical staff is the replacing or reconditioning of the oil filling in oil filled transformers. This is a very exacting process, and entails some danger when the oil is an older type known as PCB. Regardless of the danger, there is also the problem of maintaining the oil in completely moisture-free condition during handling and charging into the transformer. The process requires special handling and dehydrating equipment which outside contracting firms have available, along with the evacuating equipment needed to maintain a vacuum on the transformer while the oil is being charged into the jacket. It is true that you can probably rent the equipment in most areas, but there is always the possibility of accidents, which involves the matter of liability for personnel health and safety. It is far better to engage a responsible firm which is equipped and insured to do the job safely, quickly, and with a guarantee of final results.

There are several other maintenance jobs which could be assigned to contract firms, and we give a list below of a few of them. The degree to which this is done depends upon the size and complexity of your equipment and the composition of the plant staff. They are:

motor rewinding

rotor rebalancing

bearings replaced

realigning motors to equipment

check-out and repair of electronic panels

repair of specialized electric process equipment

updating plant electrical drawings.

As always, it is a good idea to use competitive bidding to obtain the best job for the money paid out. Some firms, after accepting bids, will select the one most qualified, then continue to rehire them for subsequent jobs, without further bidding. This has the advantage of saving money only if the plant equipment they are working on is of a specialized nature unique to the plant or the process. Engaging the same firm time after time thus saves the expense of the learning curve being required each time a new firm works on the job. However, if the job and the equipment are purely routine, run-of-the-mill projects, there is no real sense in not taking bids for each job. Taking competitive bids helps to ensure that the firm is getting the best price, and tends to keep all bidders on an equal level. It also reduces the possibility of under-the-table payments and gouging on the part of the contractor.

Regarding updating of the plant's electrical drawings mentioned above, the contracting firm may not be in a position to provide this service, as it requires a qualified designer/draftsman to either update the plant drawings, or provide completely new ones. Either way, the process is usually expensive, and few industrial plants are willing to pay for the service and convenience. Too often, all that can be expected is (hopefully) for the best trained member of the electrical staff to alter the existing drawings. This is better than no attempt at all.

Another item to insist on, when the maintenance is being performed by an outside force, is updating and recording data on the plant maintenance records and logs. At least, they should provide the engineering staff with a complete record of all services performed, including date, time, and the signature of the man doing the work, with some other responsible authority signing also. This completed record then becomes part of the plant maintenance records and should be filed with them.

As a cost-saving step, one of the plant's electrical maintenance staff may be assigned to assist the contractor's electrician in performance of the work. Also, this same staff electrician may be given the responsibility of recording the proceedings, which is an excellent way to become acquainted with the plant electrical system.

We mentioned above that outside contractors may be called in to service electronic control panels. There are many pieces of process equipment in today's plants that are controlled by such panels. In some plants, one or more of the operators of that equipment may be qualified to perform the service and testing functions required to keep their equipment operating. This condition may also exist in one of the other service or utility departments, such as in the boiler plant, where the boilers are controlled by electronic control panels.

This situation may lead to constant haggling over who has the final responsibility for the maintenance and integrity of such control equipment. This condition should not exist for long, and it is in only the poorly managed maintenance departments where it may be found. The problem entails not only the testing and repair of the equipment involved, but also in the supply of all necessary parts and in the keeping of all service records.

It goes without saying that there should be a well-established line of demarcation regarding who has the responsibility for maintenance of this equipment. There should be a written memo or directive on record calling out this decision, so that as personnel changes in the plant, the policy will be perpetuated.

This matter may seem to be elementary; however, it is surprising what conditions exist in some plant maintenance departments, most often, but not always, in the smaller plants. In some larger plants, where competition for plant maintenance funds are involved and are a constant source of irritaiton, this situation may lead to ridiculuous antagonisms.

C. Establishing a Preventive Maintenance Program

The plant electrical equipment is very well suited to conventional preventive maintenance procedures. With few exceptions, it is usually well established in a particular location, is confined to specific functions of a repetitive nature, and is usually crucial to the plant's processes and purpose.

Before launching into the procedure for organizing a plant preventive maintenance (PM) program, some remarks of a general nature are in order.

First of all, assuming that there has been no such program in the past in the firm, the extent to which one is to be implemented depends upon several very important features. There is a lot at stake, as the set-up costs may be higher than management anticipates, which is usually the case when something new is proposed by those who are not directly involved in the implementation.

Probably the first decision to be settled upon, after the decision to start a PM program is made, is the determination of maintenance functions for which the plant staff is to be responsible, and which ones are to be handled by outside contractors, as mentioned in Section B. This decision may have to wait until well into the organization of the PM program, as some very startling shortages in maintenance items and functions will no doubt be discovered as the organizing proceeds. This could lead to a reevaluation of the entire program, as could other unforeseen requirements.

Another budget-buster could be the need for special instruments and tools, as the plant may have been limping along for months without adequate equipment for proper maintenance.

And, finally, the manpower requirements may change, or the skills required may not match the available personnel. This could require retraining, at the least. The end result is that the plant's staff, from Management down, should be prepared for some changes. Obviously, Management must be sold completely on the need for the PM program, as must also those who are to implement it.

What are the goals, and how are they to be achieved? The purpose is, in brief, as follows.

a. Maintain the electrical equipment in such condition as to ensure uninterrupted operations for as long as possible, without disastrous plant shut-downs for repairs.

b. Maintain the electrical equipment in such condition that it will always operate at the highest possible efficiency, at the least practical cost.

Basically, the PM program accomplishes this by the following general procedures, performed on an established, recommended frequency:

a. equipment inspection

b. cleaning the equipment

c. tightening of all connections

d. adjusting and lubricating all moving parts

e. testing the critical portions of the circuits involved

f. correcting any deficiencies discovered

g. recording and signing the PM record sheets in the proper manner, adding all pertinent comments

h. updating the plant equipment and spare parts records as required.

We shall now present one method of organizing a PM program for the average industrial plant. We do not intend that it be considered the only method, but it should at least give the enterprising plant electrical worker the basic procedures, which may be altered to suit the individual plant size, complexity, and manpower availability.

To start, assemble all available literature covering the manufacturer's recommended maintenance procedures for the electrical equipment. Add to this a supply of standard size ruled, three-hole punched notebook paper, with a suitable binder.

Assigning one complete sheet to each item of equipment to be maintained, place the name of the equipment at the top, along with the following data.

a. Plant equipment number. If not already numbered, this is the time to start this very important function.

b. Location of the equipment, by floor, column number, etc.

c. From the maintenance manuals, list all items requiring service, with the frequency (weekly, monthly, etc.) listed on the right side of the sheet.

d. After each maintenance function listed in (c.), list the tools required, including instruments, the manpower, and the estimated time required at the site. The tools and instruments may be separated by those normally carried on the belt, in hand-carried tool boxes, or on service carts.

The next step is to obtain a supply of standard 3×5 index cards, and make out one card for each maintenance item and frequency combination from the above data sheets. The cards should be headed by the equipment name and the frequency of service required. It should then be followed by those items a–d above, and with any spare parts data from other plant files, as described in Chapter 18. This card file may be incorporated in the plant equipment files, depending upon location and availability. See Fig. 20-1 for an example.

Once the above index cards have been completely filled out as described, they should be placed in an ordinary file box, with tabbed index dividers, separating the cards by frequency of service, such as weekly, monthly, etc. If the size and number of maintenance functions un-

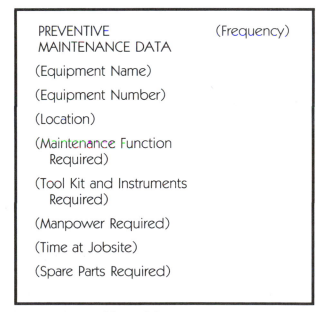

PREVENTIVE
MAINTENANCE DATA   (Frequency)

(Equipment Name)

(Equipment Number)

(Location)

(Maintenance Function
 Required)

(Tool Kit and Instruments
 Required)

(Manpower Required)

(Time at Jobsite)

(Spare Parts Required)

Fig. 20-1. Data file card for preventive maintenance program.

der each frequency heading permits it, they may be further subdivided by manpower requirements, or tool kits required for the function.

If the total plant electrical equipment main-

tenance requirements are rather minor, then the card file step may be omitted, and the PM Record Sheets, described next, may be produced directly from the basic work sheets already produced.

<u>PREVENTIVE MAINTENANCE RECORD</u>

(Equipment Description)   (Location)
(Equipment Number)   (Frequency)
(Maintenance To Be Performed)   (Manpower)
(Tools, instruments required)   (Spare Parts)

Date	By	Remarks	Date	By	Remarks

Fig. 20-2. Service record sheet, preventive maintenance program.

Finally, the last paper form to be developed is the "PREVENTIVE MAINTENANCE RECORD" sheet. This should be on standard notebook paper, three-hole punched, lined, and similar in form to Fig. 20-2.

In performing their regular PM functions, the crew should have with them a three-ring binder, containing the following material.

a. A schedule or routing sheet, giving the locations and sequence of the maintenance functions to be performed under the existing frequency schedule being followed.

b. The PM Record sheets described above, in the same order as shown on the routing sheet.

c. Photocopies of the manufacturer's maintenance instructions for the function to be performed.

As the PM program is placed into operation, it is to be expected that some alterations to the first approach may have to be made. As the crews involved progress along the learning curve, the times and schedules may have to be adjusted, and the paper work, records, and plant files altered to bring the entire program up to date.

C H A P T E R 21
SOME FINAL ADVICE

A. There's More to It Than Skill

By now it should be quite obvious that there is a lot more to your job than just perfecting your skills, for which you are being paid the going wage in the industry. The firm paying your wages is made up of people, many of whom go about their own prescribed job functions every day, with their own particular skills.

The one guiding instrument for this entire force is what is known as "PLANT POLICY," formulated usually by unknown individuals in the higher echelons of the firm, and split into lesser documents for the guidance of the departments down through the hierarchy to the lowliest position in the plant. Each of those lesser plant policy statements are interpreted by individuals, with their own ideas and understanding of the terms of the policy. That, coupled with the variations in temperaments of the plant personnel from top to bottom can, and often does, lead to conflicts throughout the plant. These little differences in personality and drives have to be understood and considered by any plant worker who is interested in advancement and fulfillment of his objectives before he can hope to reach his maximum potential in the plant.

What does it take for today's electrician to function within that jungle of personalities, other than perfecting his job skills?

B. Personal Factors

Probably the most important item which will shape the worker's success or failure on the job is his actual attitude towards the firm, his fellow employees, his supervisor, and his chosen career. A negative attitude in any of those areas will seriously hamper a worker's rise up the ladder.

So the first rule is: STRIVE TO MAINTAIN A POSITIVE, UPBEAT ATTITUDE TOWARDS YOUR JOB AND THE PLANT PERSONNEL.

The next thing affecting your survivability is whether or not your supervisor can rely on you to perform consistently at your very best. When you are assigned to a particular task, will you perform it willingly and skillfully? Or will your performance vary considerably from day to day, depending upon how you feel each day?

Can your supervisor rely on you being on the job on time, not extending your lunch break over the prescribed time, and working your coffee breaks into your daily schedule?

How about your personal behavior and grooming? Are they such as to inspire confidence in your fellow workers and your supervisor? In Chapter 2 we pointed out that you are working with one of the forces that holds the universe together, and thus it must be treated with respect if you are to survive. This rules out any thought of "horseplay" while on the job or on company premises. Save it for when you are on your own time, if you are inclined to hilarity and playfulness.

Remember! ELECTRICITY IS SERIOUS BUSINESS!

One more admonition: ALCOHOL, DRUGS, AND ELECTRICITY DO NOT MIX!

If you are addicted to any form of these, we strongly urge you to change your occupation to one less dangerous, then get treatment immediately for your habit, before you kill yourself and others on the job.

What this all reduces to is: any personal habits which reduce a worker's alertness and caution on the job will place himself and others at high risk.

C. Plant-Related Factors

Keep in mind that your job depends upon the firm remaining in operation. In order for the firm to remain in operation, it must make a profit. For both of these to happen, the employees must perform their tasks with a reasonable degree of skill, diligently applied. Among other things, this requires that the employees show a sense of loyalty to the firm or plant.

One of the reasons why West Germany and Japan have been able to rise to prominence since destruction in World War II is the basic loyalty of the workers to their firm. And loyalty to the firm means a feeling of respect for those in authority, among other things.

A feeling of respect for those in authority may, at times, be difficult to muster by an employee. Regardless of what you may think of those in authority, and specifically, your immediate supervisor, the fact remains that the firm has seen fit to appoint him to that position, and you are expected to take direction from him.

As an example, we cite the principle behind the military salute. The salute is a mark of respect to the uniform and the rank, and not the personality behind that uniform and rank.

In short, you do not have to like your supervisor, but you should show him the respect due his position and authority.

In most cases, the electrician works with tools and equipment supplied by the plant. There are very good reasons why the plant worker should show a healthy respect for the firm's property, and especially for the tools they have supplied.

First, of course, is the attitude displayed when an employee mistreats his tools supplied by the firm. Mistreatment can be caused by either carelessness or disrespect for the property of others. A worker who is careless with his tools may be equally careless in performance of his duties.

Remember: ANY DANGER FROM WORKING WITH DAMAGED TOOLS ON ELECTRICITY PLACES THE ELECTRICIAN IN DANGER FIRST!

The subject of "Teamwork" has been given much play in the business literature in recent years. A certain amount of this will be found in the plant's maintenance department, as most maintenance tasks are, or should be, carried out with at least two workers. This is usually done for reasons of safety, and for training an apprentice.

Our suggestion in this respect is to be flexible and cooperative. Do not expect that your team will always be compatible, but be willing to work with whomever is assigned to work with you. In fact, the more often the team composition is altered, the better chance for each worker to learn to cooperate with others, and thus become a better worker and more valued employee.

Should you find yourself a member of a permanent team, with a strong leader, there is danger of your career being tied too strongly into that of the leader. Strong group leaders often make enemies higher up. High echelon political shifts are often followed by a clean sweep downward. As a result, faithful team followers may be discharged through no apparent fault of their own.

Avoid this situation if you possibly can, and strive to be independant enough to be able to survive on your own ability.

D. Plant Politics and Gossip

In the beginning of this chapter, we mentioned the diverse personalities of the plant personnel, and how this often leads to conflicts within the organization. The solution to these conflicts leads to political maneuvers, gossip, and power-plays of all types.

There is a general pattern to these political movements and activities. Generally speaking, the higher up in the hierarchy, the more intense are the politics and the gossip that goes with it. This follows from the higher salary brackets with more money and power at stake.

Fortunately, the electrical maintenance worker will probably not find these political fights too intense at the maintenance department level. They will often exist, but to a much lesser extent than on the higher plant levels.

The best advice we can offer should you find the political atmosphere permeating your level is: keep your eyes, ears, and above all, your mind open at all times, and your mouth shut! Be aware of what is going on in the plant and the department, but do not take any position openly with one faction or another. Any maneuvering on your part should be quiet, cautious, and subtle, and aimed only at protecting your own position to prevent any reflection or fall-out on you.

In short—PLAY IT COOL AND SMART!

Much the same can be said about the gossip that often circulates in plants at all levels. This is often in connection with the politics being practiced in the plant, and should be regarded in the same light. When you are made a party to gossip, be completely noncommittal in your response, and do not encourage further embellishment. Absorb it into your mind, evaluate it as to source, the person being maligned, your own knowledge of all parties, and your sense of fairness and logical behavior, then store it away in your mind.

Above all—DO NOT REPEAT IT!

If you then find yourself passing judgement at a later date on the recipient of that gossip, discount it considerably, based upon your own evaluation.

E. Conflicts

We have mentioned already in this chapter the possibility of conflicts developing between individuals or groups in any plant or organization. With the primary goal of the firm being to assemble the many and various skills required to perform the plant operations, it is inevitable that there will at times be clashes between personalities. These are often a direct result of the political power-plays.

Where do these clashes come from? They come from differences in the following areas;

striving for power

personality differences

approach to the job

interpretation of the job description

indefinable subconscious biases.

The one thing to keep in mind is that any conflict between two people or groups does not always reduce to one being right and the other being wrong. Once this fact is recognized, a compromise may be easier to find, and your position in the plant made more secure.

When confronted with a clash with someone else in the plant, attempt to ascertain which of the above differences is causing the confrontation. Be fair and objective, then govern yourself with that idea in mind.

F. Plant-Sponsored Activities

Many plants initiate a complete program of outside activities among the employees of a leisure-time nature. This is done to promote cooperation among the employees, provide relaxation and education, and in general to promote the morale of the employees. They also provide a release of excess energy in a safe manner while accomplishing all of the above.

Generally, the larger the plant, the more plant-sponsored activities will be found. The degree of participation in those activities may be used as an approximate measure of the morale among the employees. They usually are not, and should not, be used as an evaluation of the merits of an employee, and the more progressive firms will take steps to see that is not done.

Whether or not you, as an electrical maintenance worker, take part in any of these activities is entirely at your own discretion. Do not feel that it is expected of you, and that therefore you should sign up for at least one of these activities. That is all we shall say concerning plant-sponsored activities, other than to point out that they are an excellent way to get to know your fellow workers and enlarge your circle of acquaintances. You work with these people during the day, and getting to know them in a relaxed atmosphere can help considerably in smoothing your work load on the job. Also, remember that many jobs are obtained through the help of acquaintances—directly and through references.

G. Moving Up

By applying the principles and procedures laid down in this book, after months—and possibly years—of hard work, the time may come when you will be offered a promotion. The new position, whether Group Leader, Department Supervisor, or whatever is needed to fulfil the plant's requirement, will no doubt carry increased responsibilities, with an increase in monetary compensation.

Congratulations!

But before you accept the offer, stop and consider the full impact on your life and the lives of your "buddies" in the electrical maintenance fraternity. We shall present here a few things to be considered.

Can you leave them and become their leader?

Will you still feel at ease in their backyard parties, their outside activities, such as the firm's bowling team? Or does the firm expect you to change—"grow" will probably be the term used? You may be expected to change

your circle of friends and acquaintances, as the new position may put you in the category known as "lower management." The new position, if accepted, may find you drifting automatically into a new peer group, with different aspirations, interests, and outlook towards the plant and its employees.

Consider, also, how much you have learned about your fellow electricians. You probably know all of their habits, attitudes, and other little quirks. For instance, "good ol' Joe." Good electrician—likable, outgoing, fun at a party, knows his job well. But he has one fault; he has a habit of walking off the property with the firm's small tools, and sometimes the plant materials, for use in his workshop or home. This is a very common problem in industry. Do you feel you can now become the plant watch-dog? How would you handle such a situation?

Consider, also, the effect on your future in the firm should you not accept the offer of a promotion. What has been the past practice of the firm when other employees have rejected a promotion? Any reason to believe your situation is any different? You will probably be kept in your present position, someone else on the crew will be offered the promotion, or someone from another department may be brought in to fill it. You may be compensated partially with steady increases in wages, but that is only possible up to the lower level of the position you just rejected. This places a definite ceiling on your future income from this firm.

So, after all of this, you have decided to take the offer, and will do your best to fulfill your new responsibilities. Fine, you are now off your old plateau, and onto a new one, with increased opportunities.

Go for it! And give it your best.

H. Moving On

Our society is known for its mobility, including the movement of workers between plants. In the past, plant-sponsored pension plans have tended to keep employees tied to one job for life. Fortunately, that is changing for the better, so that now workers are feeling free to move from job to job, and do so quite readily. This results in a steady rise in the individual's experience and worth to the employer, as well as an improvement in the worker's feelings and personal life.

Regardless of the real reason you are leaving a firm, keep in mind that the reasons given will be recorded on your personnel file and will forever be used when any new prospective employer checks back for your reasons for leaving.

Some careful thought should be given when you leave your present firm for a new one. Therefore, we shall list here some of the more acceptable reasons, with the suggestion that you choose one, and be very careful how you use it, so that it will not come back to haunt you later. Pick one, and stay with it!

Your new job opens up a new field with unlimited potential, not available in your present position.

You have been offered a promotion in another firm which opens up a new opportunity to you.

You wish to broaden your experience, and you feel you have reached the limit of your possibilities in this firm.

Transportation is a problem and you have a job offer closer to home.

A friend or a relative needs your help and skills in a new venture.

You may feel frustrated, cheated, washed-out, stressed-out, and brow-beaten by your supervisor, but resist the urge to list any such negative reasons for leaving. They will leave the impression that you were unable to get along with others in the plant, and this is what the prospective employer will think when he is given those reasons on record in your personnel file. If it is obvious to all that you are leaving for one of those reasons listed in the first sentence of this paragraph, and it cannot be covered up, simply say that you do not appear to be as effective an employee as you should be, and you have doubts about your ability to continue to function in the present conditions. You plan on taking a vacation, taking further stock of yourself, and reevaluating your future before trying again at another firm. None of these reasons places the blame on either yourself or your supervisor. There simply was a conflict of personalities.

Remember, that supervisor which you would like to "tell off" may be asked by a prospective employer to give a reference for you at some future date.

Rule Number 1: DO NOT BURN BRIDGES BEHIND YOU! Leave the firm on a smooth note if possible.

Rule Number 2: Before leaving, contact as many of your fellow workers as possible, smoothing out any past difficulties or differences.

Rule Number 3: Assemble as many promises of references as possible, with all names and addresses.

Rule Number 4: In the exit interview, do not unburden your gripes. Be sure and give only positive and constructive comments about the firm and your opinion of it and the employees. Soft-pedal any known conflicts existing between you and others in the firm.

That is all the advice we can offer. The rest is up to you. Play it by ear from day to day, and Good Luck!

APPENDICES

Table of Contents

A P P E N D I X A

SYMBOLS, VOCABULARY, AND LEGEND

A B B R E V I A T I O N S

a-c	alternating-current	E_p	primary voltage
a-f	audio-frequency	E_s	secondary voltage
B	flux density	f	frequency
C	capacitance	f_r	frequency at resonance
cemf	instantaneous counter electro-motive force	H	magnetizing flux
		h	henry
cm	centimeters	I	current
cps	cycles per second	i	instantaneous current
C_T	total capacitance	I_c	capacitive current
d	distance between points	i_c	instantaneous capacitive current
DC, d-c	direct-current	I_{eff}	effective current
de	change in voltage	i-f	intermediate-frequency
di	change in current	I_L	inductive current
dq	change in charge	i_L	instantaneous inductive current
dt	change in time	I_m	maximum current
E	voltage	I_{max}	maximum current
e	instantaneous voltage	I_p	plate current
E_c	capacitive voltage	I_R	current through resistance
e_c	instantaneous capacitive voltage	i_R	instantaneous current through resistance
E_L	inductive voltage	I_s	secondary current
e_L	instantaneous inductive voltage	I_T	total current
E_m	maximum voltage	$I\emptyset$	phase current
E_{max}	maximum voltage	K	coefficient of coupling
emf	electromotive force	kc	kilocycle

L	inductance	r-f	radio-frequency
L-C	inductance-capacitance	R_G	grid resistance
L-C-R	inductance-capacitance-resistance	R_o	load resistance
L_T	total inductance	rpm	revolutions per minute
mh	millihenry	sq cm	square centimeters
N	revolutions per minute	t	time constant
N_p	primary turns	t	time (seconds)
N_s	secondary turns	μf	microfarad
P	power	$\mu\mu f$	micromicrofarad
p	instantaneous power	V, v	volt
P_{ap}	apparent power	X_c	capacitance reactance
P_{av}	average power	X_L	inductive reactance
P_p	primary power	Z	impedance
P_s	secondary power	Z_o	load impedance
Q	charge or quality	Z_p	primary impedance
q	instantaneous charge	Z_s	secondary impedance
		Z_T	total impedance

ELECTRICAL TERMS

AGONIC—An imaginary line to the earth's surface passing through points where the magnetic declination is 0 degrees; that is, points where the compass points to true north.

AMMETER—An instrument for measuring the amount of electron flow in amperes.

AMPERE—The basic unit of electrical current.

AMPERE-TURN—The magnetizing force produced by a current of one ampere flowing through a coil of one turn.

AMPLIDYNE—A rotary magnetic or dynamo-electric amplifier used in servomechanism and control applications.

AMPLIFICATION—The process of increasing the strength (current, power, or voltage) of a signal.

AMPLIFIER—A device used to increase the signal voltage, current, or power, generally composed of a vacuum tube and associated circuit called a stage. It may contain several stages in order to obtain a desired gain.

AMPLITUDE—The maximum instantaneous value of an alternating voltage or current, measured in either the positive or negative direction.

ARC—A flash caused by an electric current ionizing a gas or a vapor.

ARMATURE—The rotating part of an electric motor or generator. The moving part of a relay or vibrator.

ATTENUATOR—A network of resistors used to reduce voltage, current, or power delivered to a load.

AUTOTRANSFORMER—A transformer in which the primary and secondary are connected together in one winding.

BATTERY—Two or more primary or secondary cells connected together electrically. The term does not apply to a single cell.

BREAKER POINTS—Metal contacts that open and close a circuit at timed intervals.

BRIDGE CIRCUIT—The electrical bridge circuit is a term referring to any one of a variety of electric circuit networks, one branch of which, the "bridge" proper, connects two points of equal potential and hence carries no

G R E E K   A L P H A B E T

Name	Capital	Lower Case	Designates
Alpha	A	α	Angles.
Beta	B	β	Angles, flux density.
Gamma	Γ	γ	Conductivity.
Delta	Δ	δ	Variation of a quantity, increment.
Epsilon	E	ϵ	Base of natural logarithms (2.71828).
Zeta	Z	ζ	Impedance, coefficients, coordinates.
Eta	H	η	Hysteresis coefficient, efficiency, magnetizing force.
Theta	Θ	θ	Phase angle.
Iota	I	ι	
Kappa	K	κ	Dielectric constant, coupling coefficient, susceptibility.
Lambda	Λ	λ	Wavelength.
Mu	M	μ	Permeability, micro, amplification factor.
Nu	N	ν	Reluctivity.
Xi	Ξ	ξ	
Omicron	O	o	
Pi	Π	π	3.1416
Rho	P	ρ	Resistivity.
Sigma	Σ	σ	
Tau	T	τ	Time constant, time-phase displacement.
Upsilon	Y	υ	
Phi	Φ	ϕ	Angles, magnetic flux.
Chi	X	χ	
Psi	Ψ	ψ	Dielectric flux, phase difference.
Omega	Ω	ω	Ohms (capital), angular velocity (2π f).

current when the circuit is properly adjusted or balanced.

BRUSH—The conducting material, usually a block of carbon, bearing against the commutator or sliprings through the current flows in or out.

BUSBAR—A primary power distribution point connected to the main power source.

CAPACITOR—Two electrodes or sets of electrodes in the form of plates, separated from each other by an insulating material called the dielectric.

CHOKE COIL—A coil of low ohmic resistance and high impedance to alternating current.

CIRCUIT—The complete path of an electric current.

CIRCUIT BREAKER—An electromagnetic or thermal device that opens circuit when the current in the circuit exceeds a predetermined amount. Circuit breakers can be reset.

CIRCULAR MIL—An area equal to that of a circle with a diameter of 0.001 inch. It is used for measuring the cross section of wires.

COAXIAL CABLE—A transmission line consisting of two conductors concentric with and insulated from each other.

COMMUTATOR—The copper segments on the armature of a motor or generator. It is cylindrical in shape and is used to pass power into or from the brushes. It is a switching device.

CONDUCTANCE—The ability of a material to conduct or carry an electric current. It is the reciprocal of the resistance of the material, and is expressed in mhos.

CONDUCTIVITY—The ease with which a substance transmits electricity.

CONDUCTOR—Any material suitable for carrying electric current.

CORE—A magnetic material that affords an easy path for magnetic flux lines in a coil.

COUNTER EMF—Counter electromotive force; an emf induced in a coil or armature that opposes the applied voltage.

CURRENT LIMITER—A protective device similar to a fuse, usually used in high amperage circuits.

CYCLE—One complete positive and one complete negative alternation of a current or voltage.

DIELECTRIC—An insulator; a term that refers to the insulating material between the plates of a capacitor.

DIODE—Vacuum tube—a two element tube that contains a cathode and plate; semi-conductor—a material of either germanium or silicon that is manufactored to allow current to flow in only one direction. Diodes are used as rectifiers and detectors.

DIRECT CURRENT—An electric current that flows in one direction only.

EDDY CURRENT—Induced circulating currents in a conducting material that are caused by a varying magnetic field.

EFFICIENCY—The ratio of output power to input power, generally expressed as a percentage.

ELECTROLYTE—A solution of a substance which is capable of conducting electricity. An electrolyte may be in the form of either a liquid or a paste.

ELECTROMAGNET—A magnet made by passing current through a coil of wire wound on a soft iron core.

ELECTROMOTIVE FORCE (emf)—The force that produces an electric current in a circuit.

ELECTRON—A negatively charged particle of matter.

ENERGY—The ability or capacity to do work.

FARAD—The unit of capacitance.

FEEDBACK—A transfer of energy from the output circuit of a device back to its input.

FIELD—The space containing electric or magnetic lines of force.

FIELD WINDING—The coil used to provide the magnetizing force in motors and generators.

FLUX FIELD—All electric or magnetic lines of force in a given region.

FREE ELECTRONS—Electrons which are loosely held and consequently tend to move randomly among the atoms of the material.

FREQUENCY—The number of complete cycles per second existing in any form of wave motion; such as the number of cycles per second of an alternating current.

FULL-WAVE RECTIFIER CIRCUIT—A circuit which utilizes both the positive and the negative alternations of an alternating current to produce a direct current.

FUSE—A protective device inserted in series with a circuit. It contains a metal that will melt or break when current is increased beyond a specific value for a definite period of time.

GAIN—The ratio of the output power, voltage, or current to the input power, voltage, or current, respectively.

GALVANOMETER—An instrument used to measure small DC currents.

GENERATOR—A machine that converts mechanical energy into electrical energy.

GROUND—A metallic connection with the earth to establish ground potential. Also, a common return to a point of zero potential.

HERTZ—A unit of frequency equal to one cycle per second.

HENRY—The basic unit of inductance.

HORSEPOWER—The English unit of power, equal to work done at the rate of 550 foot-pounds per second. Equal to 746 watts of electrical power.

HYSTERESIS—A lagging of the magnetic flux in a magnetic material behind the magnetizing force which is producing it.

IMPEDANCE—The total opposition offered to the flow of an alternating current. It may consist of any combination of resistance, inductive reactance, and capacitive reactance.

INDUCTANCE—The property of a circuit which tends to oppose a change in the existing current.

INDUCTION—The act or process of producing voltage by the relative motion of a magnetic field across a conductor.

INDUCTIVE REACTANCE—The opposition to the flow of alternating or pulsating current caused by the inductance of a circuit. It is measured in ohms.

IN PHASE—Applied to the condition that exists when two waves of the same frequency pass through their maximum and minimum values of like polarity at the same instant.

INVERSELY—Inverted or reversed in position or relationship.

ISOGONIC LINE—An imaginary line drawn through points on the earth's surface where the magnet's variation is equal.

JOULE—A unit of energy or work. A joule of energy is liberated by one ampere flowing for one second through a resistance of one ohm.

KILO—A prefix meaning 1,000.

LAG—The amount one wave is behind another in time; expressed in electrical degrees.

LAMINATED CORE—A core built up from thin sheets of metal and used in transformers and relays.

LEAD—The opposite of LAG. Also, a wire or connection.

LINE OF FORCE—A line in an electric or magnetic field that shows the direction of the force.

LOAD—The power that is being delivered by any power-producing device. The equipment that uses the power from the power-producing device.

MAGNETIC AMPLIFIER—A saturable reactor type device that is used in a circuit to amplify or control.

MAGNETIC CIRCUIT—The complete path of magnetic lines of force.

MAGNETIC FIELD—The space in which a magnetic force exists.

MAGNETIC FLUX—The total number of lines of force issuing from a pole of a magnet.

MAGNETIZE—To convert a material into a magnet by causing the molecules to rearrange.

MAGNETO—A generator which produces alternating current and has a permanent magnet as its field.

MEGGER—A test instrument used to measure insulation resistance and other high resistances. It is a portable, hand-operated DC generator used as an ohmmeter.

MEGOHM—A million ohms.

MICRO—A prefix meaning one-millionth.

MILLI—A prefix meaning one-thousandth.

MILLIAMMETER—An ammeter that measures current in thousandths of an ampere.

MOTOR-GENERATOR—A motor and a generator with a common shaft used to convert line voltages to other voltages or frequencies.

MUTUAL INDUCTANCE—A circuit property existing when the relative position of two inductors causes the magnetic lines of force from one to link with the turns of the other.

NEGATIVE CHARGE—The electrical charge carried by a body which has an excess of electrons.

NEUTRON—A particle having the weight of a proton but carrying no electric charge. It is located in the nucleus of an atom.

NUCLEUS—The central part of an atom that is mainly comprised of protons and neutrons. It is the part of the atom that has the most mass.

NULL—Zero.

OHM—The unit of electrical resistance.

OHMMETER—An instrument for directly measuring resistance in ohms.

OVERLOAD—A load greater than the rated load of an electrical device.

PERMALLOY—An alloy of nickel and iron having an abnormally high magnetic permeability.

PERMEABILITY—A measure of the ease with which magnetic lines of force can flow through a material as compared to air.

PHASE DIFFERENCE—The time in electrical degrees by which one wave leads or lags another.

POLARITY—The character of having magnetic poles, or electric charges.

POLE—The section of a magnet where the flux lines are concentrated; also where they enter and leave the magnet. An electrode of a battery.

POLYPHASE—A circuit that utilizes more than one phase of alternating current.

POSITIVE CHARGE—The electrical charge carried by a body which has become deficient in electrons.

POTENTIAL—The amount of charge held by a body as compared to another point or body. Usually measured in volts.

POTENTIOMETER—A variable voltage divider; a resistor which has a variable contact arm so that any portion of the potential applied between its ends may be selected.

POWER—The rate of doing work or the rate of expending energy. The unit of electrical power is the watt.

POWER FACTOR—The ratio of the actual power of an alternating or pulsating current, as measured by a wattmeter, to the apparent power, as indicated by ammeter and voltmeter readings. The power factor of an inductor, capacitor, or insulator is an expression of their losses.

PRIME MOVER—The source of mechanical power used to drive the rotor of a generator.

PROTON—A positively charged particle in the nucleus of an atom.

RATIO—The value obtained by dividing one number by another, indicating their relative proportions.

REACTANCE—The opposition offered to the flow of an alternating current by the inductance, capacitance, or both, in any circuit.

RECTIFIERS—Devices used to change alternating current to unidirectional current. These may be vacuum tubes, semiconductors such as germanium and silicon, and dry-disk rectifiers such as selenium and copper-oxide.

RELAY—An electromechanical switching device that can be used as a remote control.

RELUCTANCE—A measure of the opposition that a material offers to magnetic lines of force.

RESISTANCE—The opposition to the flow of current caused by the nature and physical dimensions of a conductor.

RESISTOR—A circuit element whose chief characteristic is resistance; used to oppose the flow of current.

RETENTIVITY—The measure of the ability of a material to hold its magnetism.

RHEOSTAT—A variable resistor.

SATURABLE REACTOR—A control device that uses a small DC current to control a large AC current by controlling core flux density.

SATURATION—The condition existing in any circuit when an increase in the driving signal produces no further change in the resultant effect.

SELF-INDUCTION—The process by which a circuit induces an emf into itself by its own magnetic field.

SERIES-WOUND—A motor or generator in which the armature is wired in series with the field winding.

SERVO—A device used to convert a small movement into one of greater movement or force.

SERVOMECHANISM—A closed-loop system that produces a force to position an object in accordance with the information that originates at the input.

SOLENOID—An electromagnetic coil that contains a movable plunger.

SPACE CHARGE—The cloud of electrons existing in the space between the cathode and plate in a vacuum tube, formed by the electrons emitted from the cathode in excess of those immediately attracted to the plate.

SPECIFIC GRAVITY—The ratio between the density of a substance and that of pure water at a given temperature.

SYNCHROSCOPE—An instrument used to indicate a difference in frequency between two AC sources.

SYNCHRO SYSTEM—An electrical system that gives remote indications or control by means of self-synchronizing motors.

TACHOMETER—An instrument for indicating revolutions per minute.

TERTIARY WINDING—A third winding on a transformer or magnetic amplifier that is used as a second control winding.

THERMISTOR—A resistor that is used to compensate for temperature variations in a circuit.

THERMOCOUPLE—A junction of two dissimilar metals that produces a voltage when heated.

TORQUE—The turning effort or twist which a shaft sustains when transmitting power.

TRANSFORMER—A device composed of two or more coils, linked by magnetic lines of force, used to transfer energy from one circuit to another.

TRANSMISSION LINES—Any conductor or system of conductors used to carry electrical energy from its source to a load.

VARS—Abbreviation for volt-ampere, reactive.

VECTOR—A line used to represent both direction and magnitude.

VOLT—The unit of electrical potential.

VOLTMETER—An instrument designed to measure a difference in electrical potential in volts.

WATT—The unit of electrical power.

WATTMETER—An instrument for measuring electrical power in watts.

APPENDIX B
WIRE AND CONDUIT DATA

1987 NATIONAL ELECTRICAL CODE* TABLE 310-16 — WIRE & CONDUIT TABLES

AMPACITIES OF INSULATED CONDUCTORS RATED 0-2000 VOLTS, BASED ON AMBIENT AIR TEMPERATURE OF 30°C (86°F) IN RACEWAY IN FREE AIR

TRADE SIZE OF METALLIC CONDUIT OR TUBING BASED ON NEC CHAPTER 9, TABLE 1 FOR 40% FILL AND TABLES 3A, 3B, 3C & 4 — REFER TO NEC CHAPTER 9 FOR MAXIMUM NUMBER OF CONDUCTORS IN TRADE SIZES OF METALLIC CONDUIT OR TUBING. PROPERTIES OF CONDUCTORS MUST AGREE WITH NEC CHAPTER 9 TABLE 8. DIMENSIONS OF INSULATED CONDUCTORS PER NEC CHAPTER 9 (TABLE 5 FOR COPPER — TABLE 5A FOR ALUMINUM THW, THHN, XHHW)

(UNDERLINED INSULATION TYPE INDICATES DRY AND DAMP LOCATION ONLY RATINGS)

Table 310-16 Ampacity Insulated COPPER ‡	CU WIRE SIZE	75°C (167°F) ☆THW ☆RHW-USE Without Outer Covering		CU WIRE SIZE	75°C (167°F) ☆THWN, ☆XHHW		CU WIRE SIZE	▲ 90°C (194°F) FEP(14-2), ☆THHN FEPB(14-8), ☆XHHW	
		Conduit 3W	‡Conduit 3Ø4W		Conduit 3W	‡Conduit 3Ø4W		Conduit 3W	‡Conduit 3Ø4W
20	■14†	½	½	#14†	½	½	...	...	...
20	...	...	...	...	...	...	#14†	½	½
25	■12†	½	½	#12†	½	½	...	...	...
25	...	...	...	...	...	...	#12†	½	½
35	■10†	½	½	#10†	½	½	...	...	...
40	...	...	...	...	...	...	#10†	½	½
50	■8	¾	1	#8	½●	¾	...	...	...
55	...	...	...	...	...	...	#8	½●	¾
65	#6	1	1	#6	¾	¾◆	...	...	...
75	...	...	...	...	...	...	#6	¾	¾◆
85 (100*)	#4	1	1¼	#4	1	1	...	...	...
95 (100*)	...	...	...	...	...	...	#4	1	1
100 (110*)	#3	1¼	1¼	#3	1	1¼	...	...	...
110	...	...	...	...	...	...	#3	1	1¼
115 (125*)	#2	1¼	1¼	#2	1	1¼	...	...	...
130 (125*)	...	...	...	...	...	...	#2	1	1¼
130 (150*)	#1	1¼	1½	#1	1¼	1½	...	...	...
150 (150*)	...	...	...	...	...	...	#1	1¼	1½
150 (175*)	1/0	1½	2	1/0	1¼	1½	...	...	...
170 (175*)	...	...	...	...	...	...	1/0	1¼	1½
175 (200*)	2/0	1½	2	2/0	1½	2	...	...	...
195 (200*)	...	...	...	...	...	...	2/0	1½	2
200	3/0	2	2	3/0	1½	2	...	...	...
225	...	...	...	...	...	...	3/0	1½	2
230	4/0	2	2½	4/0	2	2	...	...	...
255	250M	2½	2½	250M	2	2½	...	...	...
260	...	...	...	...	...	...	4/0	2	2
285	300M	2½	3	300M	2	2½	...	...	...
290	...	...	...	...	...	...	250M	2	2½
310	350M	2½	3	350M	2½	3	...	...	...
320	...	...	...	...	...	...	300M	2	2½
335	400M	3	3	400M	2½	3	...	...	...
350	...	...	...	...	...	...	350M	2½	3
380	...	...	...	...	...	...	400M	2½	3
380	500M	3	3½	500M	3	3	...	...	...
420	600M	3	3½	600M	3	3½	...	...	...
430	...	...	...	...	...	...	500M	3	3
460	700M	3½	4	700M	3	3½	...	...	...
475	750M	3½	4	750M	3½	4	600M	3	3½
490	800M	3½	4	800M	3½	4	...	...	...
520	900M	4	5	900M	3½	4	700M	3	3½
535	...	...	...	...	...	...	750M	3½	4
545	1000M	4	5	1000M	3½	5	...	...	...
555	...	...	...	...	...	...	800M	3½	4
585	...	...	...	...	...	...	900M	3½	4
615	...	...	...	...	...	...	1000M	3½	4

Table 310-16 Ampacity Insulated ALUMINUM ‡	AL WIRE SIZE	75°C (167°F) ☆THW		AL WIRE SIZE	75°C (167°F) ☆XHHW		AL WIRE SIZE	▲ 90°C (194°F) ☆THHN ☆XHHW	
		Conduit 3W	‡Conduit 3Ø4W		Conduit 3W	‡Conduit 3Ø4W		Conduit 3W	‡Conduit 3Ø4W
...	...	...	...	...	...	...	...	...	...
...	...	...	...	...	...	...	...	...	...
...	...	...	...	...	...	...	...	...	...
...	...	...	...	...	...	...	...	...	...
...	...	...	...	...	...	...	...	...	...
35	...	...	...	...	...	...	...	...	...
40	■8	¾	¾	#8	½	¾	...	...	...
45	...	...	...	...	...	...	...	...	...
50	#6	¾	1	#6	¾	¾	...	...	...
60	...	...	...	...	...	...	#6	¾	¾
65	#4	1	1	#4	¾	1	...	...	...
75	...	...	...	...	...	...	#4	¾	1
75	...	...	...	...	...	...	...	...	...
85	...	...	...	...	...	...	...	...	...
90 (100*)	#2	1	1¼	#2	1	1¼	...	...	...
100 (100*)	...	...	...	...	...	...	#2	1	1¼
100 (110*)	#1	1¼	1½	#1	1¼	1¼	...	...	...
115 (110*)	...	...	...	...	...	...	#1	1¼	1¼
120 (125*)	1/0	1¼	1½	1/0	1¼	1½	...	...	...
135 (125*)	...	...	...	...	...	...	1/0	1¼	1½
135 (150*)	2/0	1½	2	2/0	1¼	1½	...	...	...
150 (150*)	...	...	...	...	...	...	2/0	1¼	1½
155 (175*)	3/0	1½	2	3/0	1½	2	...	...	...
175 (175*)	...	...	...	...	...	...	3/0	1½	2
180 (200*)	4/0	2	2	4/0	1½	2	...	...	...
205	250M	2	2½	250M	2	2	...	...	...
205 (200*)	...	...	...	...	...	...	4/0	1½	2
230	300M	2	2½	300M	2	2½	...	...	...
230	...	...	...	...	...	...	250M	2	2
250	350M	2½	3	350M	2	2½	...	...	...
255	...	...	...	...	...	...	300M	2	2½
270	400M	2½	3	400M	2½	2½	...	...	...
280	...	...	...	...	...	...	350M	2	2½
305	...	...	...	...	...	...	400M	2½	2½◇
310	500M	3	3	500M	2½	3	...	...	...
340	600M	3	3½	600M	3	3	...	...	...
350	...	...	...	...	...	...	500M	2½	3
375	700M	3	3½	700M	3	3½	...	...	...
385	750M	3	3½	750M	3	3½	600M	3	3
420	...	...	...	...	...	...	700M	3	3½
435	...	...	...	...	...	...	750M	3	3½
445	1000M	3½	4	1000M	3½	4	...	...	...
500	...	...	...	...	...	...	1000M	3½	4

AMPACITY CORRECTION FACTORS

For ambient temperatures other than 30°C (86°F), multiply the ampacities shown above by the appropriate factor shown below.

Ambient Temp. °C	THW-RHW-VSE	THWN-XHHW	FEP-THHN-XHHW	Ambient Temp. °F	THW	XNNW	THHN-XHHW
21-25	1.05	1.04	1.04	70-77	1.05	1.04	1.04
26-30	1.00	1.00	1.00	79-86	1.00	1.00	1.00
31-35	.94	.96	.96	88-95	.94	.96	.96
36-40	.88	.91	.91	97-104	.88	.91	.91
41-45	.82	.87	.87	106-113	.82	.87	.87
46-50	.75	.82	.82	115-122	.75	.82	.82
51-55	.67	.76	.76	124-131	.67	.76	.76
56-60	.58	.71	.71	133-140	.58	.71	.71
61-70	.33	.58	.58	142-158	.33	.58	.58
71-80	...	.41	.41	160-176	...	.41	.41

☆ Ratings for three-wire, single-phase residential service entrance conductors and three-wire, single-phase feeder that carries the total current supplied by that service. (NEC Note 3 to Table 310-16 for conductor types RHH-RHW-THW-THWN-THHN-XHHW.) The grounded service entrance conductor (neutral) may be two AWG sizes smaller than the ungrounded conductors provided the requirements of NEC 230-42(c) and 250-23(b) are met.

‡ On a 4-wire, 3-phase wye circuit where the major portion of the load consists of electric discharge lighting, data processing, or similar equipment, derate ampacities to 80% per notes 8 & 10 to NEC Table 310-16.

◆ #6 XHHW copper requires 1" conduit for 3Ø4W.

● #8 XHHW copper requires ¾" conduit for 3W.

◇ 400 MCM THHN Aluminum requires 3" conduit for 3Ø4W.

■ Special 90°C rating for THW (14-8) within electric discharge lighting equipment per NEC Article 410-31 and Table 310-13.

▲ For panelboard circuits, wire size should be no smaller than shown above for 75°C wire. 90°C wire should be used for locations where ambient temperature above 30°C is expected, and then derated per correction factors in NEC Table 310-16.

† The overcurrent protection for conductor types marked with a dagger (†) shall not exceed 15 amperes for 14 AWG, 20 amperes for 12 AWG, and 30 amperes for 10 AWG copper; after any correction factors for ambient temperatures and number of conductors have been applied. (These are the values shown in this table.)

NEC 220-3 (a) Continuous and Noncontinuous Loads.

The branch-circuit rating shall not be less than the noncontinuous load plus 125 percent of the continuous load. (See Exceptions in NEC)

NEC 220-10 (b) Continuous and Noncontinuous Loads.

Where a feeder supplies continuous loads or any combination of continuous and noncontinuous loads, the rating of the overcurrent device shall not be less than the noncontinuous load plus 125 percent of the continuous load.

The ampacity of the ungrounded service conductor shall not be less than the noncontinuous load plus 125 percent of the continuous load. (See Exceptions in NEC)

NEC 430-22 (a) Single Motor Circuit Conductors.

Branch circuit conductors supplying a single motor shall have an ampacity not less than 125 percent of the motor full-load current rating. (See Exceptions in NEC)

★ NEC is a Registered Trademark of the National Fire Protection Association.

Courtesy of Square D Co.

Conductor Insulation

Trade Name	Type Letter	Temp. rating	Application Provisions
Rubber-Covered Fixture Wire	*RF–1	60°C 140°F	Fixture wiring. Limited to 300 V.
Solid or 7–Strand	*RF–2	60°C 140°F	Fixture wiring.
Rubber-Covered Fixture Wire	*FF–1	60°C 140°F	Fixture wiring. Limited to 300 V.
Flexible Stranding	*FF–2	60°C 140°F	Fixture wiring.
Heat-Resistant Rubber-Covered Fixture Wire	*RFH–1	75°C 167°F	Fixture wiring. Limited to 300 V.
Solid or 7–Strand	*RFH–2	75°C 167°F	Fixture wiring.
Heat-Resistant Rubber-Covered Fixture Wire	*FFH–1	75°C 167°F	Fixture wiring. Limited to 300 V.
Flexible Stranding	*FFH–2	75°C 167°F	Fixture wiring.
Thermoplastic-Covered Fixture Wire—Solid or Stranded	*TF	60°C 140°F	Fixture wiring.
Thermoplastic-Covered Fixture Wire—Flexible Stranding	*TFF	60°C 140°F	Fixture wiring.
Cotton-Covered, Heat-Resistant, Fixture Wire	*CF	90°C 194°F	Fixture wiring. Limited to 300 V.
Asbestos-Covered Heat-Resistant, Fixture Wire	*AF	150°C 302°F	Fixture wiring. Limited to 300 V. and Indoor Dry Location.
Silicone Rubber Insulated Fixture Wire	*SF–1	200°C 392°F	Fixture wiring. Limited to 300 V.
Solid or 7 Strand	*SF–2	200°C 392°F	Fixture wiring
Silicone Rubber Insulated Fixture Wire	*SFF–1	150°C 302°F	Fixture wiring. Limited to 300 V.
Flexible Stranding	*SFF–2	150°C 302°F	Fixture wiring.
Code Rubber	R	60°C 140°F	Dry locations.
Heat-Resistant Rubber	RH	75°C 167°F	Dry locations.
Heat Resistant Rubber	RHH	90°C 194°F	Dry locations.
Moisture-Resistant Rubber	RW	60°C 140°F	Dry and wet locations. For over 2000 volts, insulation shall be ozone-resistant.
Moisture and Heat Resistant Rubber	RH-RW	60°C 140°F	Dry and wet locations. For over 2000 volts, insulation shall be ozone-resistant.

*Fixture wires are not intended for installation as branch circuit conductors nor for the connection of portable or stationary appliances.

Conductor Insulation—Continued.

Trade Name	Type Letter	Temp. rating	Application Provisions
		75°C	Dry locations.
Thermoplastic and Fibrous Outer Braid	TBS	90°C 194°F	Switchboard wiring only.
Synthetic Heat-Resistant	SIS	90°C 194°F	Switchboard wiring only.
Mineral Insulation (Metal Sheathed)	MI	85°C 185°F	Dry and wet locations with Type O termination fittings. Max. operating temperature for special applications 250°C.
Silicone-Asbestos	SA	90°C	Dry locations—max. operating temperature for special application 125°C.
Fluorinated Ethylene Propylene	FEP or FEPB	90°C 194°F 200°C 392°F	Dry locations. Dry locations—special applications.
		167°F	For over 2000 volts, insulation shall be ozone-resistant.
Moisture and Heat Resistant Rubber	RHW	75°C 167°F	Dry and wet locations. For over 2000 volts, insulation shall be ozone-resistant.
Latex Rubber	RU	60°C 140°F	Dry locations.
Heat Resistant Latex Rubber	RUH	75°C	Dry locations.
Moisture Resistant Latex Rubber	RUW	60°C 140°F	Dry and wet locations.
Thermoplastic	T	60°C 140°F	Dry locations.
Moisture-Resistant Thermoplastic	TW	60°C 140°F	Dry and wet locations.
Heat-Resistant Thermoplastic	THHN	90°C 194°F	Dry locations.
Moisture and Heat-Resistant Thermoplastic	THW	75°C 167°F	Dry and wet locations.
Moisture and Heat-Resistant Thermoplastic	THWN	75°C 167°F	Dry and wet locations.
Thermoplastic and Asbestos	TA	90°C 194°F	Switchboard wiring only.
Varnished Cambric	V	85°C 185°F	Dry locations only. Smaller than No. 6 by special permission.
Asbestos and Varnished Cambric	AVA	110°C 230°F	Dry locations only.
Asbestos and Varnished Cambric	AVL	110°C 230°F	Dry and wet locations.
Asbestos and Varnished Cambric	AVB	90°C 194°F	Dry locations only.

Conductor Insulation—Continued.

Trade Name	Type Letter	Temp. rating	Application Provisions
Asbestos	A	200°C 392°F	Dry locations only. In raceways, only for leads to or within apparatus. Limited to 300 V.
Asbestos	AA	200°C 392°F	Dry locations only. Open wiring. In raceways, only for leads to or within apparatus. Limited to 300 V.
Asbestos	AI	125°C 257°F	Dry locations only. In raceways, only for leads to or within apparatus. Limited to 300 V.
Asbestos	AIA	125°C 257°F	Dry locations only. Open wiring. In raceways, only for leads to or within apparatus.
Paper		85°C 185°F	For underground service conductors, or by special permission.

A P P E N D I X   C
MOTOR DATA

FULL LOAD CURRENTS OF MOTORS

[The following data are approximate full-load currents for motors of various types, frequencies, and speeds. They have been compiled from average values for representative motors of their respective classes. Variation of 10 percent above or below the values given may be expected.]

Amperes—Full-load current

The columns below the "Direct-current motors" group are direct-current; all remaining columns fall under "Alternating-current motors."

| Hp. of motor | Direct-current motors | | | Single-phase motors | | Squirrel-cage induction motors | | | | | | | | | | Slip-ring induction motors | | | | | | | | | |
| | | | | | | Two-phase | | | | | Three-phase | | | | | Two-phase | | | | | Three-phase | | | | |
	115-volt	230-volt	550-volt	110-volt	220-volt	110-volt	220-volt	440-volt	550-volt	2,200-volt	110-volt	220-volt	440-volt	550-volt	2,200-volt	110-volt	220-volt	440-volt	550-volt	2,200-volt	110-volt	220-volt	440-volt	550-volt	2,200-volt
¼	—	—	—	4.8	2.4	4.3	2.2	1.1	0.9	—	5.0	2.5	1.3	1.0	—	—	—	—	—	—	—	—	—	—	—
½	4.5	2.8	—	7	3.5	4.7	2.4	1.2	1.0	—	5.4	2.8	1.4	1.1	—	—	—	—	—	—	—	—	—	—	—
¾	6.5	3.3	1.4	9.4	4.7	5.7	2.9	1.4	1.2	—	6.6	3.3	1.7	1.3	—	—	—	—	—	—	—	—	—	—	—
1	8.4	4.2	1.7	11	6.5	7.7	4.0	2	1.6	—	9.4	4.7	2.4	2.0	—	6.2	3.1	1.6	1.2	—	7.2	3.6	1.8	1.5	—
1½	12.5	6.3	2.6	16.2	7.6	10.4	5	3	2.0	—	12.0	6	3.0	2.4	—	6.7	3.4	1.7	1.4	—	7.8	3.9	2.0	1.6	—
2	16.1	8.3	3.4	20	10	—	8	4	3.0	—	—	9	4.5	4.0	—	12.5	6.2	3.1	2.5	—	14.4	7.2	3.6	2.9	—
3	28	12.3	5.0	28	14	—	13	7	6	—	—	15	7.5	6.0	—	17.3	8.7	4.3	3.5	—	20.2	10	5.0	4	—
5	40	19.8	8.2	46	23	—	19	9	7	—	—	22	11	9.0	—	—	13.0	6.5	5.2	—	—	15	7.5	6	—
7½	58	28.7	12	68	34	—	24	12	10	—	—	27	14	11	—	—	20.0	10.0	7.6	—	—	25	13	10	—
10	75	38	16	86	43	—	33	16	13	—	—	38	19	15	—	—	24.3	12.1	10.0	—	—	28	14	11	—
15	112	56	23	—	—	—	45	23	19	—	—	52	26	21	5.7	—	39	19.5	15.6	—	—	45	23	18	—
20	140	74	30	—	—	—	55	28	22	6	—	64	32	26	7	—	49	24.7	19.8	—	—	56	28	22	—
25	185	92	38	—	—	—	67	34	27	7	—	77	39	31	8	—	60	30.0	24.0	6.4	—	67	34	27	7.5
30	220	110	45	—	—	—	88	44	35	9	—	101	51	40	10	—	72	36.0	28.8	7.8	—	82	41	33	9
40	294	146	61	—	—	—	108	54	43	11	—	125	63	50	13	—	93	46.5	37.3	9.5	—	106	53	42	11
50	364	180	75	—	—	—	129	65	52	13	—	149	75	60	15	—	113	57	45	12.1	—	128	64	51	14
60	436	215	90	—	—	—	156	78	62	16	—	180	90	72	19	—	135	68	54	14.0	—	150	75	60	16
75	540	268	111	—	—	—	212	106	85	22	—	246	123	98	25	—	164	82	65	17.3	—	188	94	75	19
100	—	357	146	—	—	—	268	134	108	27	—	310	155	124	32	—	214	108	87	21.7	—	246	123	99	25
125	—	443	184	—	—	—	311	155	124	31	—	360	180	144	36	—	267	134	108	27	—	310	155	124	31
150	—	—	220	—	—	—	415	208	166	43	—	480	240	195	49	—	315	158	127	32	—	364	182	145	37
175	—	—	—	—	—	—	—	—	—	—	—	—	—	—	—	—	—	—	—	—	—	—	—	—	—
200	—	—	295	—	—	—	—	—	—	—	—	—	—	—	—	—	430	216	173	44	—	490	245	196	52

Conductor Sizes and Overcurrent Protection for Motors

These values are in accordance with the National Electrical Code. They can be used for all installations except those intended for commercial or industrial use. For commercial or industrial use, consult pertinent sections of the National Electrical Code.

(1)	(2)	(3)	(5)	(6)	Maximum allowable rating or setting of branch circuit protective devices			
					(7) With code letters	(8) With code letters	(9) With code letters	(10) With code letters
Full load current rating of motor in amperes.	Minimum-sized conductor in raceways. For conductors in air, or for other insulations, see tables B–2 and B–3. AWG and MCM		For running protection of motors.ᵃ		Single-phase and squirrel-cage and synchronous. Full voltage, resistor, or reactor starting. Code letters F to V inc. Without code letters Same as above.	Single-phase and squirrel-cage and synchronous. Full voltage, resistor, or reactor starting. Code letters B to E inc. Without code letters Squirrel-cage and synchronous, auto-transformer starting. High reactance squirrel-cage.ᵇ Both not more than 30 amperes.	Squirrel-cage and synchronous. Auto-transformer starting. Code letters B to E inc. Without code letters Squirrel-cage and synchronous, auto-transformer starting. High reactance squirrel-cage.ᵇ Both more than 30 amperes.	All motors, Code letter A Without code letters DC and wound-rotor motors.
	Rubber Type R Type T	Rubber Type RH	Maximum rating of non-adjustable protective devices. Amperes	Maximum setting of adjustable protective devices. Amperes				
1	14	14	2	1.25	15	15	15	15
2	14	14	3	2.50	15	15	15	15
3	14	14	4	3.75	15	15	15	15
4	14	14	6	5.00	15	15	15	15
5	14	14	8	6.25	15	15	15	15
6	14	14	8	7.50	20	15	15	15
7	14	14	10	8.75	25	20	15	15
8	14	14	10	10.00	25	20	20	15
9	14	14	12	11.25	30	25	20	15
10	14	14	15	12.50	30	25	20	15
11	14	14	15	13.75	35	30	25	20
12	14	14	15	15.00	40	30	25	20

Conductor Sizes and Overcurrent Protection for Motors—Continued

(1)	(2)	(3)	(5)	(6)	Maximum allowable rating or setting of branch circuit protective devices			
					(7) With code letters	(8) With code letters	(9) With code letters	(10) With code letters
13	12	12	20	16.25	40	35	30	25
14	12	12	20	17.50	45	35	30	25
15	12	12	20	18.75	45	40	30	25
16	12	12	20	20.00	50	40	35	25
17	10	10	25	21.35	60	45	35	30
18	10	10	25	22.50	60	45	40	30
19	10	10	25	23.75	60	50	40	30
20	10	10	25	25.00	60	50	40	30
22	10	10	30	27.50	70	60	45	35
24	10	10	30	30.00	80	60	50	40
26	8	8	35	32.50	80	70	60	40
28	8	8	35	35.00	90	70	60	45
30	8	8	40	37.50	90	70	60	45
32	8	8	40	40.00	100	80	70	50
34	6	8	45	42.50	110	90	70	60
36	6	8	45	45.00	110	90	80	60
38	6	6	50	47.50	125	100	80	60
40	6	6	50	50.00	125	100	80	60
42	6	6	50	52.50	125	110	90	70
44	6	6	60	55.00	125	110	90	70
46	4	6	60	57.50	150	125	100	70
48	4	6	60	60.00	150	125	100	80
50	4	6	60	62.50	150	125	100	80
52	4	6	70	65.00	175	150	110	80
54	4	4	70	67.50	175	150	110	90
56	4	4	70	70.00	175	150	120	90
58	3	4	70	72.50	175	150	120	90
60	3	4	80	75.00	200	150	120	90
62	3	4	80	77.50	200	175	125	100
64	3	4	80	80.00	200	175	150	100
66	2	4	80	82.50	200	175	150	100
68	2	4	90	85.00	225	175	150	110
70	2	3	90	87.50	225	175	150	110
72	2	3	90	90.00	225	200	150	110
74	2	3	90	92.50	225	200	150	125
76	2	3	100	95.00	250	200	175	125
78	1	3	100	97.50	250	200	175	125
80	1	3	100	100.00	250	200	175	125
82	1	2	110	102.50	250	225	175	125
84	1	2	110	105.00	250	225	175	150
86	1	2	110	107.50	300	225	175	150
88	1	2	110	110.00	200	225	200	150
90	0	2	110	112.50	300	225	200	150
92	0	2	125	115.00	300	250	200	150
94	0	1	125	117.50	300	250	200	150
96	0	1	125	120.00	300	250	200	150
98	0	1	125	122.50	300	250	200	150
100	0	1	125	125.00	300	250	200	150
105	00	1	150	131.50	350	300	225	175
110	00	0	150	137.50	350	300	225	175
115	00	0	150	144.00	350	300	250	175
120	000	0	150	150.00	400	300	250	200
125	000	00	175	156.50	400	350	250	200

Conductor Sizes and Overcurrent Protection for Motors—Continued

(1)	(2)	(3)	(5)	(6)	Maximum allowable rating or setting of branch circuit protective devices			
					(7) With code letters	(8) With code letters	(9) With code letters	(10) With code letters
130	000	00	175	162.50	400	350	300	200
135	0000	00	175	169.00	450	350	300	225
140	0000	00	175	175.00	450	350	300	225
145	0000	000	200	181.50	450	400	300	225
150	0000	000	200	187.50	450	400	300	225
155	0000	000	200	194.00	500	400	350	250
160	250	000	200	200.00	500	400	350	250
165	250	0000	225	206.00	500	450	350	250
170	250	0000	225	213.00	500	450	350	300
175	300	0000	225	219.00	600	450	350	300
180	300	0000	225	225.00	600	450	400	300
185	300	0000	250	231.00	600	500	400	300
190	300	250	250	238.00	600	500	400	300
195	350	250	250	244.00	600	500	400	300
200	350	250	250	250.00	600	500	400	300
210	400	300	250	263.00		600	450	350
220	400	300	300	275.00		600	450	350
230	500	300	300	288.00		600	500	350
240	500	350	300	300.00		600	500	400
250	500	350	300	313.00			500	400
260	600	400	350	325.00			600	400
270	600	400	350	338.00			600	450
280	600	400	350	338.00			600	450
290	700	500	350	363.00			600	450
300	700	500	400	375.00			600	450
320	750	600	400	400.00				500
340	900	600	450	425.00				600
360	1000	700	450	450.00				600
380	1250	750	500	475.00				600
400	1500	900	500	500.00				600
420	1750	1000	600	525.00				
440	2000	1250	600	550.00				
460		1250	600	575.00				
480		1500	600	600.00				
500		1500		625.00				

ᵃ For running protection of motors, notify values in columns 5 and 6, if nameplate-full-load current values different than those shown in table. Reduce current values shown in columns 5 and 6 by 8 percent for all motors other than open-type motors marked to have a temperature rise of over 40° C. (72° F.)

ᵇ High-reactance, squirrel-cage motors are those designed to limit the starting current by means of deep-slot secondaries or double-wound secondaries and are generally started on full voltage.

DC Motor Characteristics and Selection Chart

Typical applications		Type of motor	Speed		Starting torque in percent of full load torque	Maximum torque
			Classification	Regulation		
Fans	Saws	Shunt-wound	Constant	5 to 10 Percent	150 Percent	Limited by Commutation
	Band					
Blowers	Circular					
Positive	Joiners					
Pressure	Molders					
	Planners					
Laundry	Line shafts					
Washers	Motor-generator sets					
Flat work	Buffers					
Ironers	Drill presses					
	Grinders					
Stokers	Lathes					
Dough mixers						
Fans	Boring mill		Adjustable by field control	5 to 15 Percent Dependent upon field setting	150 Percent	
Blowers	Drills					
Ironer	Milling machines					
Flat work	Stokers					
Lathe	Dough mixers					
Elevators	Rubber Calendars	Series-wound	Adjustable by variable voltage	Slight	150 Percent	
Passenger	Mine hoist					
Paper mills						
Cranes	Bridges		Varying	Dependent upon the load; will run away if unloaded.	300 to 400 Percent	
Hoists	Coal					
Valves	Ore					
Turntable	Vehicles					
Pumps	Crushers	Compound-wound	Varying	10 to 25 Percent	175 to 200 Percent	
Centrifugal	Large bandsaws					
Displacement						
Presses	Sanders					
Printing						
Rotary	Rolls					
	Bending					
Elevators	Straightening					
Passenger						
Freight	Dough mixers					
	Laundry					
Conveyors	Extractors					
Car pullers	Power hammers					

AC Motor Characteristics and Selection Chart

Applications	Class	Group	Designation	Starting torque in percent of full load torque	Starting current in percent of full load torque	Pull-in torque in percent of full load torque	Pull-out torque in percent of full load torque	Slip in percent
Fans Centrifugal Propeller — Lathes Shapers Drill presses	Squirrel-cage induction—constant speed.	1	Normal torque.	135 to 200 Percent	Full voltage applied. 500 to 650 Percent. Usually started on reduced voltage.	....	Not less than 200 Percent.	3 to 6 Percent
Pumps Centrifugal Rotary Turbine — Circular saws Small and Medium		1	Normal starting current.					
Positive Pressure Blowers — Joiners Molders Sanders		2	Normal torque.	135 to 175 Percent	Full voltage applied 400 to 500 Percent. Meets E.E.I. requirements up to 30 hp. Above 30 hp starting currents may not be within E.E.I. limits.	....	Not less than 200 Percent.	3 to 5 Percent
Line shafts Small stokers Metal grinders Planer[3] — Laundry Washers Job printing		2	Low starting current.					
Pumps Reciprocating Displacement Foundry Tumbling Barrels Conveyors Starting Loaded — Compressors Air Refrigerating		3	High torque.	200 to 250 Percent	E.E.I. limits up to 30 hp 1800 rpm 450 to 550 Percent.	....	Not less than 200 Percent.	5 to 7 Percent
Crushers Without flywheels — Bucket-type Elevators Grain elevator legs		3	Low starting current.					
Dough mixers Ball mills Large bandsaw Turntables Passenger and freight elevators		4	High torque.	300 to 400 Percent	300 to 500 Percent	....	Not less than 250 Percent.	12 to 18 Percent
		4	High slip.					
Cranes, hoists, lifts, valves					Same as for elevators except motor will not meet all conditions as to voltage fluctuations and quiet operation.			

Type	No.	Application	Characteristic					
	5	Punch presses, shears, laundry extractors and drives with flywheels or high inertia to accelerate.	High torque. Medium slip.	300 to 400 Percent		375 to 500 Percent	300 to 400 Percent	7 to 12 Percent
Slipring induction	6	Drives requiring large starting torques and minimum starting currents, as conveyors, hoist, fans, pumps, compressors, etc. Reduced speed operations of fans, pumps, etc.	Wound rotor induction motor.	Depends on external rotor resistance.			Not less than 200 Percent.	3 to 5 Percent
[b] Synchronous	7	Reciprocating air compressors and other machines which can be started light.	Low speed 100 Percent P.F.[c]	40 Percent	275 to 300 Percent	40 Percent	150 Percent	-----
	8		Low speed 80 Percent P.F.	40 Percent	250 to 275 Percent	40 Percent	225 Percent	-----
	9	Drives requiring 200 hp or less and 514 rpm or over.	General purpose 100 Percent P.F.	110 Percent	500 Percent	110 Percent	150 or 175 Percent	-----
	10	Drives requiring 150 hp or less and 514 rpm or over.	General Purpose 80 Percent P.F.	125 Percent	575 Percent	125 Percent	200 or 250 Percent	-----

Above values will serve as a guide. [a] Larger values of torque and current apply to the higher speed motors. [b] Synchronous motors are also made to meet other requirements of torque, speed, etc. [c] PF = Power factor. EEI = Edison Electrical Institute.

Selection Chart for DC Motor Controllers

Type of motor			Constant speed	Adjustable speed
Method of control			Type of controllers	
For manual control.	Non-reversing	Across-the-line starting (with OL protection).	Manual starters, up to 2 hp. Generally toggle type quick acting with thermal overload.	No
		Reduced current starting (with LV protection).	Rheostat	No
		Speed adjustment (with LV protection).	Rheostat	Rheostat
	Reversing	Across-the-line starting	Drum switch up to 2 hp.	No
		Speed adjustment	Drum controller and armature circuit resistor.	Drum controller and field rheostat.
For magnetic control (remote pushbutton operating)*	Non-reversing	Across-the-line starting (OL and LV protection).	Line starter	No
		Reduced current starting (OL and LV protection).	Time starter	Time starter and field rheostat.
	Reversing	Across-the-line starting (OL and LV protection).	Line starter	No
		Reduced current starting (OL and LV protection).	Time starter	Time starter and field rheostat.

OL = Overload   LV = Low voltage
*Other remote control devices can be used but low voltage release may be obtained instead of low voltage protection.

Selection Chart for AC Motor Controllers

Type of motor	For manual control				
	Nonreversing			Reversing	
	Across-the-line starting (with OL protection).	Reduced-voltage starting (with LV protection).	Speed adjustment (with LV protection).	Across-the-line starting.	Speed adjustment.
Squirrel-cage. General purpose. Normal starting torque. Normal starting current.	Manual starter up to 7½ hp. Generally toggle type, quick acting.	(1) Auto-transformer starter with OL protection. (2) Linestarter with OL protection, and primary rheostat.	No	Drum switch up to 15 hp.	No
Squirrel cage. Normal starting torque. Low starting current.	Manual starter up to 7½ hp. Generally toggle type, quick acting.	(1) Auto-transformer starter with OL protection. (2) Linestarter with OL protection, and primary rheostat.	No	Drum switch up to 15 hp.	No
Squirrel-cage. High starting torque. Low starting current.	Manual starter up to 7½ hp. Generally toggle type quick acting.	No	No	Drum switch up to 15 hp.	No
High slip	Manual starter. Generally toggle type, quick acting.	No	No	Drum switch	No
Synchronous	Circuit breaker with field control.	Auto-transformer starter with OL protection and field control.	No	Special	No

Selection Chart for AC Motor Controllers—Continued

Type of motor	For manual control				
	Nonreversing			Reversing	
	Across-the-line starting (with OL protection).	Reduced-voltage starting (with LV protection).	Speed adjustment (with LV protection).	Across-the-line starting.	Speed adjustment.
Wound-rotor or slip-ring.	No	Linestarter for primary control. Rheostat or drum controller for secondary control.	Linestarter for primary control. Rheostat or drum controller for secondary control.	No	Linestarter for primary control. Faceplate, rheostat or drum control for secondary control.
Single phase ½ hp, maximum.	Manual starter. Generally toggle type, quick acting.	Rheostat	No	No	No

OL = Overload.   LV = Low voltage. Drum controller (not switch) include secondary registors.

Type of motor	For magnetic control (Remote pushbutton operation)*				
	Nonreversing			Reversing	
	Across-the-line starting (with OL and LV protection)		Reduced voltage starting (with OL and LV protection)	Across-the-line starting (with OL and LV protection)	Reduced current starting (with OL and LV protection)
Squirrel-cage. General purpose. Normal starting torque. Normal starting current.	Linestarter	Combination of linestarter and circuit breaker or safety switch.	Resistance-type or auto-transformer starter.	Linestarter	Special
Squirrel-cage. Normal starting torque. Low starting current.	Linestarter	Combination of linestarter and circuit breaker or safety switch.	Resistance-type or auto-transformer starter.	Linestarter	Special
Squirrel-cage. High starting torque. Low starting current.	Linestarter	Combination of linestarter and circuit breaker or safety switch.	No	Linestarter	No
High-slip	Linestarter	Combination of linestarter and circuit breaker or safety switch.	Special	Linestarter	Special
Synchronous	Linestarter with field control.	No	Resistance-type or auto-transformer starter.	Special	Special
Wound-rotor or slip-ring.	Linestarter in combination with secondary controller.	Combination line starter in combination with secondary controller.	Resistance-type starter.	No	Special
Singlephase 7½ hp. maximum.	Linestarter	No	Special	No	No

*Other remote controls can be used but low voltage release may be obtained instead of low voltage protection.

Tri-onic®–Class RK5

Time Delay Fuses

TR

SINGLE PHASE MOTOR FUSE SELECTION

115 Volt Single Phase
UL Class RK5 TR Fuses

MOTOR HP	FULL LOAD AMPERES	RECOMMENDED FUSE AMPERE RATING			
		MINIMUM		TYPICAL	HEAVY LOAD
		1.0 S.F.	1.15 S.F.		
⅙	4.4	5	5	6¼	9
¼	5.8	6¼	7	9	10
⅓	7.2	8	9	10	12
½	9.8	10	12	15	17½
¾	13.8	15	15	20	25
1	16	17½	20	25	30
1½	20	20	25	30	35
2	24	25	30	35	40
3	34	35	40	50	60
5	56	60	70	80	100
7½	80	90	100	125	150
10	100	110	125	150	175

230 Volt Single Phase
UL Class RK5 TR Fuses

MOTOR HP	FULL LOAD AMPERES	RECOMMENDED FUSE AMPERE RATING			
		MINIMUM		TYPICAL	HEAVY LOAD
		1.0 S.F.	1.15 S.F.		
⅙	2.2	2½	2½	3½	4
¼	2.9	3²⁄₁₀	3½	4½	5
⅓	3.6	4	4½	5⁶⁄₁₀	6¼
½	4.9	5⁶⁄₁₀	6	7	9
¾	6.9	7	8	10	12
1	8	9	10	12	15
1½	10	10	12	15	17½
2	12	12	15	17½	20
3	17	17½	20	25	30
5	28	30	35	40	50
7½	40	45	50	60	70
10	50	50	60	70	90

Minimum
Highest fuse rating which will provide both overload and short circuit protection per the NEC. Choosing this fuse rating eliminates the need for an overload relay. Nuisance fuse opening may occur if motor is loaded to its rating.

Typical
Suggested rating when fuse is used in conjunction with an overload relay. Fuse sized near 150% of motor full load current.

Heavy Load
Maximum size for effective short circuit protection. Not applicable for motors marked with code letter A.

Tri-onic®–Class RK5
Time Delay Fuses
TR/TRS
THREE PHASE MOTOR FUSE SELECTION

575 Volt Three Phase
UL Class RK5 — TRS Fuses

MOTOR HP	FULL LOAD AMPERES AT 575V	RECOMMENDED FUSE AMPERE RATING — MOTOR ACCELERATION TIMES		
		MINIMUM 2 SECONDS	TYPICAL 5 SECONDS	HEAVY LOAD OVER 5 SECONDS
½	.8	1	1¼	1⁴⁄₁₀
¾	1.1	1⁴⁄₁₀	1⁸⁄₁₀	1⁸⁄₁₀
1	1.4	1⁸⁄₁₀	2	2½
1½	2.1	2½	3	4
2	2.7	3²⁄₁₀	4	5
3	3.9	5	6	7
5	6.1	8	9	12
7½	9	12	15	17½
10	11	15	15	20
15	17	20	25	30
20	22	30	35	40
25	27	35	40	50
30	32	40	45	60
40	41	50	60	80
50	52	70	80	100
60	62	75	90	110
75	77	100	125	150
100	99	125	150	175
125	125	150	175	225
150	144	175	225	250
200	192	250	300	350

230 Volt Three Phase
UL Class RK5 — TR Fuses

MOTOR HP	FULL LOAD AMPERES AT 230V	RECOMMENDED FUSE AMPERE RATING — MOTOR ACCELERATION TIMES		
		MINIMUM 2 SECONDS	TYPICAL 5 SECONDS	HEAVY LOAD OVER 5 SECONDS
½	2	2½	3	4
¾	2.8	3½	4	5
1	3.6	4½	5⁶⁄₁₀	6¼
1½	5.2	6¼	8	9
2	6.8	8	10	12
3	9.6	12	15	17½
5	15.2	17½	25	30
7½	22	25	30	40
10	28	35	40	50
15	42	50	60	80
20	54	70	80	100
25	68	80	100	125
30	80	100	125	150
40	104	125	150	200
50	130	175	200	225
60	154	175	225	250
75	192	225	300	350
100	248	300	350	400
125	312	400	450	500
150	360	450	500	600
200	480	600	—	—

460 Volt Three Phase
UL Class RK5 — TRS Fuses

MOTOR HP	FULL LOAD AMPERES AT 460V	RECOMMENDED FUSE AMPERE RATING — MOTOR ACCELERATION TIMES		
		MINIMUM 2 SECONDS	TYPICAL 5 SECONDS	HEAVY LOAD OVER 5 SECONDS
½	1	1¼	1⁴⁄₁₀	1⁸⁄₁₀
¾	1.4	1⁸⁄₁₀	2	2½
1	1.8	2¼	2½	3²⁄₁₀
1½	2.6	3²⁄₁₀	4	5
2	3.4	4	5	6
3	4.8	5⁶⁄₁₀	7	8
5	7.6	9	10	15
7½	11	15	15	20
10	14	17½	20	25
15	21	25	30	40
20	27	35	40	50
25	34	40	50	60
30	40	50	60	70
40	52	70	75	90
50	65	80	100	125
60	77	90	110	150
75	96	110	150	175
100	124	150	175	225
125	156	200	225	250
150	180	225	250	350
200	240	300	350	400

Minimum

Fuses are sized near 125% of motor full load current. This sizing is not recommended if motor acceleration time exceeds 2 seconds. Minimum sizing will provide close overload relay back-up protection but may not coordinate with some NEMA Class 20 overload relays.

Typical

Suggested for most applications. Will coordinate with NEMA Class 20 overload relays. Suitable for motor acceleration times up to 5 seconds.

Heavy Load

Maximum fuse rating for effective short circuit protection. Not applicable for motors marked with code letter A.

Amp-trap®–Class J
Time Delay Fuses

AJT
230 Volt Three Phase
UL Class J

MOTOR HP	FULL LOAD AMPERES AT 230V	RECOMMENDED FUSE AMPERE RATING		
		MOTOR ACCELERATION TIMES		
		MINIMUM 2 SECONDS	TYPICAL 5 SECONDS	HEAVY LOAD OVER 5 SECONDS
½	2	3	3	4
¾	2.8	4	4	6
1	3.6	5	6	8
1½	5.2	8	8	10
2	6.8	8	10	15
3	9.6	12	15	20
5	15.2	17½	25	30
7½	22	25	30	45
10	28	35	40	60
15	42	50	60	90
20	54	70	80	110
25	68	80	100	150
30	80	100	125	175
40	104	125	150	225
50	130	175	200	250
60	154	175	225	300
75	192	225	300	400
100	248	300	350	500
125	312	400	450	600
150	360	450	500	600
200	480	600	—	—

460 Volt Three Phase
UL Class J

MOTOR HP	FULL LOAD AMPERES AT 460V	RECOMMENDED FUSE AMPERE RATING		
		MOTOR ACCELERATION TIMES		
		MINIMUM 2 SECONDS	TYPICAL 5 SECONDS	HEAVY LOAD OVER 5 SECONDS
½	1	1½	1½	2
¾	1.4	2	2	3
1	1.8	3	3	4
1½	2.6	4	4	5
2	3.4	4	5	6
3	4.8	6	8	10
5	7.6	10	10	15
7½	11	15	15	25
10	14	17½	20	30
15	21	25	30	45
20	27	35	40	60
25	34	40	50	70
30	40	50	60	80
40	52	70	80	100
50	65	80	100	125
60	77	90	110	150
75	96	110	150	200
100	124	150	175	250
125	156	200	225	300
150	180	225	250	350
200	240	300	300	450

THREE PHASE MOTOR FUSE SELECTION

575 Volt Three Phase
UL Class J

MOTOR HP	FULL LOAD AMPERES AT 575V	RECOMMENDED FUSE AMPERE RATING		
		MOTOR ACCELERATION TIMES		
		MINIMUM 2 SECONDS	TYPICAL 5 SECONDS	HEAVY LOAD OVER 5 SECONDS
½	.8	1	1½	1½
¾	1.1	1½	2	2
1	1.4	2	2	3
1½	2.1	3	3	4
2	2.7	4	4	6
3	3.9	5	6	8
5	6.1	8	10	12
7½	9	12	15	17½
10	11	15	15	20
15	17	20	25	35
20	22	30	35	45
25	27	35	40	50
30	32	40	45	60
40	41	50	60	80
50	52	70	80	100
60	62	80	90	125
75	77	100	125	150
100	99	125	150	200
125	125	150	175	250
150	144	175	225	300
200	192	250	300	400

Minimum
Minimum sizing may not be heavy enough for motors with code letter G or higher.

Typical
Suggested for most applications. Will coordinate with NEMA Class 20 overload relays. Suitable for motor acceleration times up to 5 seconds.

Heavy Load
Largest ampere rating allowed by NEC for short circuit protection. Not applicable for motors marked with code letter A.

APPENDIX D
LIGHTING

Performance Data For Philips Fluorescent Lamps

(Subject to change without notice)

Ordering Code	Nom. Lamp Watts	Description	Bulb	Nom. (2) Length Inches	Rated Avg. Life at 3 Hrs/ Start	Approx. Lumens at 40% Rated Avg. Life	Rated Initial Lumens Cool White (1)(3)
Preheat Lamps - Bipin Base							
F4T5/CW	4		T5	6	6000	93	135
F6T5/CW	6		T5	9	7500	230	295
F8T5/CW	8		T5	12	7500	300	400
F13T5/CW	13		T5	21	7500	655	820
F14T12/CW	14		T12	15	9000	560	710
F15T8/CW	15		T8	18	7500	765	870
F15T12/CW	15		T12	18	9000	695	800
F20T12/CW	20		T12	24	9000	1110	1300
F25T12/CW	25		T12	33	7500	1755	1950
F30T8/CW	30		T8	36	7500	2000	2200
F40CW/EW-PH	34	Econ-o-watt®	T12	48	15000	2510	2850
F90T12/CW/60/EW	84	Econ-o-watt	T12	60(4)	9000	5750	6250
Preheat Lamps - Bipin Base							
F40CW	40		T12	48	20000(5)	2820	3150
F40T10CW/99	40	Extended Service	T10	48	24000(5)	2720	3200
F40AX/41	40	Advantage X	T10	48	24000(5)	3150	3700
FB40CW/6	40	"U" Bent	T12	22-7/16	12000	2640	2950
Octolume™ T8 Lamps - Bipin Base							
FO25/41	25		T8	36	20000	1935	2150
FO32/41	32		T8	48	20000	2600	2900
FO40/41	40		T8	60	20000	3285	3650
Rapid Start Lamps - Bipin Base							
F30T12/CW/RS/EW-II	25	Econ-o-watt	T12	36	18000	1700	2000
F30T12/CW/RS	30		T12	36	18000	2070	2300
F40CW/RS/EW-II	34	Econ-o-watt	T12	48	20000	2500	2775
Instant Start Lamps - Bipin Base							
F40T12/CW/IS	40		T12	48	7500	2850	3100
F40T12/CW/IS/60	40		T12	60(6)	7500	2700	3400
High Output Lamps - Recessed Double Contact Base (800ma.)							
F24T12/CW/HO	30		T12	24	9000	1410	1700
F30T12/CW/HO	42		T12	30	9000	1925	2290
F36T12/CW/HO	50		T12	36	9000	2480	2900
F48T12/CW/HO	60		T12	48	12000	3740	4300
F72T12/CW/HO	85		T12	72	12000	5785	6650
F96T12/CW/HO/EW	95	Econ-o-watt	T12	96	12000	7220	8300
F96T12/CW/HO	110		T12	96	12000	8005	9200
Very High Output Lamps - Recessed Double Contact Base (1500ma.)							
F18T12/CW/135RFL/VHO-II	59	Reflector	T12	18	9000	1175	1400
F24T12/CW/135RFL/VHO-II	73	Reflector	T12	24	9000	2015	2400
F48T12/CW/VHO	110		T12	48	12000	5975	6950
F72T12/CW/VHO	160		T12	72	12000	9300	11100
F96T12/CW/VHO/EW	185	Econ-o-watt	T12	96	12000	11500	14000
F96T12/CW/VHO	215		T12	96	12000	13000	15500
F96T12/CW/VHO-O	212	Outdoor	T12	96	10000	13000	15500(7)
FJ96T12/CW/VHO-O	212	Low Temp T14½Jacket	T12	96	10000	13000	15500(7)
Slimline Lamps - Single Pin Base							
F42T8/CW/EW	23	Econ-o-watt	T8	42	7500	1400	1800
F64T8/CW/EW	35	Econ-o-watt	T8	64	7500	2350	2850
F72T8/CW	37.5		T8	72	7500	2800	3000
F96T8/CW	50		T8	96	7500	3860	4300
F48T12/CW/EW	30	Econ-o-watt	T12	48	9000	2300	2500
F48T12/CW	38.5		T12	48	9000	2760	3000
F72T12/CW	56		T12	72	12000	4280	4600
F96T12/CW/EW	60	Econ-o-watt	T12	96	12000	5150	5600
F96T12/CW	73.5		T12	96	12000	5800	6300
Circline Lamps - Four Pin Base							
FC6T9/CW	20		T9	6½" Dia.	12000	590	800
FC8T9/CW	22		T9	8¼" Dia.	12000	775	1050
FC12T9/CW	32		T9	12" Dia.	12000	1440	1800
FC16T9/CW	40		T9	16" Dia.	12000	1850	2500

(1) Rated initial lumen and wattage values apply when lamps have burned 100 hours and are measured in still air at 25°C Ambient under specified test conditions.

(2) Includes two standard lamp holders.

(3) Initial lumen values apply to Cool White, for other colors including Ultralume, refer to SG100 Specification Guide.

(4) Interchangeable with 90T17.

(5) Life on rapid start circuits at 3 hours burning per start. Life on preheat circuits at 3 hours burning per start for F40 is 15,000 hours, for F40T10/99 is 19,000 hours.

(6) Interchangeable with F40T17/IS.

(7) Peak values.

Lamp Type	No. of Lamps	Power Factor (5)	Line Current (Amps) 120V	Line Current (Amps) 277V	Input[2] Watts	Min.[2] Lamp Starting Temp.	Sound[2] Rating
Preheat							
4, 6, 8 T-5	1	N	.165		5[3]	50	A
15 T-8 - T-12, 20 T-12	1	N	.34		5[3]	50	A
30 T-8, 40 T-12	1	N	.65		10[3]	50	A
8" Circline	1	N	.34		5[3]	50	A
12" Circline	1	N	.65		10[3]	50	A
40 T-12	1	H	.50		11[3]	50	A
15 T-8 - T-12, 20 T-12	1	N	.72		10[3]	50	A
40 T-12	2	N	.85		10[3]	50	B
Trigger Start							
4, 6, 8 T-5	2	N	.48		7[3]	50	A
15 T-8 - T-12, 20 T-12	2	N	.55		10[3]	50	A
14 T-12, 15 T-8 - T-12, 20 T-12	2	H	.47	.20	10[3]	0	A
Rapid Start							
6", 8" Circline	1	N	.72		13[3]	50	A
12", 16" Circline	1	N	.40		17[3]	50	A
30 T-12	2	H	.68	.30	81	50	A
30 T-12/EW	2	H	.63	.27	73	60	A
40 T-12	2	H	.80	.35	96	50	A
40 T-12	2	H[4]	.73	.32	86	50	A
40 T-12/EW	2	H	.72	.31	82	60	A
40 T-12/EW	2	H[4]	.63	.27	72	60	A
Rapid Start 800 ma							
24, 36, 42, 48 T-12/HO	2	H	1.3	.56	145	50	B
72 T-12/HO	2	H[4]	1.8	.77	200	50	C
96 T-12/HO	2	H[4]	2.05	.88	237	50	C
96 T-12/HO/EW	2	H[4]	1.85	.80	207	60	C
Rapid Start 1500 ma							
48 T-12/VHO	2	H	2.1	.92	242	−20	D
72 T-12/VHO	2	H	2.8	1.1	290	−20	D
96 T-12/VHO	2	H	3.7	1.65	440	−20	D
Slimline 200 ma							
42 T-8	2	H	.67		70	0	B
64 T-8	2	H	1.2		125	0	D
Slimline 425 ma							
48 T-12	2	H	.85	.37	102	50	A
72, 96 T-12	2	H[4]	1.35	.60	158	50	C
96 T-12	2	H[4]	1.1	.47	123	60	C
PL* Lamps							
5, 7	1	N	.18		4[3]	0	A
9	1	N	.16		4[3]	0	A
13	1	N	.30		4[3]	0	A
Electronic Ballasts — Rapid Start							
40 T-12	2	H	.59	.25	71	50	A
40 T-12/EW	2	H	.9	.43	61	60	A
Electronic Ballasts — Instant Start							
32 T-8	2	H	.55	.24	62	50	A
Dimming Ballast							
40 T-12	2	H	1.0		96	50	A

(1) For typical ballasts. Most ballasts are listed by Underwriters Labs (UL), Certified Ballast Manufacturers (CBM) or Canadian Standards Assoc. (CSA). All CBM ballasts are UL listed. For specific data, refer to ballast manufacturers catalog.

(2) Rating is the same for 120 volt and 277 volt ballasts

(3) Watts lost (not input watts)

(4) High Efficiency ballast

(5) H: High Power Factor, N: Normal Power Factor

(6) Two-lamp ballasts should only operate pairs of standard or Econ-o-watt lamps, not one of each type.

Class P

Class P ballasts are U.L. designated ballasts employing internal thermal protection to limit operating case temperature to 110°C if malfunction of windings or power capacitor occurs.

Maintenance Hints

Philips lamps, with good auxiliaries on the proper line voltages and with normal conditions of operation, ordinarily give long trouble-free life with satisfactory illumination. However, each element (the lamp, the starter, the holder, the ballast, and the operating conditions) has an influence on the service provided by the lighting system.

The following check chart will serve as a guide to recognize an operating problem, to quickly determine the cause and to correct it.

Check Chart for Fluorescent Lamps & Equipment

Indication of trouble	System	Cause	Remedy
1. Blinking on and off. a) Accompanied by shimmering during lighted period.	All	Normal end of lamp life, emission material on electrode depleted.	Replace lamp.
b) Blinking of relatively new lamp.	All	Loose circuit contact.	Seat lamp securely. Indicator "bumps" should be directly over holder slot. Check if lampholders are rigidly mounted and properly spaced. Tighten all connections.
	All	Failed ballast, wrong ballast for application or not within specification.	Replace ballast.
	All	Cold drafts hitting lamp.	Enclose or protect lamp. May need low temperature rated ballast.
	All	Low circuit voltage.	Check and correct if possible.
	Preheat	Failed or incorrect starter.	Replace starter or lamp life will be shortened.
2. Ends of lamp remain lighted.	Preheat	Starter contacts stuck together.	Replace starter or lamp life will be shortened.
3. No starting, slow starting or erratic starting.	All	Failed lamp.	Check lamp in another fixture. Replace if failed.
	All	Open circuit.	Check wiring and possible poor contact between lamps and holders. Replace broken or cracked holders.
	All	Failed ballast, wrong ballast for application, or not within specification.	Replace ballast.
	All	Low temperature. Wrong ballast for application.	Replace with proper low temp ballast.
	All	Low circuit voltage.	Check and correct if possible.
	Preheat	Starter sluggish or at end of life.	Replace starter.
	Preheat	Reset type starter not reset.	Reset starter.
	Preheat & Rapid Start	Instant start lamps installed.	Replace with correct lamp.
	Rapid Start	No cathode heat. Lamp improperly seated in holder or poor contact in holder or open wiring of failed ballast.	Check lamp seating and contacts. Check heater voltage. Check wiring.
	Instant & Rapid Start	Series ballast. Both lamps out.	Only one may have failed. Check both lamps.
	Instant & Rapid Start	Series ballast. One lamp out. One burning dimly. Failed lamp or open wiring or failed ballast.	Check lamps. Check wiring and ballast.
	Instant, Trigger & Rapid Start	Accumulation of dirt on lamps which overcomes effect of silicone coating.	Wash lamps with water containing mild detergent and rinse with clear water.

Check Chart for Fluorescent Lamps & Equipment (continued)

Indication of trouble	System	Cause	Remedy
J. No starting, slow starting or erratic starting. (continued)	Trigger & Rapid Start	Starting aid more than ½" from lamp or fixture not grounded.	Check starting aid. Ground fixture.
4. Short Life.	All	Mortality rate of lamps (lamps of shorter life will be balanced by those of longer life to give rated average life).	See mortality curve for lamps on page 13.
	All	Too frequent starting of lamps. Cycling of ballast thermal switch.	The average life of fluorescent lamps depends on the number of starts and the hours of operation. Published life rating is generally based on operating periods of 3 hours per start.
	All	Loose circuit contact causing on-off blink.	Be sure lampholders are rigidly mounted and that lamps are securely seated. Check wiring.
	All	Too low or too high circuit voltage.	Check if within the range on the ballast nameplate.
	All	Low ambient temperature causes slow starting.	Use ballasts rated for temperatures involved.
	All	Wrong ballast for application or not within specification or failed.	Replace ballast.
	Instant Start	Preheat or Rapid Start lamps installed.	Replace with correct lamp.
	Instant & Rapid Start	Series ballast. Both lamps out or one lamp operating dimly.	Only one may have failed. Check both lamps. Replace failed lamp promptly.
	Trigger & Rapid Start	Accompanied by severe end discoloration. No cathode heat.	Check lamp seating and contacts. Check heater voltage. Check wiring.
	HO and VHO lamps	Incorrect lamps installed. HO lamps operate at 800 ma, VHO lamps at 1500 ma. Interchanging will reduce life.	Check fixture or ballast for required lamps. Replace with correct lamp.
5. End blackening — a normal sign of age in service but if unusual check the following: a) Dense blackening at one end or both, extending two to three inches from base.	All	Normal end of lamp life.	Replace lamp.
	Rapid & Trigger Start	Accompanied by short life. No cathode heat.	Check lamp seating and contacts. Check heater voltage. Check wiring.
b) Blackening generally within one inch of ends.	All	Mercury deposit.	Should evaporate as lamp is operated.
c) Blackening early in life. (Indicates emission material on electrodes being sputtered off too rapidly.)	Preheat	Starter defective, causing on-off blink or prolonged flashing at each start.	Replace starter.
	Preheat	Ends of lamp remain lighted because of starter failure.	Replace starter.
		Too low or too high voltage.	Check the line voltage to be certain it is within the range shown on the ballast nameplate.
	All	Loose circuit contact causing on-off blink.	Be sure the lamp holders are rigidly mounted and the lamp is securely seated; check circuit wiring.
	All	Ballast improperly designed or not within specifications or wrong ballast being used.	Use CBM certified ballast of correct rating for lamp size and application. Note table IV.

(continued on next page)

Check Chart for Fluorescent Lamps & Equipment (continued)

Indication of trouble	System	Cause	Remedy
6. Dense spot — black, about ½" wide, extending about one-half way around tube, centering about one inch from base.	All	Normal sign of age in service. If early in life, indicates excessive lamp starting or high operating current.	Check for off rating of ballast or unusually high circuit voltage or operating current.
7. Rings — brownish ring at one end or both, about two inches from base.	All	May develop on some lamps during operation.	Will not affect the lamp performance.
8. Dark streaks — streaks lengthwise on tube.	All	Globules of mercury condensed on lower part of tube.	The lower half of the tube is cooler than the upper half; by rotating the tube 180° these mercury globules should evaporate due to the increased warmth.
9. Pronounced swirling, spiraling or fluttering of arc stream.	All	May occur in new lamps.	Should stabilize in normal operation.
	Preheat	Starter not performing properly to preheat electrodes, or wrong starter.	Replace starter.
	All	Ballast not within specifications, or wrong ballast being used.	Use CBM certified ballasts of correct rating for lamp size and application.
	All	High circuit voltage.	Check and correct, if possible.
	All	Low temperature or cold drafts hitting lamp.	Enclose or protect lamps; may need low temp rated ballast.
10. Radio Interference causing "buzz" in radio.	All	Lamp radiation "broadcast" through radio receiver.	Capacitor in starter or ballast may be defective or not included — replace with high quality starter or ballast. Or move receiver and antenna away from lamps (at least 8' from 40 watt lamps, 20' for lamps over 100 watts). Provide ground connection for the radio. If antenna is external, connect to radio by means of a grounded shielded lead-in wire. Attach large mesh metal screen to bottom of fixture. Screen should be grounded through fixture.
	All	Line radiation and line feedback.	Apply radio-interference filter at lamp or fixture; sometimes possible to apply filters at power outlet or panel box.
11. Noise-humming sound which may be steady or come and go.	All	Fixture components vibrating, or noisy ballast.	Tighten louvers, ballast, glass, side panels, etc. If ballast continues to be noisy, replace it with new or better sound rated ballast.
	All	Damaged ballast due to over heating. Prolonged blinking of lamp plus other conditions over heat ballasts.	Replace ballast. Correct overheating conditions.
12. Dark section of tube - ⅓ to ½ of tube gives no light (tubes longer than 24") (on direct current).	All	DC operation without using reversing switches.	Install reversing switches.
13. Decreased light output. a) During the first 100 hours of use.		The lamp light output during the first 100 hours of operation is above published rating. (The published rating is based on output at the end of 100 hours of operation.)	
b) Anytime.	All	Cold drafts hitting lamp or low temperature.	Enclose or protect lamp. Econ-o-watt lamps are more susceptible to this condition than standard lamps.
	All	Heat confined around lamp results in lower light output.	Better ventilation of fixture.

Check Chart for Fluorescent Lamps & Equipment (continued)

Indication of trouble	System	Cause	Remedy
13. Decreased light output. b) Anytime. (continued)	All	Low circuit voltage.	Check voltage and correct if possible.
	All	Dust or dirt on lamp, fixture, walls, or ceiling.	Clean.
	Instant Trigger & Rapid Start	Both lamps on 2 lamp series ballast. Failed ballast.	Replace ballast.
14. Color and brightness differences. a) Lamps operated at unequal brilliancy.	All	Normal lamp lumen maintenance. See Fig. 13, Pg. 14	Replace lamp. Group relamp, see Pg. 13
	All	Variations in ambient temperature.	Protect lamps operating in drafts or cold temperatures, provide ventilation where heat is confined around lamps.
	All	Failed ballast.	Test lamps in another fixture. If operated at equal brightness, replace ballast.
b) Different color appearance in difference locations of same installation.	All	May be due to reflector finish, wall finish, other nearby lights or cooling drafts on lamp.	Interchange lamps to determine cause.
	All	Lamps of different colors.	Check lamp marking. Replace with correct color lamp.
15. Lamp tight or loose in fixture.	All	Push-Pull Holder spring weak or plunger jammed.	Replace holders.
	All	Holders loose or mounting plate bent. Holders spaced wrong.	Repair mounting. Correct fixture dimensions.

INCORRECT   CORRECT   INCORRECT

  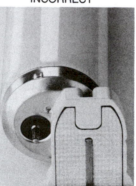 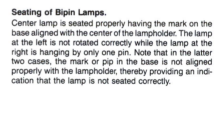

Seating of Bipin Lamps.
Center lamp is seated properly having the mark on the base aligned with the center of the lampholder. The lamp at the left is not rotated correctly while the lamp at the right is hanging by only one pin. Note that in the latter two cases, the mark or pip in the base is not aligned properly with the lampholder, thereby providing an indication that the lamp is not seated correctly.

Cathode Heat Testers. (For Rapid Start)
Testers which contain miniature lamps for recessed double contact and bipin lampholders.

Important Measurements

Preheat Current

Preheat current is only measured on preheat circuits. This is accomplished by installing an ammeter in the starting circuit as shown in **Fig. 23.** Preheat current is read during the starting cycle with a lamp in the circuit. Proper values are found in Table I.

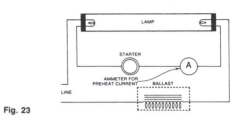

Fig. 23

Starting Voltage

In order to read starting or open circuit voltage as it is sometimes called, the lamp in question must be removed from the circuit. For Rapid Start circuits the probes of a voltmeter may be inserted across the opposing lampholders as shown in **Fig. 24.** For Slimline or Instant Start circuits the same procedure is used but care must be taken to complete the primary circuit at the lampholder having two conductors. This can be accomplished as shown in **Fig. 25** by physically shorting the two conductors.

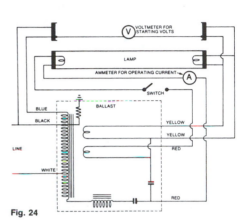

Fig. 24

Operating Current

Operating current is always measured with the lamp in the circuit. For instant start or Slimline circuits (see **Fig. 25**), the ammeter is placed on the single conductor side of the line. For Rapid Start circuits, operating current can be measured at either lampholder. For this circuit, however (note **Fig. 24**), an ammeter is inserted across one conductor and a switch across the other. The switch is closed until the lamp starts, then opened while the current is read.

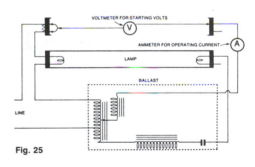

Fig. 25

Cathode Heater Voltage

This measurement only applies to Rapid Start or Trigger Start circuits. In general it is only necessary to establish the presence of voltage across the two contacts of a lampholder. This may be easily accomplished by inserting the cathode heat tester into the lampholder (note **Fig. 26**). If the tester glows at fair brightness, it can be assumed that adequate cathode heat is available at the lampholder. If such a tester is not available, a voltmeter with suitable parallel resistor may be used as shown in **Fig. 26.** For Rapid Start circuits, a 10 ohm resistor is needed and for Trigger Start the resistor should be 29 ohms. A 10-watt rating is adequate for either resistor.

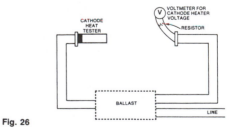

Fig. 26

End Discoloration

Blackening of the ends of the fluorescent lamp is ordinarily a sign of age in service. It is, however, aggravated by factors outside the lamp such as frequent starting, the use of starters which fail to provide adequate preheat before the arc is struck and, in the case of Rapid Start lamps, by no cathode heat. With relatively new lamps, end discoloration is also a transient mercury blackening resulting from condensed mercury depositing on the inside wall of the tube at the end. This is sometimes especially noticeable in new lamps which have been stored or shipped on end. This mercury cloud will usually disappear after a short period of normal operation.

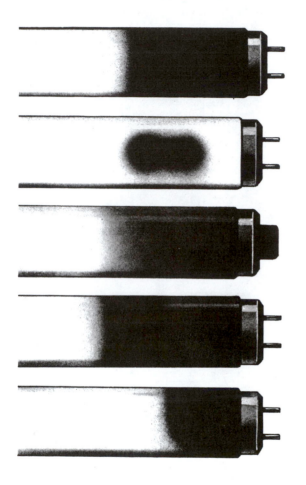

A. Blackening — May occur at end of lamp life or in Rapid Start lamps that have been operating with no cathode heat. Blackening extends from base 2 to 3 inches along lamp.

B. Mercury Condensation — May occur in new lamps, normally near the center, but sometimes in any location. Condensation has no effect on lamp performance.

C. Pressure Control Chamber — This shadow or low brightness is a characteristic of 1500 ma T12 lamps and appears only when lamps are lighted. This should not be confused with "end-of-life" discoloration above in 'A'.

D. End Banding — May develop gradually during lamp life.

E. Spotting — May develop gradually during lamp life.

SPECIFYING COLOR

COLOR SELECTION.

The choice of lamp color and space color(s) is the prerogative of the lighting designer, the interior designer and the owner. The decisions should be made jointly since both affect the success of the installation.

SELECTING THE BEST COLOR

The different colors of light sources vary only in the relative amounts of power at each wavelength in the visible spectrum. That, however, can be important because it *is* visible and people react to visible stimuli. There are certainly strong personal preference factors associated with light and color just as, for example, when people select clothing, furniture or decorations for themselves and their surroundings.

There is no one best or "true" light source color. Any spectral distribution can be said to distort object colors compared to another, whether it be from a natural source such as sunshine, north skylight, sunset, or an electric source such as incandescent, fluorescent, HID. The "right" color source for a given application depends on personal preference, custom and the tradeoffs in efficiency, cost, and color rendition. The table below suggests some appropriate choices in order of preference.

FLUORESCENT COLOR APPLICATIONS (in order of preference):	
OFFICES	
General Offices	SP35, SP41, SP30, LW, CW, WW
Color-important areas	SPX35, SPX41, SPX30, SP35, SP30
Color-critical areas	C50, CWX, WWX, SPX35, SPX41, SPX30
RETAIL STORES	
Food, Drug, Variety, Hardware	SPX35, SPX30, SPX41, SP35, SP30, SP41
Meat display	SPX30, SPX35, SPX41, CWX, WWX, SP30, SP35, N, plus incandescent
Jewelry	CWX, C50, SPX41, SP41, plus incandescent
Florist	CWX, C50, SPX41, SPX35, SP35, SP41
Shoe	SPX35, SPX30, SP35, SP41, SP30, CWX, WWX
Women's Wear	SPX30, SPX35, SPX41, SP30, SP35, WWX, CWX, SP41
Men's Wear	SPX35, SPX41, SP35, SP41, CWX
INDUSTRIAL	
General	LW, SP35, SP41, SP30, CW, WW
Printing	C50, C75, (ANSI Std. PH2.32-1972)
Textile (color checking)	C50
Paint (color checking)	C50, CWX
Meat (inspection)	CWX, C50
PUBLIC BUILDINGS	
Museums	CWX, C50, SPX35, SPX41, SPX30, WWX
Hotels/Motels, Restaurants	SPX30, SPX35, SPX41, SP30, SP35, SP41
Hospitals	SP35, SP30, SP41, SPX35, SPX30, SPX41, CWX, WWX
–Nurseries	C50
–Labs (color critical)	C50, CWX
–Treatment, intensive care, etc.	CWX, C50

SUGGESTED COLOR APPLICATIONS FOR HID LAMPS:

Clear Mercury — Landscape lighting, specialized floodlighting such as green copper roofs.
DX Mercury — Stores, public spaces–Multi-Vapor⁺ lamps however, are preferred.
WDX Mercury — Stores–Halarc lamps are preferred.
Multi-Vapor — Stores, public spaces, industrial, gymnasiums, floodlighting signs & buildings, parking areas, sports.
MV/C — Same as MV–warm color–diffuse coating reduces brightness.
Halarc — Stores, public spaces–warm color. Similar in chromaticity to incandescent.
Lucalox — Street lighting, parking areas, industrial, floodlighting, security, CCTV.
Deluxe Lucalox — Commercial exteriors, public interiors–very warm, good color rendering

Courtesy of General Electric Co., Lighting Division.

TECHNICAL DATA FOR INCANDESCENT AND QUARTZLINE® LAMPS

BULB SIZE or diameter (maximum) is expressed in eighths of an inch (⅛"). For example: an A-21 bulb is 21-eighths of an inch or 2⅝" in diameter at its maximum dimension.

BULB SHAPES (Not to Scale)

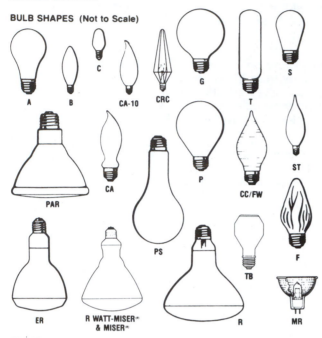

BASES (not to Scale)

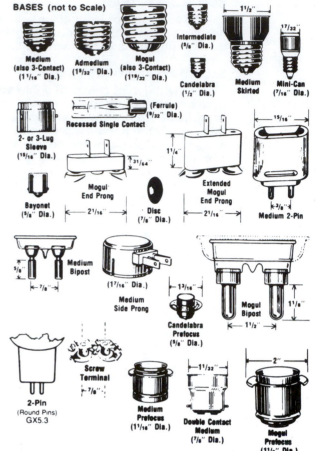

FILAMENTS are designated a letter combination where C is a coiled wire filament, CC is a coiled wire that is itself wound into a larger coil, SR is a straight ribbon filament, the number represents the type of filament support arrangement.

TYPICAL FILAMENTS (not to Scale)

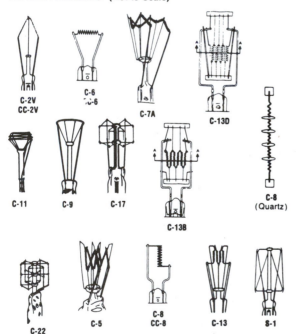

Light Center Length (L.C.L.) is the distance between the center of the filament and the L.C.L. reference plane shown below.

Base Type	LCL Reference Plane
All Screw Bases	Bottom base contact
Mini-Can	Where diameter of ceramic insulator is .531 inches
3-Contact Medium or Mogul	Bottom base contact
Medium Prefocus	Top of base fins
Mogul Prefocus	Top of base fins
Medium Bipost	Base end of bulb (Glass lamps) Bottom of ceramic base (Quartz lamps)
Mogul Bipost	Shoulder of posts (Glass lamps) Bottom of ceramic base (Quartz lamps)
2-Pin Prefocus	Bottom of ceramic base
S.C. or D.C. Bayonet Candelabra	Top of base pins
Medium Bayonet	Top of base pins
S.C. or D.C. Prefocus .	Plane of locating bosses on prefocus collar
Medium 2-Pin	Bottom of metal base shell

Burning Position—Limitations on lamp operating position are shown in the "Description" column. The following abbreviations are used:

BDTH- Burn lamp in base down to horizontal position.

BU - Burn lamp in base up position only.

Bulb Designations consist of letters and numbers, such as S-8 and PAR-46. The letters refer to the shape of the bulb and the number to its approximate diameter in eighths of an inch. **Typical** bulb shapes and sizes are shown in the lamp drawings below. For more detailed product prints consult your nearest local sales office on the back of this catalog. **Dimensions given are in Millimeters.**

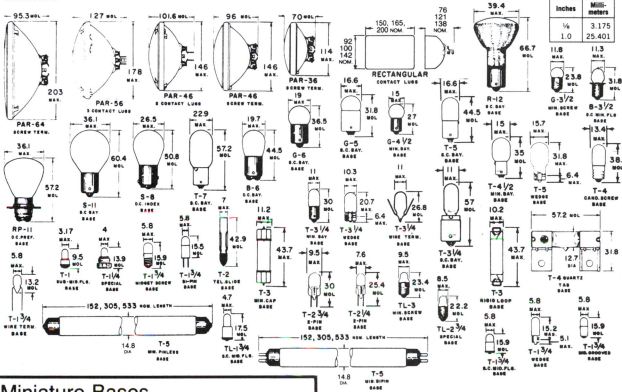

Inches	Millimeters
1/8	3.175
1.0	25.401

Miniature Bases

Bases provide electrical contact to the lamp and, in most cases, also support the lamp in the fixture. For miniature and subminiature lamps, bayonet or wedge base types are generally preferred over screw types when vibration is present. In addition, wedge bases reduce socket size and complexity. Flanged or collared types are usually associated with requirements for filament location.

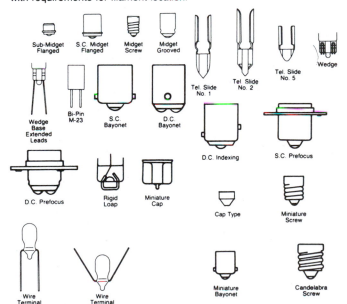

Sealed Beam Bases

Bases provide electrical contact to the lamp. The most common bases for sealed beam lamps are the screw terminal and contact lug types. Other types are also available, as illustrated.

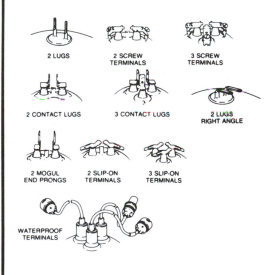

APPENDIX E

SMOKE AND FIRE DETECTORS

This section consists of material from the *Training Manual On Fire Alarm Systems*, publication No. SB 4-1985, by the National Electrical Manufacturers Association.

Section 7

PROPER MAINTENANCE OF FIRE ALARM SYSTEMS

7.1 PERIODIC MAINTENANCE

7.1.1 General

When performing periodic maintenance on the modern fire alarm system, it will be helpful to have the following available:

- The manufacturers installation and user instructions for the system control panel, smoke or flame detectors and other specialized components.
- The "as built" drawings of the system which should include location of all devices, wiring methods, and sequence connections between devices and control equipment.
- The appropriate National Fire Protection Association (NFPA) Standard if maintenance is being performed pursuant to a specific standard. NFPA standards specify frequency of testing for many devices and systems. NFPA Standard 72H entitled, *Guide for Testing Procedures for Protective Signaling Systems* is specifically recommended.
- Any record(s) of tests previously performed as well as the record from tests at system start up to allow a comparison of the electrical measurements being taken with those recently observed. Such comparisons can be a valuable aid for rapid trouble shooting. Additionally, future faults may be prevented by finding the source of a difference in an initiating device circuit resistance, voltage, or current at control unit terminals, and resistance to ground.

7.1.2 System Performance and Integrity

One way to provide a thorough test of system performance and integrity is to repeat the tests outlined in the start-up procedures published by the equipment manufacturer. If not available, Section 6 can be used as a guide.

Authorized Engineering Information 3-8-1985.

7.2 CONTROL UNITS

7.2.1 Printed Circuit Board Assemblies of Modules

Most modern control units do not require periodic adjustment or field repair other than replacement of a module, printed circuit board assembly, or adjustment of battery charging voltage. Defective printed circuit board assemblies or modules can best be serviced at the manufacturer's plant or at their authorized service facility.

When the control panel uses printed circuit boards, care should be taken to clean off excessive dust. The boards should be maintained clean and dry to ensure proper operation.

7.2.2 Relay Maintenance

The maintenance and adjustment of relays should be performed only at the manufacturer's plant or authorized service facility, or by an organization (or person) having the necessary technical experience, components, and equipment. A complete stock of spare relays should, therefore, be kept on hand so that an out-of-adjustment relay, or one with burned out, pitted contacts can be readily replaced.

Covered relays seal out dust and discourage improper adjustment by a novice.

7.2.3 Battery Charger Maintenance

. Low battery voltage is generally an indication that the battery charger or battery needs maintenance. Float-charged and trickle-charged batteries are normally kept in a fully charged condition at all times except when the battery provides power for the fire alarm system during alternating current power failures.

7.2.4 Incandescent Lamp Maintenance (Trouble lamps or lamps in Annunciators)

The life of lamps will be extended when they are illuminated only occasionally and then only for a brief period of time. Lamp life can be shortened if subjected to vibration from nearby equipment.

When lamps are used to illuminate a colored bulls-eye lens or a translucent back-lighted panel, the life of the lamp can be extended by using one with a voltage rating higher than the circuit voltage and still provide sufficient illumination for the purpose, e.g., 28- or 32-volt lamps used on 24-volt circuits, or 145-volt lamps used on 115- or 120-volt circuits. Where lamps are subject to vibration, 6- and 12-volt lamps ususally outlast 24-, 32-, and 48-volt lamps because, generally, the filaments are larger in diameter and shorter in length in the lower voltage lamps.

7.2.5 Fuses

Fuse maintenance consists of checking the fuse holders to make sure that good contact is made with the connectors on each end of the fuse and that they are not corroded. Hot fuses usually indicate either poor contact or overloaded fuses, or both. A supply of fuses of each ampere size and type should be on hand so that the equipment will not remain out of service due to a blown fuse. Fuses which exceed the ratings of those for the circuit should not be used even temporarily.

7.2.6 Circuit Breakers

Magnetic- and thermal-type circuit breakers require very little maintenance. However, they should be kept free of dust.

7.2.7 Condition of Control Unit Cabinets (Dust and Dirt Removal)

The top and bottom of the control unit cabinet and the face of the panel in which the equipment is mounted should be kept free of dust, dirt, and grime. These can cause trouble on relays or other open-contact mechanisms or be a source of corrosion of metal parts.

Neat arrangement of connecting wires from conduits, raceways, or cables at terminating points will reduce maintenance time when it is necessary to trace and disconnect a wire in the control unit from its panel terminal for isolation and test.

Authorized Engineering Information 3-8-1985.

7.3 BATTERY MAINTENANCE

Batteries should be located in a vented clean dry place, preferably on shelves or racks above the floor level, or in cabinets.

The electrolyte level of unsealed storage batteries should be checked regularly. If the electrolyte level is low, distilled water should be added to bring the level up to normal. Most tap water has sufficient metal salts or chemical to appreciably reduce battery life. Therefore, tap water should not be used instead of distilled water in the regular maintenance of the electrolyte level.

In lead-acid batteries, the colored ball floats should be checked regularly. These floats are commonly used in transparent glass or plastic battery cases to indicate the charge condition of the battery. If all balls are floating to the top of the cell, the battery is in a fully, or very nearly fully, charged condition. When all balls are "down," the

battery is in a discharge, or very nearly discharged, condition. Usually there are three colored balls in the pilot cell in a chamber wide and deep enough to float one ball above the other. Usually the red ball is on top.

The top particularly, and the sides of the cells should be kept clean and dry. The terminals between cells should be free of corrosion. The terminals should be covered with an inert acid lubricant such as white petroleum jelly, to retard future corrosion.

The instruction booklet furnished with the battery should be followed carefully and, where recommended, an equalizing charge should be given at the periods recommended. Even though the battery is equipped with ball floats, a specific gravity reading and a voltage reading should be taken at least once or twice a year, or as often as recommended in the battery instruction booklet. A lead-acid battery under a float charge should read about 2.5 volts per cell.

Authorized Engineering Information 3-8-1985.

7.4 NONCODED MANUAL BOXES

Maintenance on noncoded manual fire alarm boxes, with or without the presignal feature, should include periodic operation tests, replacement of broken "breakglass" windows or breakable elements, and checking terminal connections for loose or corroded connections. In supervised fire alarm systems, a broken connection should sound a trouble signal.

A supply of glass rods, plates, and so forth, should be kept on hand for breakglass boxes.

Authorized Engineering Information 3-8-1985.

7.5 CODED MANUAL BOXES

7.5.1 General

All coded boxes should be operated at regular intervals and a log kept showing the location of the station and the date when it was last checked. Both presignal and general alarm features should be checked. The general alarm feature is usually initiated by the operation of a key-actuated switch mounted on the box.

Since code wheels can come loose on their shafts, they should be checked about every five years.

When coded boxes are equipped with contacts for annunciation, it may be necessary to manually restore the contacts to their normal position after each alarm by means of a key or tool reset or by replacing the breakglass.

Some selective coded manual boxes are equipped with

test switches which can be used during regular maintenance. Insertion of the key, plug or special tool in the test key hole or slot and turning it in one direction will sound one tap on single-stroke bells or one blast on vibrating bells or horns; turning in the opposite direction and holding it in that position will permit operation of the station mechanism without sounding an alarm.

7.5.2 Spring-Driven Types

Maintenance of spring-driven coded manual boxes closely follows that for noncoded manual boxes. When this type of selective-coded manual box is operated, a pull handle usually winds a clock spring; when the handle is released, the spring unwinds and drives a gear train and a code wheel. The rotary motion of the code wheel makes or breaks a pair of contacts which transmit coded pulses to the control unit. The spring-actuated code mechanism is usually completely enclosed in a semi-dust-proof enclosure made of a transparent material.

In some boxes, pulling the station lever breaks a glass rod and/or immediately closes or opens one or two sets of contacts, in addition to winding the clock spring. The contacts either operate an annunciator or prevent alarm transmission from stations electrically further away from the control unit by opening or shorting the line beyond. Care should be taken to assure that these contacts function as intended.

Some boxes are wound with a key similar to a clock key. In these boxes the pull-lever mechanism only starts the code-sending mechanism or auxiliary shunt or annunciator contacts, or both. As with the types in which pulling an operating lever winds a clock spring, these prewound stations cause the code wheel to make at least three complete revolutions and sound the associated alarm signals three times.

Fully rewinding of spring-driven boxes after each operation should be a regular part of the maintenance programs.

7.5.3 Motor-Driven Coded Boxes

The maintenance of motor-driven coded boxes is similar to that for the spring-driven type.

Authorized Engineering Information 3-8-1985.

7.6 AUTOMATIC TRANSMITTERS

The maintenance problems encountered with automatically tripped transmitters are similar to those for manual coded boxes, because their construction and method of operation are similar. The automatic transmitter, however, has a remotely-controlled initiating circuit.

Prewound transmitters have either a separate pair of tripping magnets or use a relay which supervises the remote tripping wiring and generally has a built-in mechanical linkage to trip the spring-wound mechanism. When the trip coil is separate from the supervisory relay, the latter usually is mounted separately in the control unit.

Except on a direct current system, motor-driven transmitters usually are driven by a small alternating current, synchronous hysteresis motor.

Transmitters which can be tripped either manually or electrically should be tested both ways at least once every year. See the maintenance instructions for noncoded and coded boxes.

Like some manual boxes, prewound transmitters require rewinding after operation. This should be a regular part of the maintenance program. Some transmitters are equipped with a local trouble buzzer, a light and silencing switch; some have trouble contacts for connection to the main control unit; some have a combination of both. When these transmitters are checked periodically, the trouble signal features should be tested to make sure they are functioning properly.

Authorized Engineering Information 3-8-1985.

7.7 AUTOMATIC HEAT DETECTORS

7.7.1 Fixed-Temperature Detectors

Fixed-temperature-type automatic heat detectors which use fusible elements require little maintenance. This is true for most types of spot heat detectors. Loose or corroded terminal connections are, however, possible. Automatic heat detectors should not be painted.

After fifteen years, each 5th year remove at least two nonrestorable detectors out of every hundred and return them to a testing laboratory for tests similar to those prescribed and regularly performed at the facilities of Underwriters Laboratories Inc. and Factory Mutual. After low-melting-temperature detectors are tested, they should be discarded and replaced with new detectors. Since the test occurs after an initial fifteen-year no-test period and only two percent of the installed detectors are tested every fifth year after that, this is an economical way to establish the reliability of installed detectors. If a failure occurs on one of the tested detectors, additional detectors should be removed and tested until there is proven to exist either a general problem involving faulty detectors or a localized problem involving only one or two bad detectors. A record of the location of all the tested detectors should be kept.

Fixed-temperature detectors of the restorable bimetallic type (slow make or snap action) can be tested while installed on the ceiling and connected to the control unit. A heat lamp (or hair dryer) held within an inch of the detector should cause it to operate. When the lamp is moved away from the detector, the bimetal will cool and the detector contacts will reopen and again be ready for use in detecting the heat of a fire.

7.7.2 Rate-of-Rise Detectors

The tubing of line-type rate-of-rise detectors should be tested for pin-hole leaks about once a year. The manufacturer can provide an air test device to check for such leaks. The tubing system should be tested for operation at least twice a year in accordance with the manufacturer's recommendations.

Spot-type, combination rate-of-rise and fixed temperature detectors are in common use. Rate-of-rise detectors can be tested with a heat lamp. If a lamp is used to test a combination fixed temperature and rate-of-rise detector with a fusible-element, fixed-temperature feature, the lamp should be removed quickly after operation of the detector to prevent melting of the fusible element.

7.7.3 Rate-Compensation Detectors

Rate-compensation detectors are self-restoring units which can be tested with a heat lamp.

7.7.4 Explosion-Proof Detectors

Explosion-proof detectors can be of either the fixed-temperature, rate-of-rise, or rate-compensation type. They are provided within a housing which will contain any explosion causing flame or spark to the surrounding atmosphere. If these detectors are tested in an explosive atmosphere, an explosion-proof lamp unit should be used to generate the heat required to actuate the detector.

Authorized Engineering Information 3-8-1985.

7.8 SMOKE DETECTORS

Smoke detectors require periodic maintenance. All smoke detectors should be physically tested functionally at least semi-annually. Calibration tests should be con-

ducted after one year and then on alternate years thereafter if sensitivity is not changing.

Warning: Smoke detectors are sensitive electronic devices. The specific detector manufacturer's literature should be followed in performing any test or maintenance procedure. Failure to follow the manufacturer's instructions could damage the detector permanently.

Authorized Engineering Information 3-8-1985.

7.9 SPRINKLER WATERFLOW DETECTORS

Two types of sprinkler waterflow detectors are used.

7.9.1 Pressure Operated

Pressure operated detectors are pressure switches which usually are connected into the intermediate chamber of the main sprinkler valve of a sprinkler system. Generally, they have either enclosed mercury-to-metal or microswitch contacts, and they can be tested by opening the "inspector's test valve." This allows water to flow through the intermediate chamber of the main sprinkler valve and to the pressure-type waterflow detector and a hydraulically operated water motor gong, thereby closing or opening the pressure switch contacts and actuating the control unit to which the detector is connected.

Because the contacts are enclosed in a dustproof housing, cleaning is not required. Electrical connections to the switch should be checked every year or two. Some waterflow pressure switches are actuated by a reduction in pressure and are connected to the upper chamber of the main sprinkler valve. Such switches can be tested by first closing the main gate valve and then opening the inspectors test valve, thereby releasing the trapped pressure in the sprinkler piping. This causes an alarm to be sounded upon the opening of a sprinkler head even though the main gate valve is closed and only trapped pressure is released through the sprinkler head. Make certain that the main gate valve is reopened immediately after the test.

7.9.2 Vane Operated

Vane type waterflow switches are installed in insertion holes in the sprinkler risers. The whole assembly is gasketed and bolted with "U" clamps to the piping. The switches generally have either enclosed mercury or microswitch contacts which are actuated by the forward movement of the paddle lever in the riser when water flows through the pipes. Usually there is a pneumatically or electrically operated retard with these switches to prevent water hammer surges from operating the contacts. The retard is generally adjustable and can be set to delay the alarm up to 60 seconds.

Vane-type switches used as waterflow detectors can be tested by opening the inspector's test valve at the highest point in the sprinkler system piping. In this way, the retard timing can be checked.

Authorized Engineering Information 3-8-1985.

7.10 GATE-VALVE SUPERVISORY CONTACTS

Gate-valve supervisory contacts are switches which are actuated by the movement of the threaded valve stem in response to approximately two turns of the gate-valve wheel in the closing direction. The load on the contacts is usually only a fraction of the contact rating, thus, pitting and undue arcing are not likely to occur. Connections to the switch terminals should be checked every year or two, and the two-turn setting at least once a year.

Authorized Engineering Information 3-8-1985.

7.11 OPEN STEM AND YOKE (OS & Y) VALVE SUPERVISORY CONTACTS

Maintenance of OS & Y valve supervisory contacts is similar to that required for gate-valve supervisory switches.

Authorized Engineering Information 3-8-1985.

7.12 PRESSURE SWITCHES

Pressure switches on dry pipe sprinkler systems respond to high or low air pressure and require the same maintenance as those used for wet pipe systems.

Authorized Engineering Information 3-8-1985.

7.13 TANK SWITCHES FOR HIGH AND LOW ALARM SERVICE ON GRAVITY TANKS

Tank switches are actuated by a ball float which rises and falls with the level of the water in the tank. The switch mechanism may be a mercury-to-metal type, or an exposed heavy-duty, snap-action type switch. The contacts are provided in a cast, nonferrous metal housing provided with a gasket to keep moisture out. The contacts of the switch, therefore, seldom need attention and, since the exposed contacts are generally of precious metal, corrosion is not a problem.

The mechanical linkage and both the tightness and corrosive condition of the connections should be checked at least once a year.

Authorized Engineering Information 3-8-1985.

7.14 DIFFERENTIAL PRESSURE SWITCHES

Differential pressure switches are generally of the same type and construction as the high- and low-water or air-pressure switches and, therefore, require similar maintenance. They are used to start an electric pump when the pressure in the sprinkler system piping is not at least 15 pounds higher than the pressure in the water piping from the street supply. The pump builds up pressure in the system side by pumping a small quantity of water taken from the street side of the main sprinkler valve. A second differential pressure switch sounds an alarm whenever the difference in pressure between the street and system sides falls well below 15 pounds (usually due to pump failure). Pump failure may be caused by the loss of power to the motor, a burned-out motor, or a burned-out bearing on either the pump or motor.

Authorized Engineering Information 3-8-1985.

7.15 INSPECTOR'S TEST VALVES

Inspector's test valves are the conventional hydraulic, wheel-actuated valves and require maintenance similar to the valves used on regular water-supply plumbing systems.

Authorized Engineering Information 3-8-1985.

7.16 ALARM HORNS

Direct current alarm horns are usually of the vibrating-diaphragm type. An armature associated with the diaphragm makes and breaks a pair of contacts connected in series with the horn coil and alternately energizes and deenergizes the coil. Generally, a capacitor is connected across the contacts to suppress arcing. An adjusting screw controls the armature air gap and the sound level of the horn. Periodic inspection will disclose contact wear or pitting that may require smoothing with a fine file and a contact burnishing tool. When a screw and lock-nut adjustment is provided to vary the contact gap, it should be adjusted to the specification of the horn manufacturer.

Alternating current, vibrating-diaphragm horns are similar to the direct current types, except they do not require the contacts to alternately energize and de-energize the armature. The zero and peak voltage created by an ac 60-hertz sine wave provides 120 beats per second. These horns do not have contacts, and may be operated in series-connected, supervised alarm indicating circuits. Ten 12-volt horns or twenty 6-volt horns can be operated on a 120-volt circuit.

Electronic horns or sirens are either of the "trumpet" (re-entrant) type with a metal diaphragm, or the cone speaker type. If a cone is torn or warped, the speaker should be replaced.

Authorized Engineering Information 3-8-1985.

7.17 ALARM BELLS

Direct current vibrating bells have contacts which alternately energize and de-energize their coils thereby causing a striker to contact a steel gong shell and produce vibrating bell sounds. Maintenance of contacts and gap spacing are similar to that for horns. The distance between the gong striker and the steel gong shell require infrequent attention. Normally the bell movement is fastened to the base by two or four screws in elongated holes thereby permitting the striker to be adjusted closer or further away from the gong shell.

Like alternating current vibrating horns, alternating current vibrating bells usually have no contacts. They are, however, designed to operate at 60 instead of 120 beats a second by using a permanent magnet or rectifier to cut off one side of the sine wave. This reduced rate provides a cleaner bell sound. Similar to alternating current series horns, alternating current bells can be operated in a series-connected, supervised alarm indicating circuit.

Alternating and direct current single-stroke solenoid bells or chimes usually require less maintenance than other audible signals. Terminal connections need to be checked every five or ten years. When that is done, any dust or dirt which has collected between the plunger and the plunger tube should be blown free with compressed air or a bellows. These bells have no contacts and can be used in series-connected circuits.

Authorized Engineering Information 3-8-1985.

7.18 TROUBLE BELLS AND BUZZERS

Trouble bells and buzzers may be of the alternating or direct current vibrating types, and maintenance is exactly the same as for alarm horns and alarm bells.

Authorized Engineering Information 3-8-1985.

7.19 FIRE DRILL SWITCHES ON SYSTEMS

Fire drill switches are generally of the conventional key-actuated lock-switch type. These switches require no maintenance except replacement when they fail to function properly.

Authorized Engineering Information 3-8-1985.

A P P E N D I X   F
ADDITIONAL WIRING DIAGRAMS

This Appendix Courtesy of Allen-Bradley Co., Industrial Control Division

The Bulletin 609RS manual reversing starters and the Bulletin 609TS manual two-speed starters consist of two standard Bulletin 609 starters mounted in a single enclosure. Internal wiring of these starters provides the necessary connections for interchanging two motor connections in the case of the 609RS or switching to another winding in the case of the 609TS. Bulletin 609TS is for two-speed separate winding motors only. Terminal markings corresponding to those shown on the diagrams will be found on each switch.

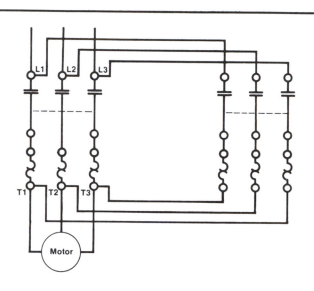

Reversing Starter

Bulletin 609RS

Sizes 0 & 1

3 Phase
2 Phase, 3 Wire

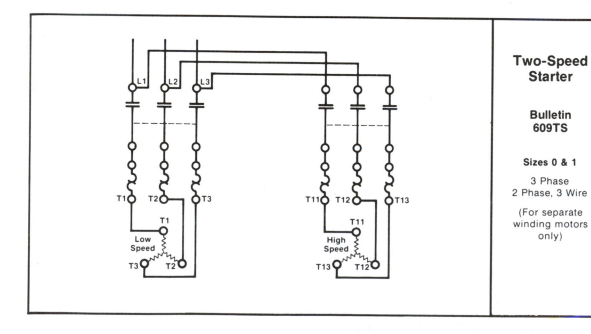

Two-Speed Starter

Bulletin 609TS

Sizes 0 & 1

3 Phase
2 Phase, 3 Wire

(For separate winding motors only)

3-Phase Starters

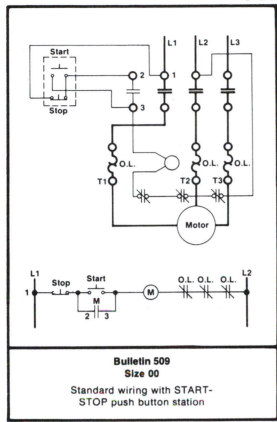

**Bulletin 509
Size 00**

Standard wiring with START-STOP push button station

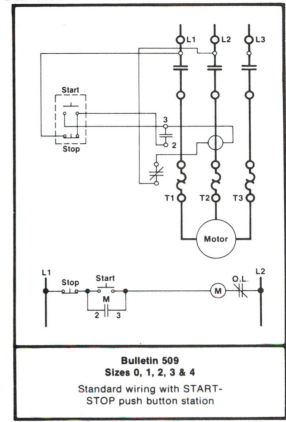

**Bulletin 509
Sizes 0, 1, 2, 3 & 4**

Standard wiring with START-STOP push button station

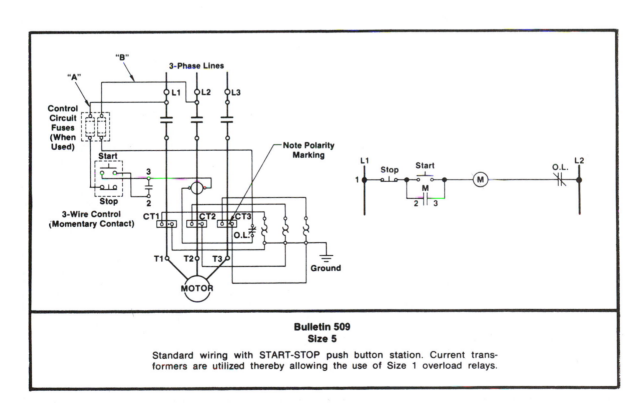

**Bulletin 509
Size 5**

Standard wiring with START-STOP push button station. Current transformers are utilized thereby allowing the use of Size 1 overload relays.

3-Phase Starters

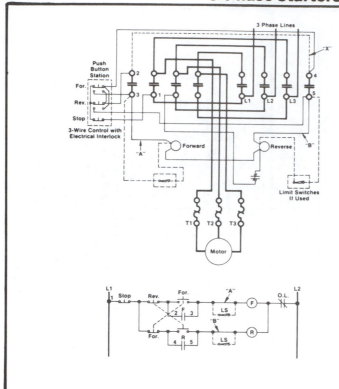

Bulletin 505
Size 00

Standard wiring with "FORWARD-REVERSE-STOP" push button station.

A mechanical interlock is provided, however electrical interlocks are not furnished on size 00 reversing starters. Electrical interlocking can be provided within the push button station, as shown in the diagram. When using this arrangement, wire "X" must be removed.

Limit switches can be added to stop the motor at a certain point in either direction. Connections "A" and "B" must be removed when limit switches are used.

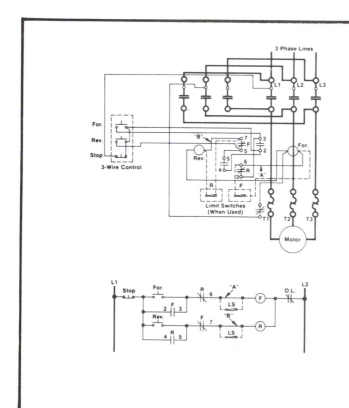

Bulletin 505
Sizes 0, 1, 2, 3 and 4

Standard wiring with "FORWARD-REVERSE-STOP" push button station. The "STOP" button must be depressed before changing directions.

A mechanical interlock and electrical interlocks are supplied as standard on all reversing starters size 0 and larger.

Limit switches can be added to stop the motor at a certain point in either direction. Connections "A" and "B" must be removed when limit switches are used.

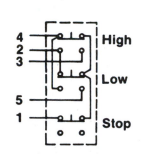

Push button connections to allow starting in either speed and changing from one speed to another without first pressing the "STOP" button.

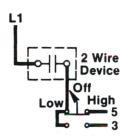

Control by an automatic "two-wire" device. A selector switch is used to determine speed.

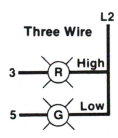

Three Wire

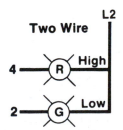

Connections for speed-indicating pilot lights. Can be added to any of the control schemes shown on this page

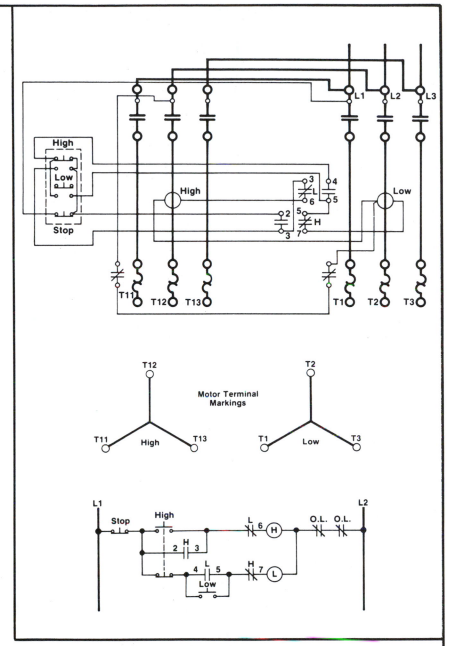

Motor Terminal Markings

For Separate Winding Motors

A typical connection for a Bulletin 520 used with a two speed separate winding motor is shown above. The wiring diagram and line diagram in the above panel illustrate connections for the following method of operation: Motor can be started in either "HIGH" or "LOW" speed. The change from LOW to HIGH can be made without first pressing STOP button. When changing from HIGH to LOW the STOP button must be pressed between speeds.

The pilot device diagrams shown in the side panel illustrate other connections that can be made to obtain different sequences and methods of operation.

Variations With START-STOP Stations

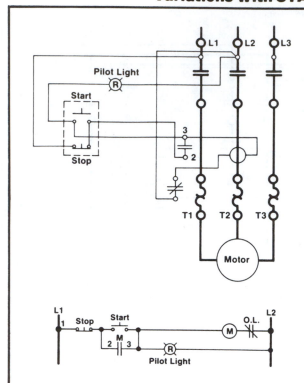

START-STOP Station with Pilot Light to Indicate When Starter is Energized.

A pilot light is to be used with a three-wire "START-STOP" station, so that it will be on when the starter is energized.

The light is shown here as separately mounted, but it can be combined in the same enclosure with the start and stop buttons. Stations combining all three are:

Type of Station	Catalog Number
Standard Duty	
120 or 240 V	800S-2SAP
Heavy Duty	
120 V	800H-2HAR
240 V	800H-2HAP
480 V, 60 Hz	800H-2HAY
600 V, 60 Hz	800H-2HAV
Oiltight	
120 V	800T-2TAR
240 V	800T-2TAP
480 V	800T-2TAY
600 V	800T-2TAV

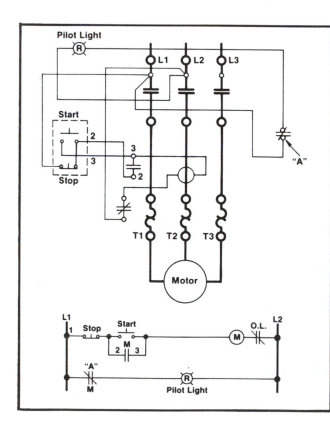

Station with Pilot Light to Indicate When Starter is Deenergized.

If it is necessary for a pilot light to show when the starter is de-energized, this requirement is most easily fulfilled by attaching a normally closed **Bulletin 595 auxiliary contact** to the starter and connecting it between L1 and L2 in series with the pilot light. "A" represents the Bulletin 595 auxiliary contact which can be added to any Allen-Bradley Bulletin 500 Line starter, sizes 0 through 4.

If the pilot light is to be included in the same enclosure with the start and stop buttons, any of the push button stations listed with drawing No. 20 can be used. The Bulletin 595 auxiliary contact has many other uses besides the ones shown here. It can also be used to operate other control circuit devices, interlock starters, etc.

Single Phase Using Standard 3-Phase Starters

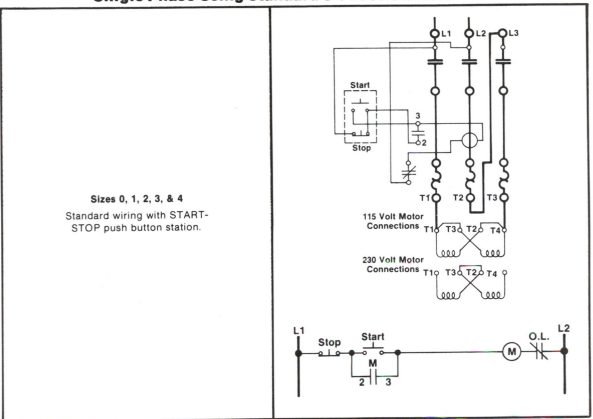

Sizes 0, 1, 2, 3, & 4

Standard wiring with START-STOP push button station.

Variations with START-STOP Stations

More Than One START-STOP Station Used to Control a Single Starter

This is a useful arrangement when a motor must be started and stopped from any of several widely separated locations.

Notice that it would also be possible to use only one "START-STOP" station and have several "STOP" buttons at different locations to serve as emergency stops.

Standard duty "START-STOP" stations are provided with the connections "A" shown in the adjacent diagram. This connection must be removed from all but one of the "START-STOP" stations used.

Heavy duty and oiltight push button stations can also be used but they do not have the wiring connection "A", so it must be added to one of the stations.

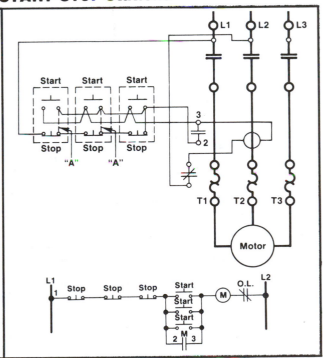

Variations With START-STOP Stations

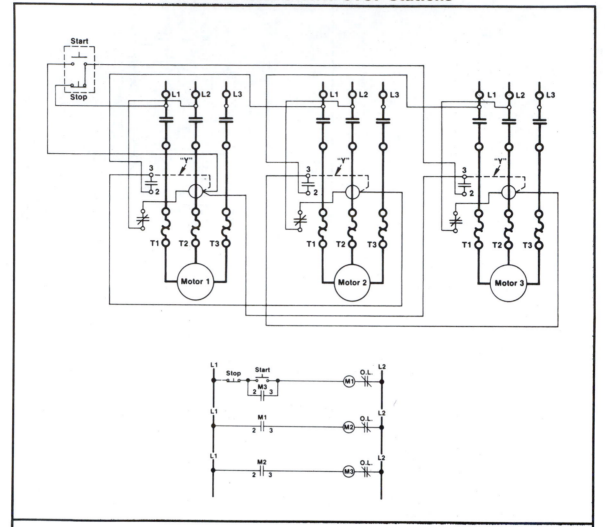

Three Starters are Operated from a Single "START-STOP" Station. An Overload on Any One of the Motors will Drop Out All Three Starters

Three Bulletin 509 solenoid type starters are to be connected so that all are controlled from a single "START-STOP" push button station. A maintained overload on any motor, tripping out the overload relays on its respective starter, will drop out all three starters disconnecting all motors from the line.

Assuming that standard Bulletin 509, Form 2 starters are to be used, then in order to obtain the desired operation, the wiring connection "Y" must be removed from each starter. The control circuits of the several starters are interconnected. It is therefore necessary to disconnect the power to the line terminals of **all** the starters in order to completely disconnect the equipment from line voltage.

Sequence Control

Sequence Control of Two Motors — One to Start and Run for a Short Time After the Other Stops

In this system it is desired to have a second motor started automatically when the first is stopped. The second motor is to run only for a given length of time. Such an application might be found where the second motor is needed to run a cooling fan or a pump.

To accomplish this an off-delay timer (TR) is used. When the start button is pressed, it energizes both M1 and TR. The operation of TR closes its time delay contact but the circuit to M2 is kept open by the opening of the instantaneous contact. As soon as the stop button is pressed, both M1 and TR are dropped out. This closes the instantaneous contact on TR and starts M2. M2 will continue to run until TR times out and the time delay contact opens.

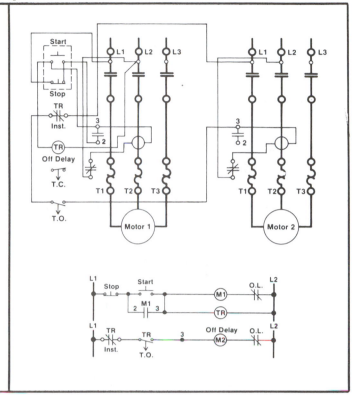

Starters Arranged for Sequence Control of a Conveyor System

The two starters are wired so that M2 cannot be started until M1 is running. This is necessary if M1 is driving a conveyor fed by another conveyor driven by M2. Material from the M2 conveyor would pile up if the M1 conveyor could not move and carry it away.

If a series of conveyors is involved, the control circuits of the additional starters can be interlocked in the same way. That is, M3 would be connected to M2 in the same "step" arrangement that M2 is now connected to M1, and so on.

The M1 stop button or an overload on M1 will stop both conveyors. The M2 stop button or an overload on M2 will stop only M2.

If standard Bul. 509 starters are used, wire "X" must be removed from M2.

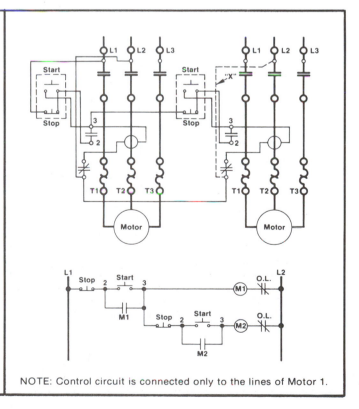

NOTE: Control circuit is connected only to the lines of Motor 1.

Jogging

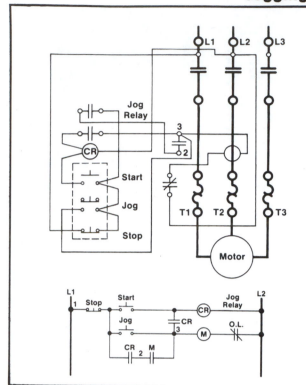

Separate "START-STOP-JOG" with Standard Push Buttons and a JOG Relay

The Bulletin 509 starter is to be operated by a "START-JOG-STOP" push button station.

The purpose of jogging is to have the motor operate only as long as the "JOG" button is held down. The starter must not "lock in" during jogging, and for this reason the "jog relay" is used.

Pushing the "START" button operates the jog relay, causing the starter to lock in through one of the relay contacts. When the "JOG" button is pressed, the starter operates, but this time the relay is not energized and thus the starter will not lock in.

CR represents the "jog relay," a Catalog Number 700-C200.

For a surface mounted heavy duty "START-JOG-STOP" station, specify Catalog Number 800H-3HG. A Catalog Number 800T-3TG can also be used.

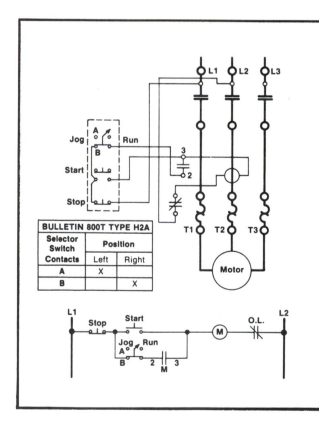

BULLETIN 800T TYPE H2A		
Selector Switch	Position	
Contacts	Left	Right
A	X	
B		X

Combined "START, JOG" and Separate "STOP" With Selector Switch Jogging With a Selector Switch

Here, a three-unit push button station with a START-STOP and selector switch is used. Heavy duty station is Catalog Number 800H-3HW14 and oiltight Catalog Number is 800T-3TW15.

The circuit to the hold-in contact "M" is broken when the selector is in the "JOG" position. The "START" button is used to "JOG" or "RUN" the motor, depending on the position of the selector switch.

Jogging

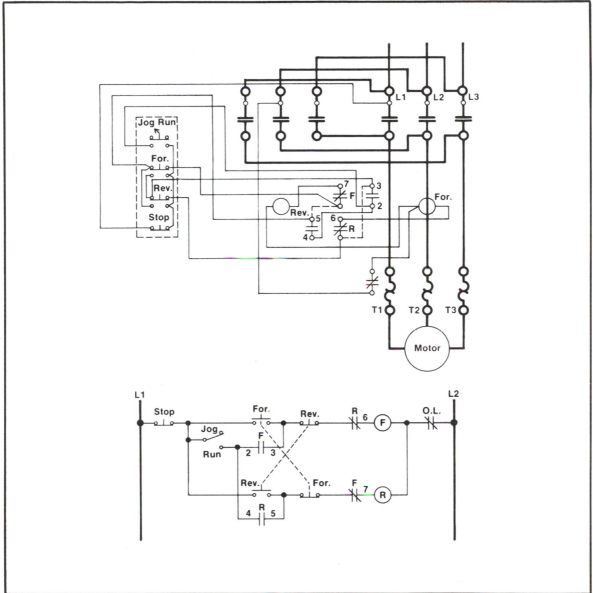

Starting, Stopping and Jogging in Either Direction. Jogging Controlled Through a Jogging Selector Switch

Here, the motor can run normally in either direction or can be jogged in either direction. With the selector in the "RUN" position, the motor can be started in either direction and will stop when the STOP button is pressed. It is not necessary to press the STOP button before changing from forward to reverse.

With the selector in the "JOG" position, the "FORWARD" and "REVERSE" buttons act as jogging buttons. The motor will run in the indicated direction when one of them is pressed but will stop as soon as the button is released.

The wiring of the standard Bulletin 505 must be modified for this type of operation. Note that the wires shown with dotted lines must be removed from the standard starter. The push button station can be either a Bulletin 800H heavy duty, or a Bulletin 800T, oiltight.

Push Button Station Variations

Limit Switch Controls Reversing

Here the direction of the motor is determined by the position of a limit switch. A START-STOP push button station is used to energize the system and the motor will start according to the position of the limit switch. The wiring of the standard Bulletin 505 need not be modified for this type of operation. Limit switch connections are made directly to the electrical interlocks.

It is necessary to use a control relay in this system such as a Catalog Number 700-C200. The limit switch can be any of several in the Bulletin 802T line having one NO and one NC contact.

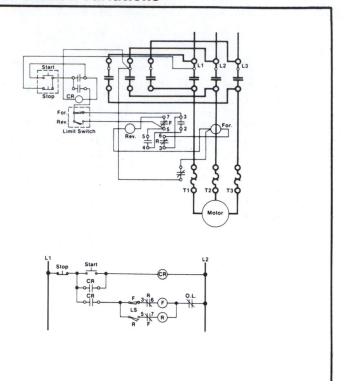

Jogging

Jogging With Relays

In this arrangement for jogging and running in either direction "jogging relays" are used to provide proper jogging. These relays guard against either the forward or reverse contactor locking in during jogging.

The push button station can be a Bulletin 800H, heavy duty, or a Bulletin 800T oiltight. Catalog Number 700-C200 relays may be used.

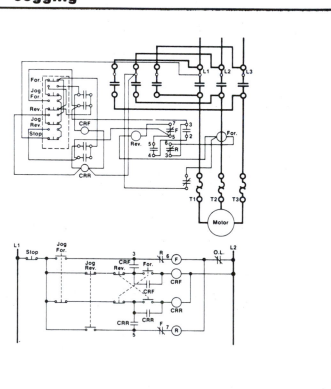

Push Button Station Variations

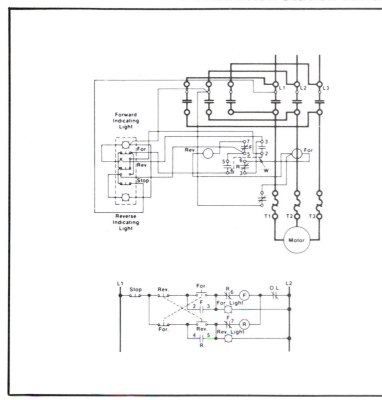

Starting and Stopping in Both Directions. Lights Indicate Direction in Which Motor is Operating

This setup provides exactly the same operation as shown in the previous diagram, except that pilot lights have been added to show which way the motor is running. Once again, standard Bulletin 505 reversing switch can be used if wire "W" is removed. The pilot lights can either be separately mounted or mounted in the push button station. If they are to be mounted in the station a Catalog Number 800H-3HA2P can be used.

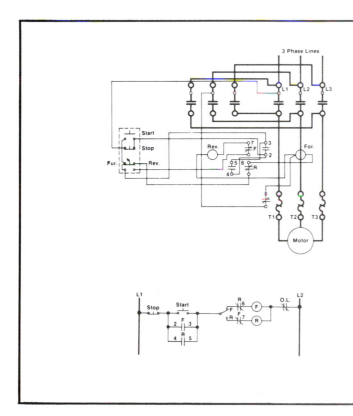

The Motor Runs in a Preselected Direction Which is Determined by The Setting of a Selector Switch

The motor can be run in either direction, but the desired direction must be set on a selector switch before starting. The motor is then operated from a "START-STOP" station as a single direction motor.

The wiring of the standard Bulletin 505 reversing starter must be modified slightly to fill this requirement. Note that the connections which normally lead from the electrical interlock contacts to points 3 and 5 have been removed and that different connections have been made to the electrical interlocks and points 3 and 5.

It is usually most convenient to include the selector switch as part of the push button station. This can be done with either a Bulletin 800H heavy duty station or a Bulletin 800T oiltight station.

A P P E N D I X G
MISCELLANEOUS DATA

NATIONAL ELECTRICAL CODE®

Any person doing electrical installations, modifications, or altering existing electrical systems should be familiar with the **NEC®**. The above-mentioned jobs should be done in compliance with the **NEC®**. To comply, a copy of the latest edition of the **NEC®** should be in every mechanics tool box. Although the entire **NEC®** is important, the following list of articles and sections are probably the ones referred to most frequently.

ARTICLE 90—Introduction including the purpose and scope of the **NEC®**.

ARTICLE 100—Definitions of frequently used terms throughout the **NEC®**.

ARTICLE 110—General requirements for electrical work, including connections, terminations, and working space around electrical equipment.

ARTICLE 210—Branch circuits.

ARTICLE 220—Branch circuit and feeder calculations.

ARTICLE 240—Overcurrent protection.

ARTICLE 250—Grounding: Parts E, F, and G.

ARTICLE 300—General requirements of wiring methods.

ARTICLE 310—Conductors for general wiring.

ARTICLE 346—Rigid metal conduit.

ARTICLE 347—Rigid nonmetallic conduit.

ARTICLE 348—Electrical metallic tubing.

ARTICLE 350—Flexible metal conduit.

ARTICLE 351—Liquid-tight flexible conduit.

ARTICLE 352—Surface raceways.

ARTICLE 362—Wireways.

ARTICLE 364—Busways.

ARTICLE 370—Outlet, device, pull and junction boxes, conduit bodies, and fittings.

ARTICLE 374—Auxiliary gutters.

ARTICLE 380—Switches.

ARTICLE 384—Switchboards and panelboards.

ARTICLE 400—Flexible cords and cables.

ARTICLE 422—Appliances.

ARTICLE 424—Fixed electrical space heating equipment.

ARTICLE 430—Motors.

ARTICLE 440—Air conditioning and refrigeration equipment.

ARTICLE 450—Transformers.

CHAPTER 5—Hazardous areas, specific section as required.

CHAPTER 6—Special equipment such as cranes, hoists, office furnishings, electric welders, data processing systems, etc.

CHAPTER 7—Special conditions such as standby systems, low voltage (less than 50 volts), fire protective signaling systems, etc.

CHAPTER 9—Tables and examples: Tables 1–9, Example #8.

Industrial Power Factor Correction Capacitors

**240 • 480 • 600 Volts
3 Phase • 60 Hertz**

CAPACITOR APPLICATION: The ICS metallized electrode capacitor assembly satisfies the continuing demand for a wide range of capacitor ratings from 5 to 400 KVAR. The ICS capacitor is intended for industrial use to improve the power factor in areas where inductive equipment such as induction motors, transformers, induction furnaces and similar equipment create low power factor conditions. Utility bill savings, improved voltage, improved motor performance, reduced line losses and increased system capacity are all features that can be made available to the industrial customer through the use of ICS capacitors.

NEW METALLIZED ELECTRODE DESIGN: The ICS metallized electrode capacitor represents a unique type of construction using a polypropylene dielectric and metallized electrodes vacuum deposited on both sides of a paper substrate. During vacuum oil processing the paper acts as a wick for thorough oil impregnation to suppress corona and to promote long capacitor life. Voids between the dielectric film

and the electrodes are eliminated with the assurance the film will operate in a complete oil environment free of gas pockets.

NON-PCB DIELECTRIC FLUID: Dykanol XND is a biodegradable, low toxicity, dielectric fluid developed by Cornell-Dubilier Group for film-paper dielectric capacitors. It meets all requirements for environmental compatibility.

SMALLER SIZE/LESS WEIGHT — LOW AS .5 WATTS/KVAR LOSSES: Size reductions are accomplished by realizing the higher voltage stress capability of the polypropylene film which also has a greater energy efficiency with respect to power losses, reducing the temperature rise within the capacitor resulting in reduced watts per KVAR, lower transportation costs and easier handling.

SELF HEALING: Self-healing properties in the event of dielectric breakdown due to excessive voltage stress result from vaporizing the electrode material around the point of failure. The self-healing operation is accomplished in milliseconds

without damage to the dielectric structure.

UL LISTED: The ICS capacitor assembly is UL listed with or without blown fuse indicating lights. The capacitor assembly is also suitable for both indoor and outdoor use. Operation between −40°C to +46°C (−40°F to +115°F) is permissible.

INDICATING LIGHTS/BLOWN FUSE INDICATION: Indicating lights (neon) are also available as an option. The lights provide a quick external inspection by maintenance personnel for blown fuse operation.

EXTERNAL CURRENT LIMITING FUSES: Each capacitor is externally fused with current limiting replaceable fuses rated at 600V with an interrupting capacity of 200,000 amps.

PRESSURE ACTIVATED CIRCUIT INTERRUPTERS: Pressure activated circuit interrupters are built into the capacitors to give maximum assurance the cases will not rupture at end-of-life.

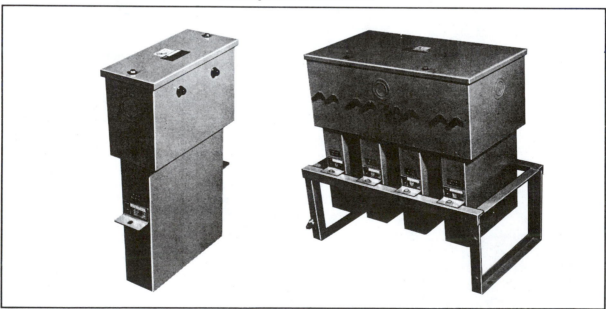

INDIVIDUAL UNIT ASSEMBLY
WITH LIGHTS

MULTIPLE UNIT ASSEMBLY
WITH LIGHTS

Courtesy of Cornell-Dubilier Electronics, Inc.

Industrial Power Factor Correction Capacitors

TYPE ICS

240 • 480 • 600 Volts
3 Phase • 60 Hertz

CLASSES 8114/8119

FEATURES

Indicating Lights: (Optional): A glowing neon light allows for quick external blown fuse operation.

Service Entrance Knockouts: Provisions for 1.25, 1.50, 2.00 and 2.50 inch conduit.

Bus Bar: Aluminum solid bus, with all aluminum hardware, provides connections between capacitors.

Phase Connectors: Multiple Units incorporate three aluminum solderless connectors securely bolted to the aluminum bus. For the individual unit each fuse has an integral solderless connector along with a copper solderless connector for the third phase.

Discharge Resistors: Each capacitor unit has carbon film resistors which reduces the residual voltage to 50 volts or less within one minute after removal of the capacitor from the energized circuit.

Indicating Fuses: Each capacitor unit is protected with two current limiting replaceable fuses. The fuses are rated at 600V with an interrupting capacity of 200,000 amps.

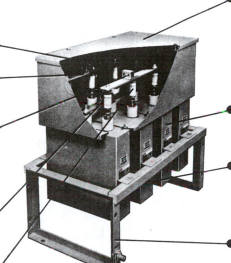

MULTIPLE UNIT ASSEMBLY

Case: The ICS capacitor enclosure uses 18 gauge steel and is jig welded, allowing for quick and easy customer floor or wall mounting. The assembly is completely gasketed providing indoor or outdoor operation. The enclosure is finished ASA gray with a UL recognized paint providing superior corrosion protection.

Capacitor Unit: Each capacitor unit is hermetically sealed in a steel case and incorporates a pressure activated circuit interrupter which virtually eliminates case rupture.

Capacitor Life: The ICS assembly provides maximum air space between capacitor units which enhances heat transfer and capacitor life. The design of ICS capacitors is based on a 20-year life expectancy.

Frame: The capacitor units are supported by a heavy steel angle, completely jig welded, assuring proper alignment and structural integrity of the entire assembly. The welded frame is also painted ASA-70 light gray, for indoor or outdoor installation.

TABLE 4
SUGGESTED MAXIMUM CAPACITOR RATINGS FOR T-FRAME NEMA CLASS B. MOTORS*

Induction Motor Rating (HP)	3600 R/MIN Capacitor Rating (KVAR)	Line Current Reduction (%)	1800 R/MIN Capacitor Rating (KVAR)	Line Current Reduction (%)	1200 R/MIN Capacitor Rating (KVAR)	Line Current Reduction (%)	900 R/MIN Capacitor Rating (KVAR)	Line Current Reduction (%)	720 R/MIN Capacitor Rating (KVAR)	Line Current Reduction (%)	600 R/MIN Capacitor Rating (KVAR)	Line Current Reduction (%)
3	1.5	14	1.5	23	2.5	28	3	38	3	40	4	40
5	2	14	2.5	22	3	26	4	31	4	40	5	40
7.5	2.5	14	3	20	4	21	5	28	5	38	6	45
10	4	14	4	18	5	21	6	27	7.5	36	8	38
15	5	12	5	18	6	20	7.5	24	8	32	10	34
20	6	12	6	17	7.5	19	9	23	10	29	12	30
25	7.5	12	7.5	17	8	19	10	23	12	25	18	30
30	8	11	8	16	10	19	14	22	15	24	22.5	30
40	12	12	13	15	16	19	18	21	22.5	24	25	30
50	15	12	18	15	20	19	22.5	21	24	24	30	30
60	18	12	21	14	22.5	17	26	20	30	22	35	28
75	20	12	23	14	25	15	28	17	33	14	40	19
100	22.5	11	30	14	30	12	35	16	40	15	45	17
125	25	10	36	12	35	12	42	14	45	15	50	17
150	30	10	42	12	40	12	52.5	14	52.5	14	60	17
200	35	10	50	11	50	10	65	13	68	13	90	17
250	40	11	60	10	62.5	10	82	13	87.5	13	100	17
300	45	11	68	10	75	12	100	14	100	13	120	17
350	50	12	75	8	90	12	120	13	120	13	135	15
400	75	10	80	8	100	12	130	13	140	13	150	15
450	80	8	90	8	120	10	140	12	160	14	160	15
500	100	8	120	9	150	12	160	12	180	13	180	15

Applies to three-phase, 60 Hz motors when switched with capacitors as a single unit.
*Taken from IEEE STD 141-1976.

Percent AR is the percent reduction in full-load line current due to capacitors: A capacitor located on the motor side of the overload relay reduces current through the relay. Therefore, a smaller relay may be necessary. The motor-overload relay should be selected on the basis of the motor full-load nameplate current reduced by the percent reduction in line current (percent AR) due to capacitors.

The capacitor size specified in the above table will increase the full load power factor to 95% and larger sizes should not be used without consulting the factory.

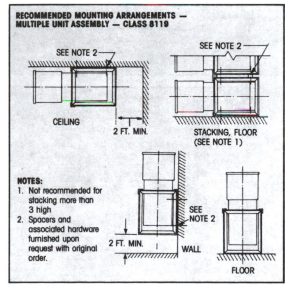

RECOMMENDED MOUNTING ARRANGEMENTS — MULTIPLE UNIT ASSEMBLY — CLASS 8119

SEE NOTE 2
SEE NOTE 2
CEILING
2 FT. MIN.
STACKING, FLOOR (SEE NOTE 1)
SEE NOTE 2
2 FT. MIN.
WALL
FLOOR

NOTES:
1. Not recommended for stacking more than 3 high
2. Spacers and associated hardware furnished upon request with original order.

UP YOUR POWER FACTOR!

Buying power (KW's) from your local power company is not as simple or as cheap as it used to be. The way energy costs have climbed over the last several years, most industrial power users with power systems more than a few years old find themselves paying a monthly bill for KVAR's also, where the power company used to charge only for KW's.

Kilovars are magnetizing power. They are needed for motors and other inductive equipment, *but they do no useful work.*

Instead of "renting" unproductive power from your Utility by the month, why not install and own your own source of KVAR's for a one time charge. Power Factor Correction Capacitors allow you to do this.

The cost is moderate and payback surprisingly short.

Think of it! No more power company excess monthly charges for low power factor or extra KVA demand.

Use this example to measure the actual savings available to you. A simple multiplication will do it.

EXAMPLE:
KW/KVA_1 = .7 OR 70% POWER FACTOR (P.F.)
KW/KVA_2 = .9 OR 90% POWER FACTOR (P.F.)

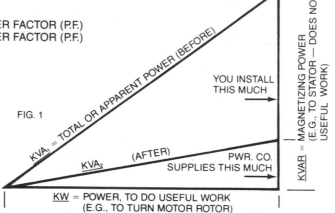

FIG. 1

YOU INSTALL THIS MUCH

PWR. CO. SUPPLIES THIS MUCH

KVA_1 = TOTAL OR APPARENT POWER (BEFORE)

KVA_2 (AFTER)

$KVAR$ = MAGNETIZING POWER (E.G., TO STATOR — DOES NO USEFUL WORK)

KW = POWER, TO DO USEFUL WORK (E.G., TO TURN MOTOR ROTOR)

POWER FACTOR, P.F. $= \dfrac{KW}{KVA}$ EQ 1

KVAR You Install $= \underbrace{\sqrt{KVA_1{}^2 - KW^2}}_{\text{Before}} - \underbrace{\sqrt{KVA_2{}^2 - KW^2}}_{\text{After}}$ EQ 2

For example, suppose you had a connected load of 500 KW, including lots of motors. You require magnetizing power (KVAR's) such that your (measured) power factor is .7, or 70%. The apparent or total power you require is 500/.7 = 714 KVA.

The power company charges you "rent" for all KVAR's below .9, or 90% P.F. How many KVAR's do you need to install to eliminate this monthly penalty charge that does not stop and will likely increase?

From EQ. 1, KVA_1 = 500/.7 = 714 KVA

KVA_2 = 500/.9 = 556 KVA

From EQ. 2, KVAR's you install

$= \sqrt{714^2 - 500^2} - \sqrt{556^2 - 500^2} = 266.5$ KVAR

or, from quick-reference table,

500 x .536 = 268 KVAR

We recommend you install 275 KVAR (closest standard package) of CDE magnetizing power to eliminate the monthly penalty charge.

To estimate your installed cost, use about $13/KVAR, installed cost estimate

= 275 x $13 = $3575

For a utility excess charge for low power factor or extra KVA demand of as little as $150/month, payback would be less than two years and you keep the excess charge after this.

How much excess charge are you paying to your power company?

To determine capacitor KVAR requirements, select your desired power factor and then read the multiplication factor from the Quick Reference Table on page 2.

To determine capacitor KVAR requirements, select your desired power factor and then read the multiplication factor from the Quick Reference Table.

QUICK REFERENCE TABLE

Multiply KW by value shown to determine capacitor KVAR needed to raise power factor to desired level.

Desired Power Factor in Percentage

Original PF	80	81	82	83	84	85	86	87	88	89	90	91	92	93	94	95	96	97	98	99	1.0
56	0.730	0.756	0.782	0.808	0.834	0.860	0.887	0.913	0.940	0.968	0.996	1.024	1.054	1.085	1.117	1.151	1.188	1.229	1.277	1.337	1.480
57	0.692	0.718	0.744	0.770	0.796	0.822	0.849	0.875	0.902	0.930	0.958	0.986	1.016	1.047	1.079	1.113	1.150	1.191	1.239	1.299	1.442
58	0.655	0.681	0.707	0.733	0.759	0.785	0.812	0.838	0.865	0.893	0.921	0.949	0.979	1.010	1.042	1.076	1.113	1.154	1.202	1.262	1.405
59	0.619	0.645	0.671	0.697	0.723	0.749	0.776	0.802	0.829	0.857	0.885	0.913	0.943	0.974	1.006	1.040	1.077	1.118	1.166	1.226	1.369
60	0.583	0.609	0.635	0.661	0.687	0.713	0.740	0.766	0.793	0.821	0.849	0.877	0.907	0.938	0.970	1.004	1.041	1.082	1.130	1.190	1.333
61	0.549	0.575	0.601	0.627	0.653	0.679	0.706	0.732	0.759	0.787	0.815	0.843	0.873	0.904	0.936	0.970	1.007	1.048	1.096	1.156	1.299
62	0.516	0.542	0.568	0.594	0.620	0.646	0.673	0.699	0.725	0.754	0.782	0.810	0.840	0.871	0.903	0.937	0.974	1.015	1.063	1.123	1.266
63	0.483	0.509	0.535	0.561	0.587	0.613	0.640	0.666	0.693	0.721	0.749	0.777	0.807	0.838	0.870	0.904	0.941	0.982	1.030	1.090	1.233
64	0.451	0.474	0.503	0.529	0.555	0.581	0.608	0.634	0.661	0.689	0.717	0.745	0.775	0.806	0.838	0.872	0.909	0.950	0.998	1.068	1.201
65	0.419	0.445	0.471	0.497	0.523	0.549	0.576	0.602	0.629	0.657	0.685	0.713	0.743	0.774	0.806	0.840	0.877	0.918	0.966	1.026	1.169
66	0.388	0.414	0.440	0.466	0.492	0.518	0.545	0.571	0.598	0.626	0.654	0.682	0.712	0.743	0.775	0.809	0.846	0.887	0.935	0.995	1.138
67	0.358	0.384	0.410	0.436	0.462	0.488	0.515	0.541	0.568	0.596	0.624	0.652	0.682	0.713	0.745	0.779	0.816	0.857	0.905	0.965	1.108
68	0.328	0.354	0.380	0.406	0.432	0.458	0.485	0.511	0.538	0.566	0.594	0.622	0.652	0.683	0.715	0.749	0.786	0.827	0.875	0.935	1.078
69	0.299	0.325	0.351	0.377	0.403	0.429	0.456	0.482	0.509	0.537	0.565	0.593	0.623	0.654	0.686	0.720	0.757	0.798	0.846	0.906	1.049
70	0.270	0.296	0.322	0.348	0.374	0.400	0.427	0.453	0.480	0.508	0.536	0.564	0.594	0.625	0.657	0.691	0.728	0.769	0.817	0.877	1.020
71	0.242	0.268	0.294	0.320	0.346	0.372	0.399	0.425	0.452	0.480	0.508	0.536	0.566	0.597	0.629	0.663	0.700	0.741	0.789	0.849	0.992
72	0.214	0.240	0.266	0.292	0.318	0.344	0.371	0.397	0.424	0.452	0.480	0.508	0.538	0.569	0.601	0.635	0.672	0.713	0.761	0.821	0.964
73	0.186	0.212	0.238	0.264	0.290	0.316	0.343	0.369	0.396	0.424	0.452	0.480	0.510	0.541	0.573	0.607	0.644	0.685	0.733	0.793	0.936
74	0.159	0.185	0.211	0.237	0.263	0.289	0.316	0.342	0.369	0.397	0.425	0.453	0.483	0.514	0.546	0.580	0.617	0.658	0.706	0.766	0.909
75	0.132	0.158	0.184	0.210	0.236	0.262	0.289	0.315	0.342	0.370	0.398	0.426	0.456	0.487	0.519	0.553	0.590	0.631	0.679	0.739	0.882
76	0.105	0.131	0.157	0.183	0.209	0.235	0.262	0.288	0.315	0.343	0.371	0.399	0.429	0.460	0.492	0.526	0.563	0.604	0.652	0.712	0.855
77	0.079	0.105	0.131	0.157	0.183	0.209	0.236	0.262	0.289	0.317	0.345	0.373	0.403	0.434	0.466	0.500	0.537	0.578	0.626	0.686	0.829
78	0.052	0.078	0.104	0.130	0.156	0.182	0.209	0.235	0.262	0.290	0.318	0.346	0.376	0.407	0.439	0.473	0.510	0.551	0.599	0.659	0.802
79	0.026	0.052	0.078	0.104	0.130	0.156	0.183	0.209	0.236	0.264	0.292	0.320	0.350	0.381	0.413	0.447	0.484	0.525	0.573	0.633	0.776
80	0.000	0.026	0.052	0.078	0.104	0.130	0.157	0.183	0.210	0.238	0.266	0.294	0.324	0.355	0.387	0.421	0.458	0.499	0.547	0.609	0.750
81		0.000	0.026	0.052	0.078	0.104	0.131	0.157	0.184	0.212	0.240	0.268	0.298	0.329	0.361	0.395	0.432	0.473	0.521	0.581	0.724
82			0.000	0.026	0.052	0.078	0.105	0.131	0.158	0.186	0.214	0.242	0.272	0.303	0.335	0.369	0.406	0.447	0.495	0.555	0.698
83				0.000	0.026	0.052	0.079	0.105	0.132	0.160	0.188	0.216	0.246	0.277	0.309	0.343	0.380	0.421	0.469	0.529	0.672
84					0.000	0.026	0.053	0.079	0.106	0.134	0.162	0.190	0.220	0.251	0.283	0.317	0.354	0.395	0.443	0.503	0.646
85						0.000	0.027	0.053	0.080	0.108	0.136	0.164	0.194	0.225	0.257	0.291	0.328	0.369	0.417	0.477	0.620
86							0.000	0.026	0.053	0.081	0.109	0.137	0.167	0.198	0.230	0.264	0.301	0.342	0.390	0.450	0.593
87								0.000	0.027	0.055	0.083	0.111	0.141	0.172	0.204	0.238	0.275	0.316	0.364	0.424	0.567
88									0.000	0.028	0.056	0.084	0.114	0.145	0.177	0.211	0.248	0.289	0.337	0.397	0.540
89										0.000	0.028	0.056	0.086	0.117	0.149	0.183	0.220	0.261	0.309	0.369	0.512
90											0.000	0.028	0.058	0.089	0.121	0.155	0.192	0.233	0.281	0.341	0.484
91												0.000	0.030	0.061	0.093	0.127	0.164	0.205	0.253	0.313	0.456
92													0.000	0.031	0.063	0.097	0.134	0.175	0.223	0.283	0.426
93														0.000	0.032	0.066	0.103	0.144	0.192	0.252	0.395
94															0.000	0.034	0.071	0.112	0.160	0.220	0.363
95																0.000	0.037	0.079	0.126	0.186	0.329

Original Power Factor in Percentage

EXAMPLE: Raise power factor of 500 KW load from 70% to 90%, installed KVAR = 500 x .536 = 268 KVAR, use 275 KVAR.

NOTE: Applying power capacitors to power systems which include solid-state DC drives or other SCR components can be complex, and should be referred to our application engineers.

Full load current charts

FULL LOAD CURRENTS IN AMPERES SINGLE PHASE DRY-TYPE TRANSFORMERS

FORMULA

$$\text{Single Phase KVA} = \frac{\text{Volts} \times \text{Load Amperes}}{1000}$$

KVA Rating	RATED LINE VOLTAGE									
	120	240	277	480	600	2400	4160	7200	7620	13200
.25	2.08	1.04	.9	0.52	0.42					
.5	4.16	2.08	1.8	1.04	0.84	0.21	0.12			
.75	6.24	3.12	2.7	1.56	1.2	0.3	0.18			
1.0	8.33	4.16	3.6	2.08	1.6	0.4	0.24			
1.5	12.5	6.24	5.4	3.12	2.4	0.6	0.36	.21	.20	.114
2.0	16.66	8.33	7.2	4.16	3.2	0.8	0.48	.28	.26	.151
3.0	25	12.5	10.8	6.1	4.8	1.2	0.72	.42	.39	.23
5.0	41	21	18	10.4	8.3	2.0	1.2	.70	.66	.38
7.5	62	31	27	15.6	12.5	3.1	1.8	1.04	.98	.57
10.0	83	42	36	21	16.5	4.1	2.4	1.39	1.31	.76
15.0	124	62	54	31	25	6.2	3.6	2.10	1.97	1.14
20.0	166	83	72	42	33	8.2	4.8	2.78	2.62	1.5
25.0	208	104	90	52	42	10.4	6	3.48	3.28	1.9
30.0	249	125	108	62	50	12.5	7	4.18	3.94	2.3
37.5	312	156	135	78	62	15.6	9	5.2	4.92	2.8
50	416	208	180	104	84	21	12	6.9	6.56	3.8
75	624	312	270	156	124	31	18	10.4	9.85	5.6
100	830	415	360	207	168	42	24	13.9	13.1	7.5
125	1040	520	450	260	208	52	30	17.3	16.4	9.5
150	1248	624	540	312	248	62	36	20.8	19.7	11.8
167	1390	695	601	348	278	70	40	23.2	21.9	12.6
200	1660	833	720	416	336	84	48	27.8	26.2	15.0
250	2080	1040	900	520	420	105	60	34.8	32.8	19
333	2780	1390	1199	695	555	139	80	46.0	43.6	25
400	3320	1660	1440	830	672	168	96	55.6	52.5	30
500	4160	2080	1800	1040	840	210	120	69.5	65.5	38
600	5000	2500	2160	1250	1000	250	144	83.6	78.7	45
750	6240	3120	2700	1560	1240	310	180	104	98.5	57
1000	8300	4150	3600	2075	1680	420	240	139	131	76
1500	12480	6240	5400	3120	2480	620	360	208	197	113
2000	16600	8300	7200	4150	3360	840	480	278	262	152

FULL LOAD CURRENTS IN AMPERES THREE PHASE DRY-TYPE TRANSFORMERS

FORMULA

$$\text{Three Phase KVA} = \frac{\text{Volts} \times \text{Load Amperes} \times 1.73}{1000}$$

KVA Rating	RATED LINE VOLTAGE										
	120	208	240	480	600	2400	4160	7200	7620	12470	13200
6	28.8	16.6	14.4	7.2	5.8	1.4	.83	.48	.45	.28	.26
9	43.2	25.0	21.6	10.8	8.7	2.2	1.2	.72	.68	.42	.39
10	48.0	27.7	24	12	9.6	2.4	1.4	.8	.76	.46	.44
15	72.0	41.6	36	18	14.4	3.6	2.08	1.2	1.1	.69	.65
20	96	55.5	48	24	19.0	4.8	2.8	1.6	1.5	.92	.9
25	120	69.5	60	30	24.0	6.0	3.5	2.0	1.9	1.16	1.1
30	144	83.0	72	36	28.8	7.2	4.2	2.4	2.3	1.39	1.3
37.5	180	104	90	45	36	9.0	5.2	3.0	2.8	1.74	1.6
45	216	125	108	54	43	10.8	6.2	3.6	3.4	2.08	2.0
50	240	138	120	60	48	12	7.0	4	3.8	2.3	2.2
75	360	208	180	90	72	18	10.4	6	5.7	3.5	3.3
100	480	278	240	120	96	24	14.0	8	7.6	4.6	4.3
112.5	540	312	270	135	108	27	15.6	9	8.5	5.2	4.9
150	720	415	360	180	144	36	21.0	12	11.4	6.9	6.6
200	960	554	480	240	192	48	28.0	16	15.2	9.2	8.6
225	1080	625	540	270	216	54	31.2	18	17.1	10.4	9.8
250	1200	695	600	300	240	60	35.0	20	18.9	11.6	10.8
300	1440	830	720	360	288	72	42.0	24	22.8	13.9	13.2
400	1920	1110	960	480	384	96	55.6	32	30.4	18.5	17.5
500	2400	1380	1200	600	480	120	70	40	38	23.1	22.0
600	2880	1660	1440	720	576	144	84	48	45.6	27.7	26.2
750	3600	2080	1800	900	720	180	104	60	57	34.7	33
1000	4800	2780	2400	1200	960	240	140	80	76	46.2	44
1500	7200	4150	3600	1800	1440	360	208	120	114	69.4	66
2000	9600	5540	4800	2400	1920	480	278	160	151	92.4	87

Temperature Conversion Centi. vs. Fahr.

0 to 100

C	*	F	C	*	F
—17.8	0	32	10.0	50	122.0
—17.2	1	33.8	10.6	51	123.8
—16.7	2	35.6	11.1	52	125.6
—16.1	3	37.4	11.7	53	127.4
—15.6	4	39.2	12.2	54	129.2
—15.0	5	41.0	12.8	55	131.0
—14.4	6	42.8	13.3	56	132.8
—13.9	7	44.6	13.9	57	134.6
—13.3	8	46.4	14.4	58	136.4
—12.8	9	48.2	15.0	59	138.2
—12.1	10	50.0	15.6	60	140.0
—11.7	11	51.8	16.1	61	141.8
—11.1	12	53.6	16.7	62	143.6
—10.6	13	55.4	17.2	63	145.4
—10.0	14	57.2	17.8	64	147.2
—9.44	15	59.0	18.3	65	149.0
—8.89	16	60.8	18.9	66	150.8
—8.33	17	62.6	19.4	67	152.6
—7.78	18	64.4	20.0	68	154.4
—7.22	19	66.2	20.6	69	156.2
—6.67	20	68.0	21.1	70	158.0
—6.11	21	69.8	21.7	71	159.8
—5.56	22	71.6	22.2	72	161.6
—5.00	23	73.4	22.8	73	163.4
—4.44	24	75.2	23.3	74	165.2
—3.89	25	77.0	23.9	75	167.0
—3.33	26	78.8	24.4	76	168.8
—2.78	27	80.6	25.0	77	170.6
—2.22	28	82.4	25.6	78	172.4
—1.67	29	84.2	26.1	79	174.2
—1.11	30	86.0	26.7	80	176.0
0.56	31	87.8	27.2	81	177.8
0	32	89.6	27.8	82	179.6
0.56	33	91.4	28.3	83	181.4
1.11	34	93.2	28.9	84	183.2
1.67	35	95.0	29.4	85	185.0
2.22	36	96.8	30.0	86	186.8
2.78	37	98.6	30.6	87	188.6
3.33	38	100.4	31.1	88	190.4
3.89	39	102.2	31.7	89	192.2
4.44	40	104.0	32.2	90	194.0
5.00	41	105.8	32.8	91	195.8
5.56	42	107.6	33.3	92	197.6
6.11	43	109.4	33.9	93	199.4
6.67	44	111.2	34.4	94	201.2
7.22	45	113.0	35.0	95	203.0
7.78	46	114.8	35.6	96	204.8
8.33	47	116.6	36.1	97	206.6
8.89	48	118.4	36.7	98	208.4
9.44	49	120.2	37.2	99	210.2
			37.8	100	212.0

100 to 1000

C	*	F	C	*	F
38	100	212	260	500	932
43	110	230	266	510	950
49	120	248	271	520	968
54	130	266	277	530	986
60	140	284	282	540	1004
66	150	302	288	550	1022
71	160	320	293	560	1040
77	170	338	299	570	1058
82	180	356	304	580	1076
88	190	374	310	590	1094
93	200	392	316	600	1112
99	210	410	321	610	1130
100	212	413	327	620	1148
104	220	428	332	630	1166
110	230	446	338	640	1184
116	240	464	343	650	1202
121	250	482	349	660	1220
127	260	500	354	670	1238
132	270	518	360	680	1256
138	280	536	366	690	1274
143	290	554	371	700	1292
149	300	572	377	710	1310
154	310	590	382	720	1328
160	320	608	388	730	1346
166	330	626	393	740	1364
171	340	644	399	750	1382
177	350	662	404	760	1400
182	360	680	410	770	1418
188	370	698	416	780	1436
193	380	716	421	790	1454
199	390	734	427	800	1472
204	400	752	432	810	1490
210	410	770	438	820	1508
216	420	788	443	830	1526
221	430	806	449	840	1544
227	440	824	454	850	1562
232	450	842	460	860	1580
238	460	860	466	870	1598
243	470	878	471	880	1616
249	480	896	477	890	1634
254	490	914	482	900	1652
			488	910	1670
			493	920	1688
			499	930	1706
			504	940	1724
			510	950	1742
			516	960	1760
			521	970	1778
			527	980	1796
			532	990	1814
			538	1000	1832

1000 to 2000

C	*	F	C	*	F
538	1000	1832	816	1500	2732
543	1010	1850	821	1510	2750
549	1020	1868	827	1520	2768
554	1030	1886	832	1530	2786
560	1040	1904	838	1540	2804
566	1050	1922	843	1550	2822
571	1060	1940	849	1560	2840
577	1070	1958	854	1570	2858
582	1080	1976	860	1580	2876
588	1090	1994	866	1590	2894
593	1100	2012	871	1600	2912
599	1110	2030	877	1610	2930
604	1120	2048	882	1620	2948
610	1130	2066	888	1630	2966
616	1140	2084	893	1640	2984
621	1150	2102	899	1650	3002
627	1160	2120	904	1660	3020
632	1170	2138	910	1670	3038
638	1180	2156	916	1680	3056
643	1190	2174	921	1690	3074
649	1200	2192	927	1700	3092
654	1210	2210	932	1710	3110
660	1220	2228	938	1720	3128
666	1230	2246	943	1730	3146
671	1240	2264	949	1740	3164
677	1250	2282	954	1750	3182
682	1260	2300	960	1760	3200
688	1270	2318	966	1770	3218
693	1280	2336	971	1780	3236
699	1290	2354	977	1790	3254
704	1300	2372	982	1800	3272
710	1310	2390	988	1810	3290
716	1320	2408	993	1820	3308
721	1330	2426	999	1830	3326
727	1340	2444	1004	1840	3344
732	1350	2462	1010	1850	3362
738	1360	2480	1016	1860	3380
743	1370	2498	1021	1870	3398
749	1380	2516	1027	1880	3416
754	1390	2534	1032	1890	3434
760	1400	2552	1038	1900	3452
766	1410	2570	1043	1910	3470
771	1420	2588	1049	1920	3488
777	1430	2606	1054	1930	3506
782	1440	2624	1060	1940	3524
788	1450	2642	1066	1950	3542
793	1460	2660	1071	1960	3560
799	1470	2678	1077	1970	3578
804	1480	2696	1082	1980	3596
810	1490	2714	1088	1990	3614
			1093	2000	3632

2000 to 3000

C	*	F	C	*	F
1093	2000	3632	1371	2500	4532
1099	2010	3650	1377	2510	4550
1104	2020	3668	1382	2520	4568
1110	2030	3686	1388	2530	4586
1116	2040	3704	1393	2540	4604
1121	2050	3722	1399	2550	4622
1127	2060	3740	1404	2560	4640
1132	2070	3758	1410	2570	4658
1138	2080	3776	1416	2580	4676
1143	2090	3794	1421	2590	4694
1149	2100	3812	1427	2600	4712
1154	2110	3830	1432	2610	4730
1160	2120	3848	1438	2620	4748
1166	2130	3866	1443	2630	4766
1171	2140	3884	1449	2640	4784
1177	2150	3902	1454	2650	4802
1182	2160	3920	1460	2660	4820
1188	2170	3938	1466	2670	4838
1193	2180	3956	1471	2680	4856
1199	2190	3974	1477	2690	4874
1204	2200	3992	1482	2700	4892
1210	2210	4010	1488	2710	4910
1216	2220	4028	1493	2720	4928
1221	2230	4046	1499	2730	4946
1227	2240	4064	1504	2740	4964
1232	2250	4082	1510	2750	4982
1238	2260	4100	1516	2760	5000
1243	2270	4118	1521	2770	5018
1249	2280	4136	1527	2780	5036
1254	2290	4154	1532	2790	5054
1260	2300	4172	1538	2800	5072
1266	2310	4190	1543	2810	5090
1271	2320	4208	1549	2820	5108
1277	2330	4226	1554	2830	5126
1282	2340	4244	1560	2840	5144
1288	2350	4262	1566	2850	5162
1293	2360	4280	1571	2860	5180
1299	2370	4298	1577	2870	5198
1304	2380	4316	1582	2880	5216
1310	2390	4334	1588	2890	5234
1316	2400	4352	1593	2900	5252
1321	2410	4370	1599	2910	5270
1327	2420	4388	1604	2920	5288
1332	2430	4406	1610	2930	5306
1338	2440	4424	1616	2940	5324
1343	2450	4442	1621	2950	5342
1349	2460	4460	1627	2960	5360
1354	2470	4478	1632	2970	5378
1360	2480	4496	1638	2980	5396
1366	2490	4514	1643	2990	5414
			1649	3000	5432

Note: — The numbers in column * refer to the temperature either in degrees Centigrade or Fahrenheit which it is desired to convert into the other scale. If converting from Fahrenheit degrees to Centigrade degrees, the equivalent temperature will be found in the left column; while if converting from degrees Centigrade to degrees Fahrenheit, the answer will be found in the column on the right.

Conversion formulas:

$$\frac{C}{100} = \frac{F-32}{180} = \frac{Kelvin-273.16}{100} = \frac{Rankine-491.7}{180} = \frac{Reaumur}{80}$$

INTERPOLATION FACTORS

C	*	F	C	*	F
0.56	1	1.8	3.33	6	10.8
1.11	2	3.6	3.89	7	12.6
1.67	3	5.4	4.44	8	14.4
2.22	4	7.2	5.00	9	16.2
2.78	5	9.0	5.56	10	18.0

Linear Conversion

Inches to Millimeters

(1 inch = 25.4 millimeters)

In.	0	1/16	1/8	3/16	1/4	5/16	3/8	7/16	1/2	9/16	5/8	11/16	3/4	13/16	7/8	15/16
0	0.0	1.6	3.2	4.8	6.4	7.9	9.5	11.1	12.7	14.3	15.9	17.5	19.1	20.6	22.2	23.8
1	25.4	27.0	28.6	30.2	31.8	33.3	34.9	36.5	38.1	39.7	41.3	42.9	44.5	46.0	47.6	49.2
2	50.8	52.4	54.0	55.6	57.2	58.7	60.3	61.9	63.5	65.1	66.7	68.3	69.9	71.4	73.0	74.6
3	76.2	77.8	79.4	81.0	82.6	84.1	85.7	87.3	88.9	90.5	92.1	93.7	95.3	96.8	98.4	100.0
4	101.6	103.2	104.8	106.4	108.0	109.5	111.1	112.7	114.3	115.9	117.5	119.1	120.7	122.2	123.8	125.4
5	127.0	128.6	130.2	131.8	133.4	134.9	136.5	138.1	139.7	141.3	142.9	144.5	146.1	147.6	149.2	150.8
6	152.4	154.0	155.6	157.2	158.8	160.3	161.9	163.5	165.1	166.7	168.3	169.9	171.5	173.0	174.6	176.2
7	177.8	179.4	181.0	182.6	184.2	185.7	187.3	188.9	190.5	192.1	193.7	195.3	196.9	198.4	200.0	201.6
8	203.2	204.8	206.4	208.0	209.6	211.1	212.7	214.3	215.9	217.5	219.1	220.7	222.3	223.8	225.4	227.0
9	228.6	230.2	231.8	233.4	235.0	236.5	238.1	239.7	241.3	242.9	244.5	246.1	247.7	249.2	250.8	252.4
10	254.0	255.6	257.2	258.8	260.4	261.9	263.5	265.1	266.7	268.3	269.9	271.5	273.1	274.6	276.2	277.8
11	279.4	281.0	282.6	284.2	285.8	287.3	288.9	290.5	292.1	293.7	295.3	296.9	298.5	300.0	301.6	303.2
12	304.8	306.4	308.0	309.6	311.2	312.7	314.3	315.9	317.5	319.1	320.7	322.3	323.9	325.4	327.0	328.6
13	330.2	331.8	333.4	335.0	336.6	338.1	339.7	341.3	342.9	344.5	346.1	347.7	349.3	350.8	352.4	354.0
14	355.6	357.2	358.8	360.4	362.0	363.5	365.1	366.7	368.3	369.9	371.5	373.1	374.7	376.2	377.8	379.4
15	381.0	382.6	384.2	385.8	387.4	388.9	390.5	392.1	393.7	395.3	396.9	398.5	400.1	401.6	403.2	404.8
16	406.4	408.0	409.6	411.2	412.8	414.3	415.9	417.5	419.1	420.7	422.3	423.9	425.5	427.0	428.6	430.2
17	431.8	433.4	435.0	436.6	438.2	439.7	441.3	442.9	444.5	446.1	447.7	449.3	450.9	452.4	454.0	455.6
18	457.2	458.8	460.4	462.0	463.6	465.1	466.7	468.3	469.9	471.5	473.1	474.7	476.3	477.8	479.4	481.0
19	482.6	484.2	485.8	487.4	489.0	490.5	492.1	493.7	495.3	496.9	498.5	500.1	501.7	503.2	504.8	506.4
20	508.0	509.6	511.2	512.8	514.4	515.9	517.5	519.1	520.7	522.3	523.9	525.5	527.1	528.6	530.2	531.8
21	533.4	535.0	536.6	538.2	539.8	541.3	542.9	544.5	546.1	547.7	549.3	550.9	552.5	554.0	555.6	557.2
22	558.8	560.4	562.0	563.6	565.2	566.7	568.3	569.9	571.5	573.1	574.7	576.3	577.9	579.4	581.0	582.6
23	584.2	585.8	587.4	589.0	590.6	592.1	593.7	595.3	596.9	598.5	600.1	601.7	603.3	604.8	606.4	608.0
24	609.6	611.2	612.8	614.4	616.0	617.5	619.1	620.7	622.3	623.9	625.5	627.1	628.7	630.2	631.8	633.4
25	635.0	636.6	638.2	639.8	641.4	642.9	644.5	646.1	647.7	649.3	650.9	652.5	654.1	655.6	657.2	658.8
26	660.4	662.0	663.6	665.2	666.8	668.3	669.9	671.5	673.1	674.7	676.3	677.9	679.5	681.0	682.6	684.2
27	685.8	687.4	689.0	690.6	692.2	693.7	695.3	696.9	698.5	700.1	701.7	703.3	704.9	706.4	708.0	709.6
28	711.2	712.8	714.4	716.0	717.6	719.1	720.7	722.3	723.9	725.5	727.1	728.7	730.3	731.8	733.4	735.0
29	736.6	738.2	739.8	714.4	743.0	744.5	746.1	747.7	749.3	750.9	752.5	754.1	755.7	757.2	758.8	760.4
30	762.0	763.6	765.2	766.8	768.4	769.9	771.5	773.1	774.7	776.3	777.9	779.5	781.1	782.6	784.2	785.8
31	787.4	789.0	790.6	792.2	793.8	795.3	796.9	798.5	800.1	801.7	803.3	804.9	806.5	808.0	809.6	811.2
32	812.8	814.4	816.0	817.6	819.2	820.7	822.3	823.9	825.5	827.1	828.7	830.3	831.9	833.4	835.0	836.6
33	838.2	839.8	841.4	843.0	844.6	846.1	847.7	849.3	850.9	852.5	854.1	855.7	857.3	858.8	860.4	862.0
34	863.6	865.2	866.8	868.4	870.0	871.5	873.1	874.7	876.3	877.9	879.5	881.1	882.7	884.2	885.8	887.4
35	889.0	890.6	892.2	893.8	895.4	896.9	898.5	900.1	901.7	903.3	904.9	906.5	908.1	909.6	911.2	912.8
36	914.4	916.0	917.6	919.2	920.8	922.3	923.9	925.5	927.1	928.7	930.3	931.9	933.5	935.0	936.6	938.2
37	939.8	941.4	943.0	944.6	946.2	947.7	949.3	950.9	952.5	954.1	955.7	957.3	958.9	960.4	962.0	963.6
38	965.2	966.8	968.4	970.0	971.6	973.1	974.7	976.3	977.9	979.5	981.1	982.7	984.3	985.8	987.4	989.0
39	990.6	992.2	993.8	995.4	997.0	998.5	1000.1	1001.7	1003.3	1004.9	1006.5	1008.1	1009.7	1011.2	1012.8	1014.4
40	1016.0	1017.6	1019.2	1020.8	1022.4	1023.9	1025.5	1027.1	1028.7	1030.3	1031.9	1033.5	1035.1	1036.6	1038.2	1039.8
41	1041.4	1043.0	1044.6	1046.2	1047.8	1049.3	1050.9	1052.5	1054.1	1055.7	1057.3	1058.9	1060.5	1062.0	1063.6	1065.2
42	1066.8	1068.4	1070.0	1071.6	1073.2	1074.7	1076.3	1077.9	1079.5	1081.1	1082.7	1084.3	1085.9	1087.4	1089.0	1090.6
43	1092.2	1093.8	1095.4	1097.0	1098.6	1100.1	1101.7	1103.3	1104.9	1106.5	1108.1	1109.7	1111.3	1112.8	1114.4	1116.0
44	1117.6	1119.2	1120.8	1122.4	1124.0	1125.5	1127.1	1128.7	1130.3	1131.9	1133.5	1135.1	1136.7	1138.2	1139.8	1141.4
45	1143.0	1144.6	1146.2	1147.8	1149.4	1150.9	1152.5	1154.1	1155.7	1157.3	1158.9	1160.5	1162.1	1163.6	1165.2	1166.8
46	1168.4	1170.0	1171.6	1173.2	1174.8	1176.3	1177.9	1179.5	1181.1	1182.7	1184.3	1185.9	1187.5	1189.0	1190.6	1192.2
47	1193.8	1195.4	1197.0	1198.6	1200.2	1201.7	1203.3	1204.9	1206.5	1208.1	1209.7	1211.3	1212.9	1214.4	1216.0	1217.6
48	1219.2	1220.8	1222.4	1224.0	1225.6	1227.1	1228.7	1230.3	1231.9	1233.5	1235.1	1236.7	1238.3	1239.8	1241.4	1243.0
49	1244.6	1246.2	1247.8	1249.4	1251.0	1252.5	1254.1	1255.7	1257.3	1258.9	1260.5	1262.1	1263.7	1265.2	1266.8	1268.4
50	1270.0	1271.6	1273.2	1274.8	1276.4	1277.9	1279.5	1281.1	1282.7	1284.3	1285.9	1287.5	1289.1	1290.6	1292.2	1293.8

LAWS OF EXPONENTS

The International Symbols Committee has adopted prefixes for denoting decimal multiples of units. The National Bureau of Standards has followed the recommendations of this committee, and has adopted the following list of prefixes:

Numbers	Powers of ten	Prefixes	Symbols
1,000,000,000,000	10^{12}	tera	T
1,000,000,000	10^9	giga	G
1,000,000	10^6	mega	M
1,000	10^3	kilo	k
100	10^2	hecto	h
10	10	deka	da
.1	10^{-1}	deci	d
.01	10^{-2}	centi	c
.001	10^{-3}	milli	m
.000001	10^{-6}	micro	u
.000000001	10^{-9}	nano	n
.000000000001	10^{-12}	pico	p
.000000000000001	10^{-15}	femto	f
.000000000000000001	10^{-18}	atto	a

To multiply like (with same base) exponential quantities, add the exponents. In the language of alegebra the rule is $a^m \times a^n = a^{m+n}$

$$10^4 \times 10^2 = 10^{4+2} = 10^6$$

$$0.003 \times 825.2 = 3 \times 10^{-3} \times 8.252 \times 10^2$$

$$= 24.756 \times 10^{-1} = 2.4756$$

To divide exponential quantities, subtract the exponents. In the language of algebra the rule is

$$\frac{a^m}{a^n} = a^{m-n} \text{ or}$$

$$10^8 \div 10^2 = 10^6$$

$$3,000 \div 0.015 = (3 \times 10^3) \div (1.5 \times 10^{-2})$$

$$= 2 \times 10^5 = 200,000$$

To raise an exponential quantity to a power, multiply the exponents. In the languague of algebra $(x^m)^n = x^{mn}$.

$$(10^3)^4 = 10^{3 \times 4} = 10^{12}$$

$$2,500^2 = (2.5 \times 10^3)^2 = 6.25 \times 10^6 = 6,250,000$$

Any number (except zero) raised to the zero power is one. In the language of algebra $x^0 = 1$

$$x^3 \div x^3 = 1$$

$$10^4 \div 10^4 = 1$$

Any base with a negative exponent is equal to 1 divided by the base with an equal positive exponent. In the language of algebra $x^{-a} = \frac{1}{x^a}$

$$10^{-2} = \frac{1}{10^2} = \frac{1}{100}$$

$$5a^{-3} = \frac{5}{a^3}$$

$$(6a)^{-1} = \frac{1}{6a}$$

To raise a product to a power, raise each factor of the product to that power.

$$(2 \times 10)^2 = 2^2 \times 10^2$$

$$3,000^3 = (3 \times 10^3)^3 = 27 \times 10^9$$

To find the nth root of an exponential quantity, divide the exponent by the index of the root. Thus, the nth root of $a^m = a^{m/n}$.

$$\sqrt{x^6} = x^{6/2} = x^3$$

$$\sqrt[3]{64 \times 10^3} = 4 \times 10 = 40$$

A P P E N D I X H
BATTERY SULPHATION*

A. Undercharging or Neglect of Equalizing Charge

Repeated partial charges that do not thoroughly mix the electrolyte are a cause of sulphation. It is extremely difficult in normal battery operation to determine when sulphation begins, and only by giving equalizing charges to individual cells and comparing specific gravity and voltage readings can the condition be detected in the early stages and corrected.

B. Standing in a Partially or Completely Discharged Condition

A battery left standing in a partially discharged condition will allow sulphate to form in the plates. Batteries should be fully charged as soon as possible after discharge, and no batteries should be allowed to stand in a completely discharged condition for more than 24 hours, or when temperatures are below freezing.

C. Low Electrolyte Level

Low electrolyte level, permitting the plates to become exposed, allows the sulphate to form and harden.

D. Adding Acid

Adding acid to a cell in which sulphation exists will aggravate the condition.

E. High Specific Gravity

Normally, the higher the specific gravity of a fully charged cell, the greater the possibility of sulphation and the more difficult it is to reduce. Cells having a specific gravity of more than 0.015 above average are likely to incur sulphation.

F. High Temperature

High temperatures accelerate sulphation, particularly in an idle, partially discharged battery.

All cells of a sulphated battery give low specific gravity and voltage readings, and the bat-

*This Appendix is Courtesy of GNB, Inc.

tery will not become fully charged after normal charging. Internal inspection will disclose that the negative plates have a slate-like appearance and feel hard and gritty. A good fully charged negative plate is spongy to the touch and gives a metallic sheen when stroked with a finger nail or knife. A sulphated positive plate is a lighter brown color than a normal plate.

The internal inspection should be made after a normal charge, since a discharged plate is always somewhat sulphated.

G. Treatment of Sulphated Batteries

Careful attention to the following procedure will often restore a sulphated battery to a good condition.

1. Thoroughly clean battery.

2. Add water to bring the electrolyte to the proper level.

3. Charge battery at the prescribed finishing rate until full ampere-hour capacity has been put in the battery, based on the six-hour rate. If the temperature rises above 110° F during these procedures, reduce the charge rate accordingly. If a cell gives low readings (0.20 volts less than the average cell voltage of the battery), remove and repair the cell.

4. Continue the charge at the finishing rate after completion of the full ampere-hour capacity charge until specfic gravity shows no change for a four-hour period with readings taken hourly. Record voltage and corrected specific gravity readings. These readings indicate the state of charge of the battery.

5. Observe the following test procedure.

   a. Give battery a test discharge and record time at which the discharge test is started.

   b. During the test, individual cell voltages and overall battery voltages are recorded at intervals. The first interval should be 15 minutes after starting the test, and then at each hour from starting time until voltage of any one cell reaches 0.05 volts above the termination value (1.70 volts).

   c. After conditions described in Step b have been reached, take voltage readings at 15 minute intervals.

   d. Record the time when each cell voltage goes below termination value.

   e. Stop the test discharge when the majority of the cells reach termination voltage and before any single cell goes into reversal.

   f. Record the specific gravity of each cell immediately after terminating the test discharge. These readings will determine whether the battery cells are uniform or if any one or more cells are low in capacity. If the battery is uniform and delivers 80 percent or more of its rated capacity, the battery can be returned to service.

6. If the battery at this point gives rated capacity, no further treatment is required other than normal recharge and equalization of specific gravity.

7. If the battery does not deliver rated capacity, continue the discharge without adjusting the discharge rate until one or more cells reach 1.0 volts.

8. Recharge the battery at the finishing rate, again charging until there is no further rise in specific gravity over a four-hour period. Take readings hourly.

9. Discharge again at the six-hour rate, and if the battery gives at least 80% capacity, recharge and put into service.

10. If this procedure does not result in at least 80% capacity, repeat immediately.

11. If the battery has not responded to steps 1–10, it is sulphated to the point where it is impractical to attempt further treatment. The battery should be replaced.

A P P E N D I X I

BASIC ANNUNCIATOR SYSTEM

This Appendix Courtesy of Panalarm Division, Ametek, Inc.

SELF-POLICING
INDUSTRIAL ANNUNCIATOR SYSTEMS

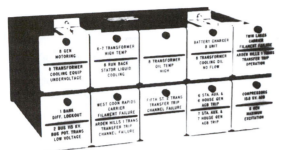

Model 1025 Twinpoint Cabinet

◄ TWINPOINT MODEL 10

With Magna-Plac nameplates, M194-W, providing nearly 5 sq. in. of engraving area for each point.

With special nameplates, A194-W. Where Sequence AF with Model SC-10 plug-in relays are specified, note that a Model ACS-PB pushbutton and Model ACSF-1 flasher may occupy one position. With Model SC-100 plug-in relays, lamp test is standard, and ACS-PB41 can be used with flasher. ▶

Model 1023 Twinpoint Cabinet

HOW ANNUNCIATORS FUNCTION

WHAT THEY ARE

The Annunciator is the central collection point of a persistent surveillance system embracing sensing devices of many types and functions. Pressure and Temperature Switches, Undervoltage Relays, Gas Analyzers, Conductivity Relays and thousands of other unrelated devices actuate the Annunciator by means of their respective electrical contacts.

WHAT THEY DO

The Annunciator *announces* to operators, supervisory personnel or casual attendants and workmen that a significant change of condition has taken place. The announcement is usually in the form of a visual and an audible signal.

HOW THEY DO IT

Annunciator designs naturally vary, but generally they call attention to an off-normal condition by sounding a horn, bell or buzzer; the condition is further identified by the lighting of a lamp with which is associated a designating nameplate. *Backlighted* types, where the nameplate itself lights up, and *Bullseye* types, where the nameplate is adjacent to the lamp, are the most common visual indications.

SEQUENCE

A description of the order of events, including the actions of the sensing device, trouble contact, horn, lamp and the attendant or operator, is called the "Sequence." Countless different sequences are possible; each may suit a particular and frequently peculiar situation, but essentially the same objective exists as follows:

1. The Annunciator alerts the attendant to an off-normal condition.

2. It enlightens the attendant as to the nature of that condition.

3. It requires acknowledgement by the attendant.

4. It advises the attendant when the condition has returned to normal.

TROUBLE CONTACTS

An Annunciator requires a normally open electrical contact that closes on off-normal conditions, or a normally closed contact that opens on off-normal conditions. These are generally referred to as "Trouble Contacts." They *make* a circuit or *break* a circuit; the electrical power for which comes from the Annunciator proper.

MISCELLANY

Horns and lamps operate in accordance with the sequence; their power originating with the Annunciator in most cases. Pushbuttons generally are employed to serve acknowledgment functions. Operational tests — a complete Annunciator check (and/or lamp tests) — are often associated with the Annunciator, and are accomplished with pushbuttons.

The following pages present a wide variety of PAN-ALARM equipment. Many physical designs and sequences are available in addition to those included. However, this catalog describes standard types available for prompt delivery. Their performance, utility and economy have been demonstrated and proved in countless applications.

Annunciator Sequences

As noted previously, the Sequence is the order of events associated with an Annunciator. The letter "A" is generally accepted as designating the following Sequence in the order indicated.

1. The trouble contact goes off-normal and
   a. The horn sounds
   b. The nameplate lamp lights up
2. The attendant responds by noting the condition and he depresses the silence pushbutton, whereupon
   a. The horn is silenced, but
   b. The nameplate lamp remains lighted
3. The trouble contact returns to normal and
   a. The nameplate lamp goes out
   b. The point involved is ready for realarming

Many variations of this basic sequence are included in this catalog. Each is designated by a second letter (F, R, M, etc.) for additional description.

Key to sequence letter code

A —Basic "A" sequence as described
O —Indicates nothing added to basic sequence
F —Flashing
R —Ringback
M —Manual lamp reset
S —Sequential or "First out" indication
D —Dimming

Typical example of Sequence AF (A with Flashing Light)

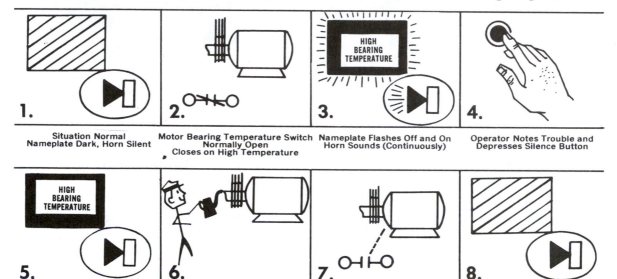

1. Situation Normal
Nameplate Dark, Horn Silent

2. Motor Bearing Temperature Switch
Normally Open
Closes on High Temperature

3. Nameplate Flashes Off and On
Horn Sounds (Continuously)

4. Operator Notes Trouble and
Depresses Silence Button

5. Nameplate Glows Steady Bright
Horn Is Silenced

6. Attendant Lubricates Bearing

7. Bearing Cools and Temperature
Switch Opens (Returns to Normal)

8. Nameplate Again Dark
Horn Remains Silent

Typical sequence chart for this order of events

	CONDITION			
DEVICE	TROUBLE CONTACT NORMAL	TROUBLE CONTACT GOES OFF-NORMAL	SILENCE BY PUSHBUTTON	TROUBLE CONTACT RETURNS TO NORMAL
LAMPS	Off	Bright Flash	Steady Bright	Off
HORN	Off	On	Off	Off

Note: After being silenced the horn is ready to sound for any other station or point going off-normal.

ANNUNCIATOR APPLICATION INFORMATION

Annunciators are subject to the same considerations associated with other electrical equipment regarding (1) Environment, (2) Power Supply, (3) Physical Mounting and (4) Function. The last of these has been discussed in the preceding pages, and the following pertinent data covers the others.

1. Environment includes pressures, temperatures and surrounding atmosphere analysis.

   a. Atmospheric pressures normally found anywhere on earth are satisfactory for annunciator operation. Special pressure chamber or high flying airborne applications must be given further consideration.

   b. Satisfactory ambient temperatures include the range of 0° F. to 110° F. Further consideration should be given to internal annunciator cabinet temperatures which normally should not exceed 165° F. Observing the above ambient range, this will not be exceeded except in cases where a majority of the alarm points are off-normal for long periods of time.

   Larger annunciator cabinets can easily have over 1000 watts to dissipate due to lamps alone when many points are off-normal. Therefore, *provision should be made for disconnecting the annunciator from its power supply during plant shutdown periods and during start-up periods.*

   c. While plug-in relays, the only moving and operating components, are hermetically sealed, all terminals, wiring, insulation and the like should be protected from excess moisture (weather exposure, splashing water, etc.) and corrosive atmospheres. Special enclosures for this purpose are shown in the various dimension sections.

   d. Ambient lighting levels higher than 80 footcandles reduce the effectiveness of back-lighted nameplates even where clear lamps and white nameplates are used in the annunciator. Colored lamps and nameplates lose some effect in ambient levels over 60 footcandles. Colored bullseyes are still less contrasting in high lighting ambients.

2. Standard power inputs are 120 volts 60 Hertz AC and 125 volts DC. Other voltages may be accommodated on special order. Commercial fluctuations in voltage of 10% are acceptable. Low energy transients even in excess of 1000 volts will not harm the annunciator nor cause it to malfunction.

3. Plug-ins and the overall annunciator will operate in any position. Conventional applications involve panel mounting. Shock loads, up to 5Gs, applied to the panel will not injure the annunciator. Special shock mounting extends this range.

4. Standard lamps for 120 volt systems are T-3¼ miniature bayonet base, 25 mA. @ 120V. They are rated, by the lamp manufacturer, for an average life of 10,000 hours at 120 volts.

   Incandescent lamps used in annunciator equipment have filament failure rates generally illustrated by the Maxwellian curve in which a few fail very early in their service history, some carry on almost indefinitely and the majority attain a so-called average life, generally indicated as rated life. The value of applied voltage significantly influences this average life.

ANNUNCIATOR ENGINEERING DATA

ANNUNCIATOR COMPONENT	POWER CONSUMPTION				CONTACT CAPACITY*
	VOLTAGE	STAND-BY	MAXIMUM	MINIMUM	
Self-Policing Plug-Ins	120 AC 125 DC	6.5 VA 3.7 Watts	8.5 VA 7.4 Watts	0 0	2.0 amps 0.5 amps
No-Drain Plug-Ins	120 AC 125 DC	0 0	7.2 VA 4.3 watts	0 0	2.0 amps 0.5 amps
Flashers, All Models	120 AC 125 DC	0 0	2.7 VA 5.0 Watts	0 0	1200 watts lamp load** 500 watts lamp load**
Auxiliary Relays (Model ACS-X1)	120 AC 125 DC	0 0	5.0 VA 3.5 Watts	0 0	10.0 amps 1.0 amps
Backlighted Nameplates 2 Lamps Per Point 1 Lamp Per Point	Any Any	0 0	6.0 Watts 3.0 Watts	0 0	— —
Bullseye Lamps	Any	0	3.0 Watts	0	—
Pushbuttons Models SW-102, SW-302	120 AC 125 DC	— —	— —	— —	6.0 amps 1.1 amps
Model SI-11	120 AC 125 DC	— —	— —	— —	20.0 amps 10.0 amps
Models 51-PB, 51-PB41 Models ACS-PB, ACS-PB41	120 AC 125 DC	— —	— —	— —	6.0 amps 2.0 amps

Horn
   Model HSA
   Model HSD
	120 AC
	125 DC
		0
		0
			20.0 VA
			13.0 Watts
				0
				0
					—
					—

*All contact ratings are for non-inductive loads at 120/60 for AC applications and at 125 volts for DC applications.

**The number of lamps handled by one flasher is a function of this rating and lamp wattage. Consider total wattage of maximum number of lamps flashing at any one time. Where lamp load exceeds flasher contact rating, an auxiliary flasher relay is required, not an additional flasher.

In most cases, plug-in relays are designed for use on either 50 or 60 cycles. Nameplates so designate, for example 120/50-60. Flashers must be ordered for exact frequency however.

TYPICAL ANNUNCIATOR DIAGRAMS

To further illustrate annunciator operation, representative schematic and general electrical and field wiring diagrams are shown below.

MODEL 10 Sequence AF with Plug-In Model ACS-8

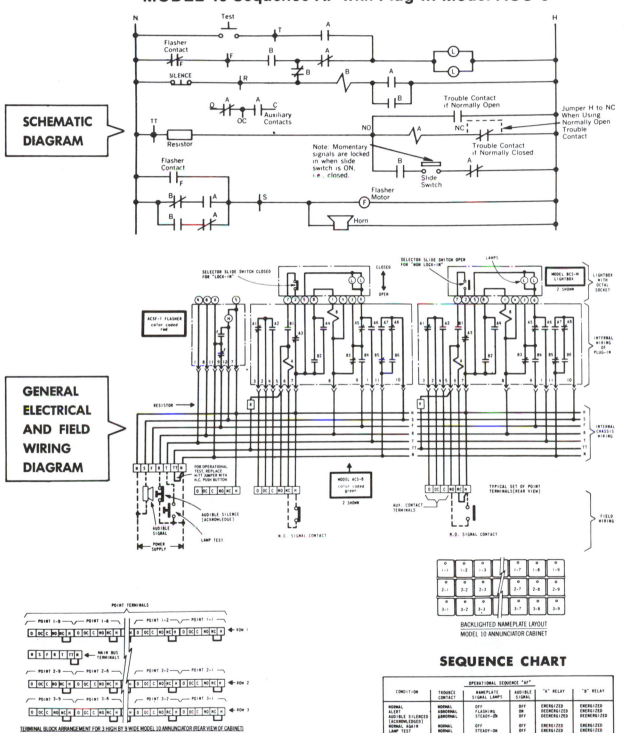

SCHEMATIC DIAGRAM

GENERAL ELECTRICAL AND FIELD WIRING DIAGRAM

BACKLIGHTED NAMEPLATE LAYOUT
MODEL 10 ANNUNCIATOR CABINET

TERMINAL BLOCK ARRANGEMENT FOR 3 HIGH BY 9 WIDE MODEL 10 ANNUNCIATOR (REAR VIEW OF CABINET)

SEQUENCE CHART

CONDITION	TROUBLE CONTACT	OPERATIONAL SEQUENCE "AF"		"A" RELAY	"B" RELAY
		NAMEPLATE SIGNAL LAMPS	AUDIBLE SIGNAL		
NORMAL	NORMAL	OFF	OFF	ENERGIZED	ENERGIZED
ALERT	ABNORMAL	FLASHING	ON	DEENERGIZED	ENERGIZED
AUDIBLE SILENCED (ACKNOWLEDGED)	ABNORMAL	STEADY-ON	OFF	DEENERGIZED	DEENERGIZED
NORMAL AGAIN	NORMAL	OFF	OFF	ENERGIZED	ENERGIZED
LAMP TEST	NORMAL	STEADY-ON	OFF	ENERGIZED	ENERGIZED

SEQUENCE AF
With Plug-In MODEL SC-10
MODEL 10

GENERAL ELECTRICAL AND FIELD WIRING DIAGRAM
(Patent Nos. 2,709,249 & 2,730,704)

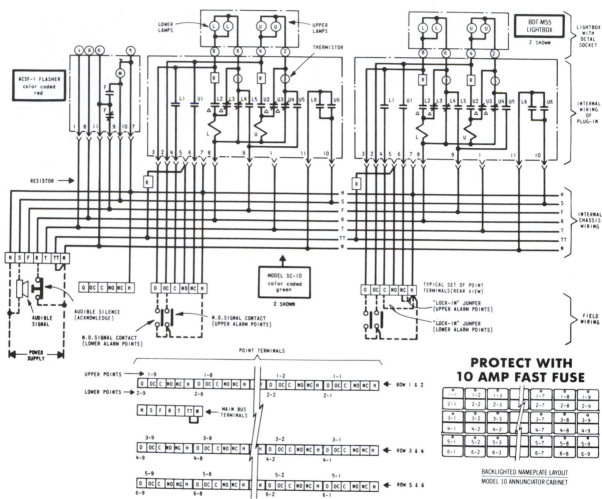

TERMINAL BLOCK ARRANGEMENT FOR 3 HIGH BY 9 WIDE MODEL 10 ANNUNCIATOR (REAR VIEW OF CABINET)

PROTECT WITH 10 AMP FAST FUSE

BACKLIGHTED NAMEPLATE LAYOUT
MODEL 10 ANNUNCIATOR CABINET

EQUIPMENT REQUIRED

One Model 10 Twinpoint Annunciator

- Model SC-10 plug-in relays and BDT-M55 lightboxes, one for each active position (2 alarm points)
- One Model ACSF-1 integral flasher, or one Model 50-F1 flasher with remote mounting receptacle WB-2.
- One audible signal, horn or bell
- One pushbutton for SILENCE

SEQUENCE CHART

CONDITION	OPERATIONAL SEQUENCE "AF"			
	TROUBLE CONTACT	NAMEPLATE SIGNAL LAMPS UPPER OR LOWER	AUDIBLE SIGNAL	"U" OR "L" RELAY
NORMAL	OPEN	OFF	OFF	DEENERGIZED
ALERT	CLOSED	FLASHING	ON	ENERGIZED
AUDIBLE SILENCED (ACKNOWLEDGED)	CLOSED	STEADY-ON	OFF	DEENERGIZED
NORMAL AGAIN	OPEN	OFF	OFF	DEENERGIZED

SEQUENCE AF
With Plug-In MODEL SC-100
MODEL 10

GENERAL ELECTRICAL AND FIELD WIRING DIAGRAM
(Patent Nos. 2,709,249 & 2,730,704)

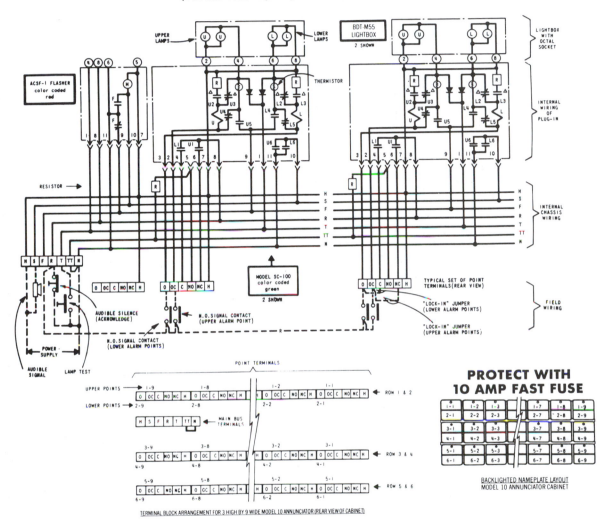

TERMINAL BLOCK ARRANGEMENT FOR 3 HIGH BY 9 WIDE MODEL 10 ANNUNCIATOR (REAR VIEW OF CABINET)

PROTECT WITH 10 AMP FAST FUSE

BACKLIGHTED NAMEPLATE LAYOUT
MODEL 10 ANNUNCIATOR CABINET

SEQUENCE CHART

CONDITION	TROUBLE CONTACT	NAMEPLATE SIGNAL LAMPS UPPER OR LOWER	AUDIBLE SIGNAL	"U" OR "L" RELAY
		OPERATIONAL SEQUENCE "AF"		
NORMAL	OPEN	OFF	OFF	DEENERGIZED
ALERT	CLOSED	FLASHING	ON	ENERGIZED
AUDIBLE SILENCED (ACKNOWLEDGED)	CLOSED	STEADY-ON	OFF	DEENERGIZED
NORMAL AGAIN	OPEN	OFF	OFF	DEENERGIZED
LAMP TEST	OPEN	DIM	OFF	DEENERGIZED

— BRIGHT ON D.C. SYSTEMS

EQUIPMENT REQUIRED

One Model 10 Twinpoint Annunciator

- Model SC-100 plug-in relays and BDT-M55 lightboxes, one for each active position (2 alarm points)
- One Model ACSF-1 integral flasher, or one Model 50-F1 flasher with remote mounting receptacle WB-2.
- One audible signal, horn or bell
- Two pushbuttons for SILENCE and TEST

A P P E N D I X   J
UNINTERRUPTIBLE POWER SOURCE SYSTEMS

This Appendix Courtesy of Emergi-Lite, Inc.

UPS Standard Features

INPUT VOLTAGE/PHASE: Either 120 or 277 single phase 2 wire (Standard); other voltages and 3 phase 4 wire available.

INPUT VOLTAGE REGULATION: ± 20%

OUTPUT VOLTAGE/PHASE: Same as input (Standard); dual voltage and 3 phase 4 wire available.

OUTPUT VOLTAGE REGULATION: ± 5%

INPUT POWER WALK-IN: 18 Cycles: limits utility inrush to 200% of normal.

INPUT FREQUENCY: 60 Hz. ± 1 Hz. (Standard); 50 Hz. optional.

OUTPUT FREQUENCY: Locked to AC input during utility "ON" condition; 60 Hz. ± 0.05% during emergency run; 50 Hz. optional.

OUTPUT WAVE FORM: Sine wave with less than 5% total harmonic distortion.

POWER FACTOR RANGE: .75 lag to .9 lead.

TRANSFER TIME UPON UTILITY FAILURE: Zero as the inverter continuously supplies power to the critical load.

OVERLOAD CAPABILITIES: 130% for 5 minutes during an emergency run.

AUDIBLE Db RATING: 55 DbA for smallest unit increasing linearly with size of units to a maximum of 78 DbA @ 75 KVA.

TEMPERATURE LIMITS: 32°F to 120°F. Temperatures higher than 80°F will shorten battery life; temperatures lower than 60°F will decrease run time. Optimum operating temperature: 77°F.

BATTERY CHARGER: 3 rate including Automatic Equalize every 90 days.

MINIMUM LOAD: 10% of rated capacity.

AUTOMATIC BYPASS SWITCH: Upon inverter failure, the AC utility automatically transfers to the load by mechanical means within approximately 1 second. Optional solid state high speed transfer (HST) device is available to reduce bypass time to 4 milliseconds.

NOTE: For 3 phase input and output, all lines to neutral performance data are the same as the single phase data shown.

SYSTEM OPERATION

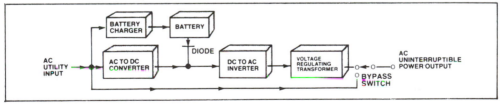

The AC Utility Input powers a Battery Charger and an AC to DC Converter. The output of the Battery Charger keeps the Battery automatically charged during the time the utility power is on. The AC to DC Converter normally provides the DC input power for the DC to AC Inverter during the time the utility power is on. The Battery is connected to the DC line feeding the Inverter through a Diode. The output of the DC to AC Inverter powers the Voltage Regulating Transformer which in turn powers the load via the Bypass Switch. Upon a utility power failure, the Battery Charger and AC to DC Converter cease to operate as their output voltages drop. The Battery is now switched automatically to the DC to AC Inverter through the Diode. The DC to AC Inverter now is being powered from the Battery. Upon the return of the AC utility, the Battery is recharged and the AC to DC Converter again powers the DC to AC Inverter. If the Inverter fails for any reason, the Automatic Bypass Switch connects the load directly to the utility until the unit can be repaired. An optional static Bypass Switch can be provided which switches within a few milliseconds when a minimal interrupt is required. The standard Automatic Bypass Switch switches in approximately 1 second.

UPS Technical Notes

The EMERGI-LITE UPS Series of Inverters is constructed of 6 modules. They are:

1. Battery Bank
2. Battery Charger
3. DC Rectifier
4. AC Present Logic and Transfer Device
5. Inverter Module
6. Bypass Switch on Inverter Failure

The battery bank and charger operate the same as the IPS series Inverters. It should be noted that EMERGI-LITE UPS Systems utilize a separate rectifier for the battery charger. Some companies use one large DC rectifier to run the Inverter continuously and charge the batteries. EMERGI-LITE, however, uses two separate transformers and rectifiers for greater reliability. The large DC rectifier provides the DC voltage and current which the Inverter needs to operate while the AC utility is up. This rectifier can provide as much as 450 Amperes at 240 volts DC.

The AC present logic is, in most UPS systems, a large power steering diode which allows the power to flow from the large DC rectifier or from the battery bank when the DC rectifier output voltage drops below a preset point. The transfer time from when the rectifier voltage drops out and the battery bank picks up the load is zero, since the diode literally reacts instantaneously.

The UPS Inverter module is similar to the IPS Inverter Module. In general, the components and the operating parameters follow the same design logic. The UPS system, however, has larger ferroresonant transformers to dissipate the heat generated by the constantly running Inverter.

Since the Inverter runs continuously and supplies the actual AC power, if the Inverter fails, the load is without power. EMERGI-LITE UPS Systems are constructed in such a way that, upon an Inverter failure, the load is automaticaly switched over to the AC feed line (if it is still available). One problem with this is the fact that the transfer device is typically an electromechanical relay which can only transfer in about one second. This is too long a time for some sensitive loads, so EMERGI-LITE offers an optional High Speed Transfer (HST) device which can connect the load to the AC feed line within four milliseconds. The HST option is a very technically advanced circuit utilizing state-of-the-art components. No matter which bypass system is used, the Inverter will attempt to return the load to the Inverter as long as the Inverter is running properly (which will be the case in the event of a momentary overload caused by a motor inrush or some other device turning on). If the Inverter is not running properly, the Inverter will lock on to the AC feed line and remain there until the Inverter is repaired. Provisions for manually transferring the Inverter are provided whether the Inverter has the HST or utilizes the relay transfer.

AC Current Requirements For UPS Systems

VA	Maximum Input (Amps)		Recommended Input Breaker Size	
	120	277	120	277
500	9	4.5	10	5
1,000	17	8	20	10
1,500	25	12.5	30	15
3,000	50	25	60	30
4,500	75	37.5	80	40
6,000	100	50	125	60

VA	Maximum Input (Amps)		Recommended Input Breaker Size	
	120	277	120	277
7,500	125	62.5	150	80
10,000	166	83	200	100
16,600	210	100	250	120
25,000	300	150	360	180

Typical Output Voltages Available

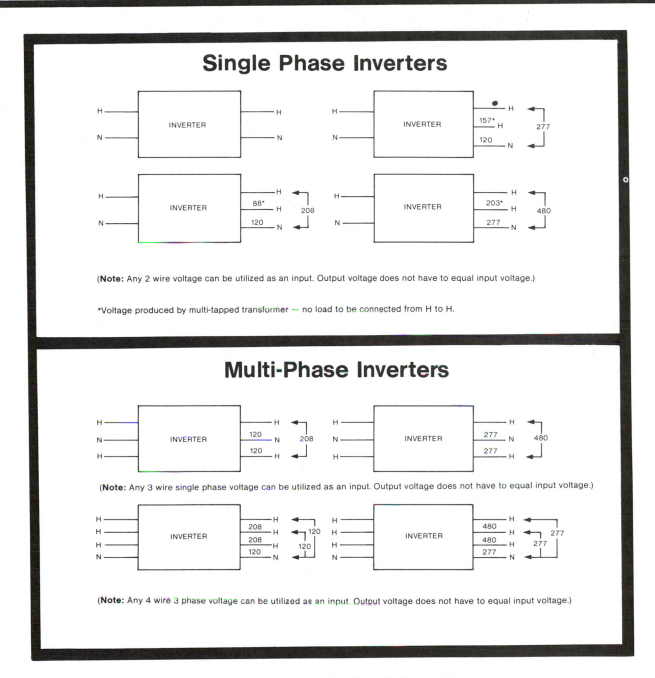

Single Phase Inverters

(**Note:** Any 2 wire voltage can be utilized as an input. Output voltage does not have to equal input voltage.)

*Voltage produced by multi-tapped transformer — no load to be connected from H to H.

Multi-Phase Inverters

(**Note:** Any 3 wire single phase voltage can be utilized as an input. Output voltage does not have to equal input voltage.)

(**Note:** Any 4 wire 3 phase voltage can be utilized as an input. Output voltage does not have to equal input voltage.)

Location of Cabinets

The Inverter cabinets should be located away from any heat-producing pipes or equipment. Avoid placing the inverters against an adjoining office wall. If this is not possible, placing the inverter on a noise-reducing pad and installing a similar pad *behind* the unit will help to reduce the noise transmitted through the floor/wall to the office area. If the inverter is to be located in an area served by air conditioning, care should be taken in sizing the air conditioning unit in order to accommodate the heat generated by the Inverter. (See table, page 15.)

Hydrogen and the Inverter

Hydrogen is lighter than air. It is colorless, tasteless and highly explosive. It is also very difficult to detect without special equipment. Hydrogen is highly concentrated under the vent cap of each cell of the battery. Smoking, sparks or flames should be prohibited in the Inverter room because of the presence of hydrogen. Flame-retardant vent caps are supplied with all EMERGI-LITE Wet Batteries to protect the cells against accidental ignition of the gas.

Every type of battery (including sealed maintenance-free types) produces hydrogen gas, which is vented through the vent caps to the air in the Inverter room or compartment. (This occurs mainly during the charging operation). The most significant period of hydrogen production occurs when the maximum system voltage is impressed on the fully charged cells. No hydrogen is produced during the float charging operation.

The volume of hydrogen generated is governed by the amount of charging current supplied to the fully charged battery by the charger. The ampere hour capacity of the cell is not a contributing factor to the production of hydrogen.

Air Exchange Parameters

When a lead acid cell is fully charged, each charging ampere supplied to the cell produces about 0.016 cubic feet (cuft) of hydrogen per hour from each cell (*not each* battery *which may consist of more than one cell*). Lead calcium and nickel cadmium batteries produce about 0.010 cuft of hydrogen for each ampere of current supplied to a fully charged cell. This volume applies at sea level, when the ambient temperature is 77°F and when the electrolyte is "gassing" or bubbling. (See accompanying KVA/Hydrogen Production Chart.)

The engineer should first calculate the cubic feet of unoccupied space in the inverter room and then arrange for sufficient ventilation to keep the hydrogen content well below 3% of that calculated volume. Most areas in which people work require 3 to 4 air changes per hour for minimum comfort. This number of air exchanges in an inverter room will usually provide a substantial margin of protection against hydrogen buildup.

Louvers located both in the doors and in the walls near the ceiling are generally adequate to provide sufficient ventilation. A small fan may be installed in the wall above the inverter to exhaust room air to the outside atmosphere if desired. If the inverter room is air conditioned as part of a general buildingwide air conditioning system, the exhaust air from the inverter room should not be returned to the air distribution system, but rather should have its own exhaust system direct to the outdoors.

KVA/HYDROGEN PRODUCTION	
KVA	HYDROGEN PRODUCTION (Approx. cu. ft. per hour)
0.5	0.2
1.0	0.4
2.0	0.5
3.0	0.8
4.0	0.9
5.0	1.3
6.0	2.0
7.5	1.9
10.0	2.4
12.5	1.2
16.6	2.9
25.0	3.8
50.0*	1.3
75.0*	2.0

*30 minute operating time

Note: No hydrogen is produced during float or discharge periods. Hydrogen is only produced during the final few hours of the recharge or equalize modes.

BIBLIOGRAPHY

General Electric Co., Lighting Division, (1) Industrial Lighting Application Bulletin, (2) Selection Guide For Quality Lighting.

GNB, Inc., Gould Industrial Battery Div., Instruction, Maintenance and Service Manual, Motive Power Batteries and Chargers.

Illuminating Engineering Society of North America, 345 East 47th St., New York, NY 10017
ED-100, IES Education Series

RP-7, American National Standard Practice For Industrial Lighting

Phillips Lighting Company, Guide To Fluorescent Lamps.

Simpson Electric Co., Simplified Electrical Appliance Servicing.

Square D Co., (1) Distribution Equipment Fundamentals, (2) Motor Control Fundamentals, (3) Wiring Diagrams, (4) Service Manual.

Sylvania Div., GTE Products Corp., Engineering Bulletins 0-330 and 0-345.

The U.S. Government Printing Office has available the following training and informational material. Their address is:

U.S. Government Printing Office
Supt. of Documents
Washington D.C. 20402

To obtain their price list, a written application must be made direct to the above address, requesting their SB-053, for the Electricity and Electronics price list. They will send you the current list, with ordering instructions.

Dept. of the Army:

(1) Electric Motor and Generator Repair, TM 5-764
(2) Interior Wiring, TM-760, S/N 008-020-00456-1
(3) Electrical Fundamentals, Alternating Current, TM-11-681, S/N 008-020-00060-4
(4) Illustrated Guide To Electrical Safety, S/N 029-015-00064
(5) Basic Electricity, Direct Current, S/N 008-047-00097-9
(6) Basic Electricity, Alternating Current, S/N 008-047-00098-7

Bureau of Naval Personnel: Basic Electricity, S/N 0502-LP-050-4300

Some of the above government publications may be out of print, or may be superseded by more recent material. In any case, it is an excellent and inexpensive source of educational material.

John Wiley & Sons, Inc., *Elements of Electrical Engineering*, by Arthur L. Cook, 4th Edition, copyright 1943.

INDEX